WIRING THE WORLD

COLUMBIA STUDIES IN INTERNATIONAL AND GLOBAL HISTORY

COLUMBIA STUDIES IN INTERNATIONAL AND GLOBAL HISTORY

The idea of "globalization" has become commonplace, but we lack good histories that can explain the transnational and global processes that have shaped the contemporary world. Columbia Studies in International and Global History encourages serious scholarship on international and global history with an eye to explaining the origins of the contemporary era. Grounded in empirical research, the titles in the series transcend the usual area boundaries and address questions of how history can help us understand contemporary problems, including poverty, inequality, power, political violence, and accountability beyond the nation-state.

Cemil Aydin, *The Politics of Anti-Westernism in Asia: Visions of World Order in Pan-Islamic and Pan-Asian Thought*

Adam M. McKeown, *Melancholy Order: Asian Migration and the Globalization of Borders*

Patrick Manning, *The African Diaspora: A History Through Culture*

James Rodger Fleming, *Fixing the Sky: The Checkered History of Weather and Climate Control*

Steven Bryan, *The Gold Standard at the Turn of the Twentieth Century: Rising Powers, Global Money, and the Age of Empire*

Heonik Kwon, *The Other Cold War*

Samuel Moyn and Andrew Sartori, eds., *Global Intellectual History*

Alison Bashford, *Global Population: History, Geopolitics, and Life on Earth*

Adam Clulow, *The Shogun and the Company: The Dutch Encounter with Tokugawa Japan*

Richard W. Bulliet, *The Wheel: Inventions and Reinventions*

SIMONE M. MÜLLER

WIRING THE WORLD

THE SOCIAL AND CULTURAL CREATION OF GLOBAL TELEGRAPH NETWORKS

COLUMBIA UNIVERSITY PRESS *NEW YORK*

Columbia University Press
Publishers Since 1893
New York Chichester, West Sussex
cup.columbia.edu

Library of Congress Cataloging-in-Publication Data

Names: Müller, Simone M.
Title: Wiring the world : the social and cultural creation of global telegraph networks / Simone M. Müller.
Description: New York : Columbia University Press, 2015. |
Series: Columbia studies in international and global history | Includes bibliographical references and index.
Identifiers: LCCN HYPERLINK "tel:2015024700" \t "_blank" 2015024700| ISBN 9780231174329 (cloth : alk. paper) | ISBN 9780231540261 (ebook)
Subjects: LCSH: Telegraph—History. | Transatlantic cables. | Telegraph—Economic aspects.
Classification: LCC HE7631 .M85 2015 | DDC 384.1—dc23
LC record available at http://lccn.loc.gov/2015024700

Columbia University Press books are printed on permanent and durable acid-free paper.

This book is printed on paper with recycled content.
Printed in the United States of America

Cover image: Copyright © Morphart Creation/Shutterstock
Cover design: Kathleen Lynch

CONTENTS

ACKNOWLEDGMENTS

"HUZZA! – The magic cable's laid!" goes the opening stance of an Atlantic cable poem by Samuel B. Sumner and Charles A. Sumner. The brothers' poem was read at the reception of Cyrus W. Field in Stockbridge, Massachusetts, in August 1858. What struck me when reading it was not the beauty of its phrasing, the ingenuity of its rhymes, or the melodic rhythm of the words assembled; I was impressed by the expression of "heartfelt thanks" that went out to the "men of tireless zeal" who had "ventured all and battled all" to accomplish the masterly task of laying an Atlantic telegraph cable. There was such authenticity to this thankfulness that it outlasted more than 150 years, pressed within the pages of a book that had seen more joyful times. The authors' thank you knew of kindness shown and noble deeds done. The poem ends with three more cheers of "Huzza! Huzza!! Huzza!!!" I was immediately smitten by the crescendo of exclamation marks and could not help smile when detecting a little footnote. There, the authors noted, how the assembly "all rose and joined in the cheer at the conclusion." Reading the little poem, I could feel the sensation of how "splendid an effect" this must have been in the summer of 1858.

Just as Cyrus W. Field did not dream up the transatlantic telegraph entirely on his own as he mused by his globe into the night, neither did this book come about entirely from my own work. Like the "hero" of the wiring of the Atlantic, I am deeply indebted to the various, oftentimes overlapping, networks that helped me to bring this book to light: my academic network, my archival network, my financial network, and finally, my social network of family and friends. Without those myriad groups of people, ideas, finances, and institutions, this book could not have been written. I would like to express my deepest gratitude to all of them—so here goes my "Huzza!!!" to you:

First and foremost, I would like to thank my two dissertation advisors. Michaela M. Hampf and Sebastian Conrad have been my lighthouses, guiding me faithfully through the stormy process of writing a dissertation and turning it into a book. They have provided tremendous support and great guidance while continuing to challenge me and keep me moving forward. It has been a great honor to work with them and I am deeply grateful for their counsel.

Many others contributed to the genesis and completion of this book. My "combatants" of the DFG-research group 955 "Actors of Cultural Globalization, 1860–1930" provided immensely helpful suggestions and support. Christina Peters, Michael Facius, Maria Moritz, Sebastian Sprute, Jessika Bönsch, as well as Frederik Schulze have more than once aided me in my struggles of putting and keeping the actors in globalization. Jochen Meissner has shepherded us all through years of writing and research, and I am truly thankful for his invaluable enthusiasm, close reading, and perceptive insights on the early versions of this manuscript. My thanks also go out to Ursula Lehmkuhl, Andreas Eckert, Stefan Rinke, Verena Blechinger-Talcott, Ulrich Mücke, and Harald Fischer-Tiné, who have generously shared their knowledge, their expertise, and their networks. It has been a wonderful experience to work with this research group. The John F. Kennedy Institute for North American Studies has been a superb home base to set out on all my travels to enquire about the wiring of the world as well as to write about and discuss everything I had found.

This book has greatly benefited from discussions and debates with colleagues and friends throughout the world. Richard Noakes, Wendy Gagen, Graeme Gooday, Léonard Laborie, Daqing Yang, Nicola Borchardt, Elisabeth Engel, and Torsten Kathke generously shared their research and ideas and provided me with feedback on my own. Particular thanks go out to Richard John for sharing with me both his enthusiasm for communications history and his encyclopedic knowledge on telegraphy, news, and postal services and to Pascal Griset for pushing me forward with his initial skepticism toward the topic as well as his astute thinking and feedback. Heather Ellis provided me with input on the history of science in Britain and helped me challenge and expand on the concept of actor networks as tools for research on the history of globalization. Finally, this book could not have been written without the continuous support of Heidi J. S. Tworek. A true congenial light in understanding global communications and news, she never tired of participating in discussions on the intricate relationship between communication and modernity, capitalism, and globalization. Moreover, with a sharp eye for both precision and narrative, she supported the process of the book's writing and revision. I would also like to thank Matthew Conelly and Adam McKeown for accepting my work into their series on international and global history. The book could not have found a better home and not a better editor than Anne Routon.

My archival network for this book stretches across more than twenty institutions and six countries. Archivists on both sides of the Atlantic were unfailingly helpful, fulfilling my requests for far-flung files and photocopies with utmost generosity. I am indebted to the staff at the National Archives in London, Paris, Washington, and Ottawa, the Bundesarchiv in Berlin, the British Library, the Library of Congress, the Bibliothèque Nationale de France, the University Archives of Glasgow and Bristol, the IET Archives London, the Royal Society Archives, the International Telecommunication Union Archives, British Telecom (BT) Archives, the Siemens Archives Munich, the North Somerset Museum Archives, the Archives des Amis des Câbles Sous-Marins, and the Museum of Communication of Berlin. Particular thanks are due to Allan Green and Charlotte Dando from the Porthcurno Telegraph Museum and Cable and Wireless Archives. This book could not have been written without the treasures harbored in these archives, and without their insight and guidance many a gemstone, including Lady Pender's diary, might have gone missing. Similarly, Alison Oswald and Bernhard Finn from the Smithsonian Institutions have shared their expert knowledge on the Western Union Archive with me and enabled me to experience firsthand a researcher's highlight when archival material, in my case the Heart's Content cable station's letter books, is "rediscovered" and opened to first use. Finally, I want to express my particular thanks to Donard de Cogan, who opened his private family archive to me.

In terms of financial networks, I gratefully acknowledge the generous support provided by the German Research Foundation, the German Academic Exchange Service, and the Smithsonian Institutions. Thanks also to the German Historical Institute Washington, which gave me the freedom offered by a fellowship to turn the dissertation into a book.

Family and friends provided me with the emotional support that kept me going, and over time, many colleagues became dear friends. All of them believed in me as much as in my project and bore generously and humorously with my curiosity and obsession over telegraphs, networks, and processes of globalization. My engineering brothers, Christopher and Thomas, were sources of influence and wisdom when it came to the technological processes of telegraph mechanics. My brother Andi, the musician, brought the telegraph polka to life for me by playing the sheet music on the piano. My parents gave me the greatest gift of all: their love and their never-ending support. Christina Brüning, Viola Eckert, Eva Hartmann, Christina Hartung, Rachel and Charles Horwitz, Björn Jesse, Marta Kozlowska, Sebastian Kunkel, and Michael Tworek have supported me in more ways than I can tell. Finally, Alexander Pohl, gave me the strength to continue when I was in doubt and the liberty to go when I needed to seek the horizon.

Over the course of writing this book, I have been very fortunate to receive advice, guidance, and support from a great number of generous individuals, not all of whom I could mention here, and my research and work owe much to the above networks. I would like to thank them all with a cheerful "Huzza!!!"

WIRING THE WORLD

0.1 Adam Weingärtner, "American Torchlight Procession Around the World," 1858. Pictures and Photograph Division, Library of Congress.

INTRODUCTION

The Class of 1866

THE CITY was in uproar. Bells were rung and guns fired. Candles and fireworks lit up New York City Hall. People swarmed to the streets with boundless enthusiasm and formed a procession along Broadway. At the grand fete at Union Square, people danced to a specially composed "Telegraph Polka" to commemorate the act of putting "a loving girdle round the earth," in the lyrics of A. Talexy.[1] Throughout "the length and breadth of the [entire] country," every tongue talked of only two things: New York businessman Cyrus W. Field, the "Columbus of America," and the "eighth wonder of the world," the Great Atlantic Cable.[2] In August 1858, an Anglo-American consortium of engineers, electricians, and entrepreneurs had finally succeeded in laying a telegraphic cable across the Atlantic. Via telegraph, Manchester cotton merchant John Pender, British engineers Charles T. Bright and William Thomson, and American philanthropist Peter Cooper congratulated themselves upon connecting the Old World and the New by electric wire. Meanwhile, newspapers all across the Euro-American hemisphere printed special editions with the "spectacular" story of the wiring of the world. On the streets of New York, blue silk ribbons with the portrait of cable entrepreneur Cyrus W. Field were peddled for women's jewelry and men's hatbands. Catering to a public clamoring for souvenirs, jewelers Tiffany and Company sold the remainder of the Atlantic cable in four-inch pieces mounted neatly with brass ferules and a certificate letter signed by Field himself.[3] That summer, these emblems of material culture created the mythological narrative of the wiring of the world: the heroic struggle of male protagonists triumphing over the "opposition of nature" through technology.[4]

Meanwhile, only some blocks from Broadway, a New York artist, Adam Weingärtner, finished his depiction of a counter-narrative to white masculine technological heroism. Just in time for the September 1 celebration of the laying of the Great Atlantic Cable, he had a lithograph ready for sale. For the artist, this lithograph, "Torchlight Procession Around the World," summed up the apotheosis of global integration in the nineteenth century. Instead of another likeness of "Cyrus the Great," "the Ship," or "the Cable," Weingärtner had sketched a parade of people from all over the world: Euro-Americans, Asians, Africans, and Native Americans dance around the globe surmounted by a cloud forming the word *Liberty*. As they hold a telegraph cable girdling the earth, the world's "civilized" and "heathens" join together in technological "revolution" that seems to hold out the prospect of "global" liberty. In the lower-left and lower-right corners, two cable ships, the American USS *Niagara* and the British HMS *Agamemnon*, frame the celebration. Ships and crews are busy laying out the Great Atlantic Cable. Nearby, an irritated Neptune sits with his mermaids in the middle of the Atlantic. The cable laying foreshadows the end of his despotic reign over maritime space. The lithograph, dedicated "To Young America," is framed by the vignette of four men who had shaped the history of the American Republic through telegraphy: Benjamin Franklin, Captain William L. Hudson, Samuel F. B. Morse, and Cyrus W. Field. After adorning these portraits, the cable follows the arch formed where the hands of two female figures meet in the upper middle part of the lithograph. Dresses and ornaments give them away as Columbia and Britannia. Between the two figures, a member of the U.S. fire brigade straddles an eagle. While he seems to be directing the group's work, he is also showering the British cable ship HMS *Agamemnon* with arrows of electrical currents.[5]

"Torchlight Procession Around the World" is unique among cable memorabilia. The lithograph departs from the iconography of the heroic Atlantic cable entrepreneur established through other portrayals of international telegraphy at the time. In his depiction of the events of the summer of 1858 when telegraphic transmissions joined the Old World and the New for the first time, Weingärtner did not focus solely on singular individuals, but on various groups of international actors. He did not foreground the North Atlantic connection, but the wider world. In the lithograph, the geopolitical ramifications of the 1858 Atlantic connection reach far beyond the telegraph plateau now joining the United

States and Great Britain as "Anglo-American brethren." Even the concept of brotherhood appears troubled, given that the American fireman is directing electrical currents toward the British cable ship. Moreover, in Weingärtner's image, this new "conduit of manifest destinarian energy" was to defibrillate the world and, possibly under the direction of the U.S. fire brigade and not Great Britain's Union Jack, spread liberty to this multiethnic and multinational procession.[6] "Torchlight Procession" in many ways symbolizes the central themes of this study: first, the centrality of international actor networks for understanding the social, cultural, economic, and technological aspects of these "wonders" of "civilization"; second, the multiple, and at times politically and socially highly charged and even contradictory, visions of a "unifying" world through electrical currents; and third, the great importance of infrastructural projects for global integration processes.[7]

Wiring the World: The Social and Cultural Creation of Global Telegraph Networks focuses on the various protagonists of the submarine telegraph system. It examines the engineers, entrepreneurs, operators, politicians, media reformers, and financiers in the pre-World War I era. It explores how these actors negotiated and battled over the very concepts that helped define an electric world in union. In their function as actors of globalization, these men, and notably also women, linked macrostructural processes of economic imperialism and geopolitics with microstructural interpretations of the means and ends of global communications. They charged economic and political processes with social and cultural meaning.[8]

ESTABLISHING A GLOBAL TELEGRAPH NETWORK

From its very beginning, ocean telegraphy spurred people's imagination of the scope and content of the world's future. When the first successful commercial submarine telegraph was laid across the English Channel in 1851, the London *Morning Herald* published an article celebrating the electric nuptials of England and France and predicting an electric union of the entire planet. In the context of submarine telegraphy, and in contrast to landline connections, entrepreneurs, journalists, politicians, and the public all thought in global dimensions. The importance of submarine telegraphy did not lay in its singularity, but in its multiplicity.

One cable never functioned in isolation but within a system of cables. Consequently, the "ultimate progress of electro-Telegraphy," the author of the article wrote, need not be confined solely to the Old World. Since the English Channel had been crossed, Ireland must follow next, "as but a matter of course." And once Ireland was reached, there lay only "a couple of thousand miles of water or so between the Old World and the New." Where, after all, lay "the practical difficulty"?[9]

The story of the Great Atlantic Cable exemplifies that there were still many "practical difficulties" to overcome. The much-celebrated cable of 1858 only functioned for a matter of weeks and transmitted just 271 messages before the connection died.[10] The group surrounding Cyrus W. Field, John Pender, and William Thomson only managed to complete a new cable across the Atlantic in 1866. In the meantime, they faced technological capriciousness, financial near-bankruptcy, and the political turmoil of the American Civil War. On top of this, rival actors discussed, scrapped, and planned alternative routes for a telegraphic connection between the Old World and the New. In the end, much cable and money were irretrievably sunk into the ocean before another submarine cable transmitted messages by electrical currents across the Atlantic.

Yet, when in 1866 ocean telegraphy finally saw its breakthrough, it initiated the rapid emergence of a system of global communication based on electric speed. After the successful transatlantic connection, contemporaries witnessed the laying of more submarine cables throughout the world. Telegraphs to India (1870), Australia (1872), and Southeast Asia (1870), as well as to the Caribbean (1873/1874), Brazil (1874), and South Africa (1879), swiftly followed. Indeed, by the late 1870s, virtually any place on earth could, at least theoretically, be reached from Europe via submarine cable through a network spanning, at that time, roughly 100,000 miles of ocean cables. In the 1880s and 1890s, popular connections were duplicated and even triplicated, and the ocean network became more densely linked with simultaneously expanding landline connections. In addition, technological developments, such as duplex telegraphy, enabled the passage of two or even four messages from both ends of the wire simultaneously. By 1900, thirteen submarine cables crossed the North Atlantic processing as many as 10,000 messages a day.[11]

Irrespective of this future success story, contemporaries already perceived the first failed attempts at submarine telegraphy as a caesura separating their age into pretelegraphic and telegraphic eras. In 1858, when the first transatlantic cable proved that long-distance communication via ocean telegraphs was feasible, Queen Victoria and the American President James Buchanan anticipated a new era opening up before them.[12] To them, the cable was a "triumph more glorious, because far more useful to mankind than was ever won by conqueror on the field of battle."[13] The Great Atlantic Cable stimulated people's imagination and hopes for a rapidly changing world synchronized to the beat of the electric telegraphs. Journalists, politicians, and merchants expected submarine telegraphy, "one of the grandest and most beneficent enterprises of the age," to "remove [political] misconceptions" and allow for transactions that would be "worth millions."[14] On top of this, President Buchanan heralded the cable as a "bond of perpetual peace and friendship between the kindred nations and an instrument destined by Divine Providence to diffuse religion, civilization, liberty, and law throughout the world."[15]

Indeed, in 1858 and even more so in 1866, the times ahead heralded great changes or "a coherent epoch of world development," according to historian Charles Maier.[16] The following decades saw a nonlinear growth of economic, political, cultural, and social global interdependence as well as the expansion of transboundary networks and spaces of interaction. The emergence of a global communication system based on electrical speed was essential for these processes. With the dematerialization of long-distance communication, message and messenger were decoupled. Through this relative increase of transmission speed compared with material carriers, the telegraph network could be used "to efficiently coordinate, control, and command . . . material movement,"[17] or as the historian Jorma Ahvenainen phrased it, only with the ocean telegraphs did world economy and world politics become possible.[18]

To this day, we continue to assume that communications hold the key to a unified world. In 2009, the scientist Charles Kuen Kao was awarded the Nobel Prize in physics. His "groundbreaking achievements" in the transmission of light in fiber-optic cables form a crucial basis of today's Internet.[19] That same year, the laying of 10,000 miles of submarine fiber-optic cable between the Arabian Peninsula and East Africa showed that the wiring of the world was not yet completed. "Closing the Final Link" was the slogan of the pan-African business consortium SEACOM that connected Kenya, Tanzania, Mozambique, Uganda, and South Africa with the world's communication network. Those formerly disconnected were finally tapping into the world communication system. In his opening address, Jakaya Kikwete, then Tanzania's president, stated that the cable connection represented "the ultimate embodiment of modernity."[20]

Despite the rhetorical analogy, history does not repeat itself, nor does the building of SEACOM's "African Internet" embody a teleology of modernity, in the sense of modernization theorists.[21] Nevertheless, these examples demonstrate that questions concerning the logic of a global media and communications system did and still do matter as modern societies depend on communication and the electronic transmission of information on a world scale. Regardless of whether a global communications system is based on ocean telegraphs or fiber-optic cables and satellites, there exist network logics inherent to any communications system that rely on its structure. Previous works have portrayed the processes driving the wiring of the world as an issue of technology, imperialism, or transcontinental business networks. With its focus on the actor networks, this book inserts social and cultural considerations alongside these political and economic issues to understand the wiring of the world and ultimately globalization processes.

ACTOR NETWORKS AND GLOBAL HISTORY

In September 1877, Emma Pender, wife of the British cable magnate John Pender, sent a letter to her son-in-law, William des Voeux. William, the husband of her daughter Marion, served as an official of the British Colonial Office. At the time, he was stationed on St. Lucia, a small island in the British Caribbean. In her letter, Emma Pender informed des Voeux about his chances of being transferred back home to London or at least to a place less remote (which in Emma Pender's imagination meant somewhere *within* the global telegraph network). The previous day, her husband had consulted about different options with Herbert from the Colonial Office. "Herbert," a frequently reappearing figure in Emma Pender's correspondence, signified Sir Robert George Wyndham Herbert, undersecretary of the British Colonial Office and an important government official. Pender and Herbert were well acquainted and met frequently, as "Mr. Pender's cable business [was] always sufficiently important to take him to the office."[22]

Accounts like these nourish the impression that the cable business and its actors were closely connected to British imperial structures. Indeed, the cable industries' excellent connections to Herbert and other imperial institutions gave them multiple advantages. It helped them gain access at relatively low cost to resources, such as the insulating material gutta percha, which was primarily found in British colonies, and to ocean sounding data. Additionally, the British government also lent its support during landing right negotiations with other governments. Furthermore, the buildup of the global communications system and its coordination and regulation were deeply entrenched in the logic of imperial power relations as well as Eurocentric notions of civilization. Cable routes primarily followed those of imperial trade and governance, and contemporaries used the technology as a distinctive marker of Euro-America's superiority over the rest of the world.[23]

Older scholarship on submarine telegraphy strongly linked cable business and imperial interests. After Harold Innis's 1950 publication *Empire and Communications*, the intricate interrelation of communication and the "rise and fall" of empires has predominantly been described under the heading of geopolitics and technology and as a struggle for global control.[24] A large majority of the existing literature on global communication and telegraphy accentuates the telegraphs' importance for imperial control and Euro-American nationalist power politics. Scholars portray telegraphs as "tools of empire" that aided the formation and consolidation of nation-states and empires or helped to generate narratives of national technological progress and development.[25] In the end, submarine telegraphy was essential for "Britain's ascendency as a world power."[26] Consequently, cables, cable manufacturers, and cable contractor companies were attributed a distinct nationality and implicit nationalist agendas. Submarine cables were considered indispensable for the nations' economic and political benefit and military security.[27]

This book reconsiders these grand narratives of imperial control and nationalist power politics. Although the actors strongly benefited from imperial and national structures and a global coloniality, i.e., a world shaped through the experience and logic of centuries of colonialism, they did not necessarily embrace imperial interests. Additionally, it is difficult to ascribe a particular nationality to distinct cables or cable companies.[28] This methodological nationalism underplays the international financing structure of the cable companies and the transnational working agreements between the companies, the nation-states, and international governance.[29] Most of all, this book argues, it underplays the actors' imagined and real spaces of action as well as their conscious instrumentalization of national institutions and their strategic nationalism reflected in their rhetoric.[30] The relationship between individual actor networks and imperial and national structures, or the micro and macro levels, was much more complex; the strong entanglements between social and cultural practices and economic and political strategies were much more important for the structure, coordination, and regulation of the global media system than so far explored.[31]

The approach of global history provides a new avenue to explore telegraphy as a historical force of globalization. Informed by postcolonial theory and subaltern studies, global history challenges Eurocentric narratives of modernization or Westernization alongside the methodological nationalism that has led scholars to assign cable connections a distinct nationality or imperial agenda.[32] Global historians of telecommunication, by contrast, use concepts of agency, technology-in-use, and space to approach telegraphy from a perspective that is not bound to the nation-state.[33] Following the global history approach, this book emphasizes, through its focus on actors and networks, the importance of other aspects of telegraphy, such as inter- and nongovernmental modes of regulation and coordination, transboundary processes of scientific and business exchanges, and alternative notions of identity formation beyond and outside of the primarily Euro-American nation-state and empire, respectively.[34] Finally, the global history approach also reminds us to analyze networks of connections against the backdrop of their disconnections and to see users in relation to nonusers.[35]

A group of people holding on to the submarine cable are at the center of Weingärtner's 1858 lithograph, "Torchlight Procession." Similarly, the individual actors of submarine telegraphy holding on to the structures shaping their world stand at the heart of this study. Yet, while the New York artist depicted a "parade" of people including Euro-Americans, Asians, Africans, and Native Americans, in reality, the actors actually constituting and contesting these cable networks formed a much more limited group. They were mainly white, male, and from the middle class. During previous centuries, the aristocracy had primarily shaped the logic and structures of politics, economy, and culture on a grand scale. Now a new, albeit very diverse, Euro-American middle class constituted the prime movers behind the globalization processes so characteristic of the nineteenth century.[36] Cyrus W. Field's portrait on New Yorker's hatbands epitomized this class. Field, the son of a clergyman from Massachusetts had passed up his chance to go to college for a career in business. At age fifteen, he started out as an errand boy for one of New York's leading dry goods stores. Quickly, Field was promoted to senior clerk and soon moved on to the paper trade, where he made a fortune within only one decade. Barely thirty, Field was one of New York's richest men before entering the telegraph business.[37] Field's British business partners, such as John Pender and Richard Glass, were successful merchants emerging from Britain's industrial middle class before they ventured into the business of submarine telegraphy. Similar to other infrastructural technologies of the time, ocean telegraphy opened up spaces for maneuvering and assigned agency to those middle-class businessmen, such as Cecil Rhodes and Ferdinand de Lesseps, in shaping the course of nineteenth-century globalization processes.

Within a matter of years after 1866, this group of cable entrepreneurs of globalization became an exclusive club. One of the quintessential legacies of the first Atlantic cable is that it not only set in motion the submarine telegraph machinery, but also formed a closed cable community. I call this group of gatekeepers the *Class of 1866*. They used concepts such as gender, class, and professionalism in addition to cosmopolitanism and Eurocentrism to define eligibility to their group and the world of telegraphy.[38] Members of this group of entrepreneurs, electricians and engineers, mariners, journalists, and cable operators dominated the scenes of the global media system until the outbreak of World War I and, in some respects, even beyond. For instance, the cotton merchant turned cable magnate John Pender, the Scottish Lord William Montague-Hay, and James Anderson, the former captain of the (in)famous cable ship, *Great Eastern*, managed to build up an empire of ocean cables. The global monopoly of their conglomerate, the Eastern and Associated Companies, lasted until 1934 when the submarine cable business gave way to technological and market forces and merged with the wireless business.

In their function as the system's gatekeepers, the Class of 1866 profited immensely from their fame as telegraphic conquerors of the Atlantic—a deliberate manipulation of public memory—as well as the business networks they had built up during the Great Atlantic Cable project. These cable actors defined not only the layout and structure of the early global media system but also the purpose and extent of global connectivity. It was their notion of *Weltcommunication*, to use a contemporary term, most notably expressed in the words of James Anderson as communication between the wealthy few, that determined the scope of global connectivity.[39]

This does not mean, however, that the Class of 1866 and their understanding of *Weltcommunication* formed a coherent representation of "Western" modernity. These Euro-American men were not a homogenous group and may not lightly be equated with a universally applicable construction of "the West" as the imperial-capitalist system of the industrial Euro-American states.[40] Socially and culturally, New York high society member Field and Scotsman Pender came from very different backgrounds. While Field was deeply religious and believed in the missionary and civilizing qualities of the ocean telegraphs, Pender, as a student of Manchester Liberalism, merely saw them as economic investment, for which, and here he agreed with Field, a concept of "universal peace" was extremely beneficial. Defining the "West" as a coherent political-economic system reduces global entanglements to a unidirectional exploitation of commodities and resources. It disregards not only social and cultural aspects but also the fact that industrial Euro-America equally adapted and responded to its entanglement with "the rest."[41] More than once, the cable entrepreneurs, engineers, and operators had to grapple with non-Western actors, cultures, and geographies surrounding their cable stations.

Historians can challenge the structural approach of a one-dimensional and linear process of modernization not just by looking at cases of friction between the West and the rest, of global fragmentation, and in extra-European regions. Rather than relabeling colonial history as global history, a study that primarily deals with the Euro-American world can also rewrite and reclaim a history of modernity in its multiplicity.[42] Until the late nineteenth century, and arguably also beyond, there was no homogeneous civilizational community of Euro-America.[43] Important cultural differences existed between the American, British, Irish, French, Canadian, Australian, and German actors involved in the wiring of the world. The first news transmissions across the Atlantic, for instance, failed, because there was no common cultural code to enable both ends of the line to understand telegraphic communications. It proved impossible for the British press to understand the significance of a telegraph message announcing the death of John Van Buren properly, and they complained bitterly about receiving "such a scrap as this" instead of the price of gold from the North American side of the Atlantic cable. John Van Buren was indicative of Anglo-American misunderstandings via telegraphy: he was son of former U.S. President Martin Van Buren to one side of the Atlantic, but insignificant and wasteful jamming of telegraphic space to the other.[44]

Approaching the history of the wiring of the world from the perspective of technology-in-use provides another angle on the plurality of visions connected to submarine telegraphy. In particular, the continuous battles between those connected and those unconnected, on the provider and the user sides, reveal vast power disparities. They illustrate the inherent differences within a Euro-American setting and counter the notion of a homogenous form of modernity. From its very beginning, ocean telegraphy served only an exclusive few. Although by the late 1870s virtually any place on earth could be reached from Europe, the network still left many gaps. Depending on the entrepreneurs' and later also on the governments' decisions about the cables' locations and purposes, many places, such as des Voeux's St. Lucia or rural areas outside of the world's leading commercial centers, remained unconnected. In addition, ocean telegraphy's exorbitant tariffs did not enable social or mass communication. In the end, the vast majority of the globe's population remained unconnected. Submarine telegraphy was by no means a *Victorian Internet*.[45] Henniker Heaton, Australian newspaper publisher, member of the British parliament, and one of this book's protagonists, fought in vain for a generally affordable, "social" cable press in the pre-World War I period. Labeled by the *New York Times* as "unofficial postmaster general of the world," Heaton attempted to secure global communication access "for the millions," on a crusade quite similar to that of human rights organizations today.[46] Finally, the global communication network also supported processes of othering by reconfiguring mental maps of the globe.[47] As Eric Hobsbawm points out, although global newsmakers at the time depended on a "shrinkage of the globe" through the instantaneity of news coverage, their reports, such as the "discovery" of David Livingstone, enforced the notion of the "dark continent" or "far-away" places that lay outside of the Euro-American system.[48]

Disputes on who could use and how to use submarine telegraphy as a global communications system generally ran along the lines of race, class, and gender. Yet, throughout the course of the nineteenth century, the protagonists of 1866 also faced challenges from inside their class in their attempt to dominate and regulate the globe as communicational space. Important figures in this struggle are the law reformers David Dudley Field, brother of the cable entrepreneur Cyrus W. Field, and Louis Renault. Both attempted to structure the global media system according to rules and regulations of *l'esprit d'internationalité,* international law, thereby trying to contribute to the promise of "universal" peace. Other cable entrepreneurs such as the Anglo-German Siemens Brothers, the Americans Jay Gould, John W. Mackay, and James Gordon Bennett, as well as the Australian and Canadian media reformers Henniker Heaton and Sandford Fleming played an important part in the processes of renegotiating global connectivity. Probably *the* key opponent of John Pender was William Siemens, German émigré, engineer, and head of Siemens Brothers London. Carl Wilhelm Siemens, often known under his Anglicized name (Charles) William, had immigrated to Great Britain in 1844 and first worked as a civil engineer before he embarked on telegraphy. By the late 1850s, he was a recognized authority in the field. In the 1870s, he challenged John Pender's established Anglo-American Telegraph Company on the North Atlantic connection setting up his own cable company, the Direct United States Cable Company. The Pender-Siemens antagonism that evolved was not merely a business rivalry but entailed a deeper divide over global communication as an economic system and the place of science in it. Read against the story of Britain's industrial "decline" and the commencement of the American century, it reveals how strongly those global visions also influenced national narratives.[49]

Finally, incorporating a spatial perspective, this book extrapolates the importance of maritime space as the ground of an alternative modernity that challenged existing notions of nationality and territoriality. These actors' identity connected far stronger with their profession and its international scientific networks and their working schemes connected with the logic of emerging global capitalism far more than nationalist interpretations of their history have shown. Almost taking on the form of a "maritime empire," the cable actors' seascape encompassed all of the world's oceans and transgressed national boundaries. Along the technical nerves of the globe, almost similar to the Socratic fable of the Greeks as "frogs around a pond," the telegraph actors constructed a distinct seascape.[50] Networks of family dynasties of telegraph operators such as the

Graves, Mackays, or Perrys ran the various cable stations throughout the world. Moreover, with the expansion of the submarine telegraph network in the nineteenth century, the relatively small group of telegraph engineers, electricians, and operators came to work and travel the world in a way that before was only open to diplomats, military, sailors, or the very rich. Laying or maintaining cables, the engineers and electricians travelled the oceans and used their time during shore leave for sightseeing in Egypt, the Caribbean, or India. Telegraph expert Charles Bright proclaimed that there was "probably no branch of engineering which len[t] itself so readily to a full sight of the world as that of telegraphy."[51]

The cable actors not only travelled the world with unprecedented freedom, but also worked indiscriminately for various different governments. This made ocean telegraphy substantially different from landlines. While the landline networks of telegraphy were run, apart from the United States, by governments, submarine telegraphy followed the logic of private enterprise.[52] Because governments were reluctant and even opposed to granting landing rights to foreign governments, it fell to the private and "neutral" cable companies to mediate between them. When wiring the Caribbean, for instance, British engineer Charles T. Bright, who had received a knighthood for his contribution to the Great Atlantic Cable, worked simultaneously for the Spanish, Dutch, British, and French. Only by employing Bright and his company as neutral mediators could a functioning ocean cable network be established within a maritime space, such as the Caribbean, where multiple governments were involved.

Finally, telegraphic engineering and science were organized transnationally.[53] The London-based Society of Telegraph Engineers, for instance, saw itself as a "cosmopolitan institution." It coordinated, institutionalized, and standardized the exchange and acquisition of knowledge among all telegraph engineers and operators in the world. Engineers and operators carried out their experiments along the ocean cable lines and so within a global maritime laboratory. The cable agents' transnationalism drew from this transboundary logic of submarine telegraphy. It was based on an ideological system where explanatory models of a universally acclaimed cosmopolitanism as well as a state-centered internationalism, at least initially, played an equal role.[54] In fact, the actors' relationship to imperial and national structures and communities is highly complex. On the one hand, they relied heavily on imperial and national structures, particularly of the British Empire. On the other hand, they primarily identified themselves through their profession and followed the necessarily transnational routine of laying and maintaining ocean cables. In the end, they found ways to employ both in the service of their worldwide system in the form of a *strategic* nationalism. As the example of the American John W. Mackay and his Commercial Cable Company shows, they were "in-betweeners" easily adapting to the varying national contexts they were operating in: American in one situation, and British or German in the next.

The Class of 1866, William Siemens and his family, the American media mogul James Gordon Bennett, and his business partner John W. Mackay are some of the more prominent names in this book but not the only ones. As networks only become visible as they emerge from the "in-between" and the intricate relationship between society and technology solely as "technology-in-use," this study extends beyond the inventors and cable heroes of the Great Atlantic Cable.[55] It also considers the shareholders who financed the system, the journalists who used the system, and the media and law reformers who attempted to shape and reform the system. Furthermore, this study also takes into consideration that direct interaction with the technology was not the only way of experiencing its global outreach. Through news agencies, newspapers, and their distribution of news, globality could be experienced by a large group of people.[56] In the end, engineers, entrepreneurs, and telegraph operators as well as media reformers, journalists, and politicians followed their own visions of an electric world in union and attempted to structure its reality accordingly. These diverse visions played themselves out in cable wars, rivalry between science and business, discourses on civilization and universal peace, and almost constant, albeit unsuccessful, challenges to the system's social scope. In sum, those establishing the structures as well as those rallying against it, users and nonusers alike, determined the history of world-spanning communication in the long nineteenth century.

LAYING OUT THE CABLE STORY

Starting with the Great Atlantic Cable enterprise, the book proceeds by topic thematically and relatively chronologically. Each chapter investigates one of the various concepts accredited to submarine telegraphy from the actors' perspective: world economy, world peace, *Weltcommunication* and world news, the global organization and codification of science, and finally, world politics. At the heart of the analysis, there is the Class of 1866, an Anglo-American group of about forty people, as the dominant mover of the cable business. However, in particular after 1880, a small group of additional actors also moved to the forefront. All in all, it is a network of about 300 known people, not counting the many faceless operators, postal administrators, or employees of cable companies, news agencies, or newspapers of whom the sources reveal only rudimentary traces. To support the flow of reading and facilitate identification, a biographical list of the main protagonists is provided in the appendix.

The study ends in 1914. The outbreak of the First World War marked the end of an era in the history of submarine telegraphy and the history of its actors.[57] By 1914, most of the key players among the cable protagonists were dead. James Gordon Bennett, newspaper publisher and cable entrepreneur, who died in 1918, was the last of a distinct generation of actors of globalization. In the decade preceding the war, not only the group of cable actors but also the technology itself experienced a distinct shift. From 1902 to 1904, the wiring of the world was ostensibly completed with two cables across the Pacific, which were eventually taken "round about the earth." Also, in 1901, Guglielmo Marconi successfully transmitted his first wireless transatlantic message. A decade later, wireless had established itself as a serious competitor to the ocean cables for global communication.[58]

Chapter 1, "Networking the Atlantic," focuses on the emergence of the Class of 1866 as gatekeepers of the global media system by retelling the story of the Great Atlantic Cable project (1854–1866) from their perspective. It shows how their transatlantic network formed, drawing on preexisting local structures, such as the cable entrepreneurs of the British national telegraph system, the group of American expatriates in London, and the elite circle of New Yorkers residing at Gramercy Park. In fact, these preexisting social networks were vital for the success of transnational infrastructure projects as neither the lone entrepreneur, like Cyrus W. Field, nor ingenious inventors and engineers, like Samuel F. B. Morse or Charles T. Bright, can be solely credited for directing globalization processes. Finally, by forging a clever and not always accurate cable memory culture, the Class of 1866 not only managed to remain the dominant force in the field of global telegraphy and dictate the contemporary cable discourse, but also, in many respects, have continued to do so until the present day.

Chapter 2, "The Battle for Cable Supremacy," examines the economic aspects of the business of submarine telegraphy with a strong focus on the 1870s when crucial structures and trends emerged. Key actors in this chapter are John Pender, James Anderson, and their Eastern and Associated Cable Companies, as well as the Siemens clan and their two cable manufacturing and operating companies, Siemens Brothers and the Direct United States Cable Company. When, in the 1870s, both entered into a fierce "cable war" over the monopoly on the Atlantic cable market, their conflict expanded beyond business rivalry into a controversy concerning the very structure of their business system. While Pender argued for a system of monopoly aiming at capital accumulation, Siemens stood for a system of competition allowing for technological progress. The nucleus of their argument has been reproduced in a larger picture with the decade-long debate among historians on the British industrial "decline" and the emergence of a multicentered economic world in the late nineteenth century. It further touches on the divisions between science and technology and a dispute on the place of the engineer in cable business. In its last part, the chapter offers an analysis of shareholder lists and gives detailed insights into the financial network structure of ocean telegraphy. From the 1870s on, the increasing popularization, affordability, and nationalization of ocean cable stock accompanied the shift from big business to small investors. Finally, the analysis sheds fresh light on the New Woman of the late nineteenth century as well as women as actors within ocean telegraphy.[59]

Chapter 3, "The Imagined Globe," explores the power relations entailed in technological progress as expressed in notions of an electric union, universal peace, and Euro-America's civilizing mission. Initially, different national and local contexts, such as the influence of Manchester Liberalism on the British actors as well as the notion of a *Societas Christiana* for the American actors, played an equally important role in the construction of concepts of an "electric union" and an imagined globe. The appearance of David Dudley Field and Louis Renault, French jurist and Nobel Peace Prize winner of 1907, as key codifiers of international law in the 1870s, however, marks the transition to the institutionalization of a third explanatory model for "universal peace," *l'esprit d'internationalité*. Finally, the chapter suggests that the idea of a unified market of morality via means of global communication encompassed not only the notion of universal peace but also the concept of "civilization." The demarcation line between the two ran not only between the "West" and the "rest" but sometimes also right through it.

Chapter 4, "*Weltcommunication*," deconstructs the notion of the telegraph as a medium of mass communication. The ocean telegraphs nourished not only a "time-space compression," but also a division of communicational space between those who could pay and those who could not.[60] Exploring the cable agents' theory of communication, the chapter shows that due to the actors' strong emphasis on revenues, the new *Weltcommunication* was from its beginning based on geopolitical structures of economic and political interests. Nevertheless, throughout the long nineteenth century, we also find repeated attempts to challenge such a materiality of global communication, first in the form of news agents in the 1880s and, later on, in Henniker Heaton's campaign for a social tariff. Finally, drawing attention to James Gordon Bennett, owner of the *New York Herald* and the Commercial Cable Company, the chapter explores yet another systemic shift in the organization of the global communication system: a market advantage was no longer obtained via control over vertical business networks concerning the manufacture, laying, and operating of cables, but via the concerted organization of the cables and their content. Bennett's fight against Jay Gould and Western Union for the independence of the Associated Press further illustrates the influence of local systems and forces on global systems. What has so far been overlooked in the literature dealing with the Commercial Cable Company and the Atlantic pool is that their conflict was as much a cable war as it was a news war.

Chapter 5, "The Professionalization of the Telegraph Engineer," focuses on the engineers, electricians, and operators and the Society of Telegraph Engineers, of which many of the Class of 1866 were members. Relating the history of the society as a cosmopolitan institution and its slow demise as a global authority, the chapter accounts for the interplay of globalization and nationalization processes. The example of the society illustrates not only a transition from transboundary cosmopolitanism to internationalism, but also the diversification of science from a London-centrism to a multicentered scientific globe. The analysis further points to the importance of the global dimension for telegraphic science and research. The society's history in its relation to the cable stations represents the interplay of local and global forces, or rather of central versus local control. In this dispute, the underlying distinction between scientific philosophy (the society) and *experientia docet* (the cable stations) related to questions of authority and discursive hegemony. This eventually furthered the notion of ocean telegraphy as a technology in stagnation.

Chapter 6, "Cable Diplomacy and Imperial Control," analyzes the interplay of globalization and nationalization processes and explores the highly complex and continuously evolving relationship between governmental representations and cable agents. As part of their job and as a side effect of landing-right regulations, the cable agents took on the role as "neutral" elements between the nations at the respective cable end. They were cable diplomats, translating and mediating between various national agendas. Events of 1898, however, with the Spanish-American War and the Fashoda Crisis, exposed as a delusion the cable agents' belief in cables solely as a means of commerce and sounded the death knell for cable neutrality. The growing world economy and the development of competing industrial powers besides Great Britain, a growing pluralism in the ocean cable system, and the period of new imperialism additionally complicated the new situation for the cable agents in the period prior to World War I. The cable actors' response to this paradigmatic shift was the employment of nationalism as a strategy: officially they tied themselves ever closer to one particular national discourse, thereby disguising and thus protecting the international financial and working setup of their company.

There is probably no image more fitting than Weingärtner's lithograph "Torchlight Procession" to capture the hopes and excitement created by the idea of global communication by means of submarine telegraphs. The painter vividly expressed one of the many visions contemporaries harbored of a rapidly changing world, girdled by a net of wires and synchronized to the beats of dots and dashes: "universal" peace, world trade, global politics, and *Weltcommunication*. Form and content of these concepts, however, varied widely, and the actors involved in the wiring of the world disputed them continuously over the course of the nineteenth century. Engineers, entrepreneurs, journalists, media reformers, politicians, and financiers harbored very distinct visions of a world in electric union as they were influenced by their respective individual, regional, national, or international structures. In the end, cultural, socioeconomic, political, and technological structures played important roles in the establishment, coordination, and regulation of the global communications system based on submarine telegraphs. Looking at the wiring of the world from the actor networks' perspective extrapolates the myriad of meanings globalization processes could take on.

1 | NETWORKING THE ATLANTIC

It was while thus studying the globe that the idea first occurred to him [Cyrus Field], that the telegraph might be carried further still, and be made to span the Atlantic Ocean.

Henry Martyn Field, *History of the Atlantic Telegraph*, 26.[1]

WITHIN THE grand narrative of the Great Atlantic Cable, this scene is one of the most oft-repeated and the picture of the lonely, yet ingenious entrepreneur one of the most powerful: it was the beginning of 1854 and the American businessman and millionaire Cyrus W. Field was standing in his library studying a globe. He had just received a nighttime visit from the Scottish-Canadian engineer Frederick N. Gisborne, who was involved in a telegraph scheme along the North American east coast. Gisborne's idea was to speed up transatlantic communication by tapping steamers coming from London at St. John's, Newfoundland, and pass messages between that point and commercial places like Boston, New York, and Philadelphia via telegraph. After a few miles of cable had been laid and the route for Newfoundland had been surveyed, work stopped for want of funds. Facing bankruptcy, Gisborne attempted to win Field as an investor.[2] The practical proximity of Newfoundland and Ireland of roughly 1,600 miles, however, impelled Field to think of a different idea: why be content with shortening communication with Europe by a mere day or two by relays of boats and carrier pigeons if one could go all the way *across* the ocean?[3]

This image of Cyrus W. Field standing by his globe and mentally reorganizing the structure of the world's networks of communication usually sets the stage for a master narrative of the wiring of the Atlantic between

1854 and 1866: Field's "epic struggle" that would require "a decade of effort, millions of dollars in capital, the solution of innumerable technological problems . . . and uncommon physical, financial, and intellectual courage."[4] However, "Cyrus the Great," as the London *Times* journalist William H. Russell later named him, was neither the first nor the only one to think of an Atlantic telegraph.[5] At the same time, four or five different schemes were seriously discussed and two undertaken. Only Field's was brought to a successful completion. Moreover, Field was not the sole and center figure of the Atlantic cable project. Notwithstanding Field's immense influence, the project's success depended on the development and cooperation of an Anglo-American group of entrepreneurs, engineers, financiers, mariners, and lawyers that later formed the network of the Class of 1866.

The wiring of the Atlantic is a story of a series of failures and a number of different undertakings. The cable is the first link in a global information and communications network that nourished the acceleration and multiplication of transnational and transcontinental interactions as one of the main characteristics of the nineteenth century.[6] Yet, the wiring of the Atlantic is also and foremost the story of its actors, denoting a crucial moment in telegraph history from which a group of white, Euro-American, middle-class entrepreneurs formed that influenced the course of globalization. Men like the seaman James Anderson, the cable manufacturer Richard A. Glass, and the merchant John Pender evolved from the project as "Don Quixotes" that had persevered, as "Cable King[s]" of the future.[7] They represented the rising middle classes of the nineteenth century, which enjoyed the benefits of a capitalist economy and an imperial Euro-American setting.[8] In their "telegraphic network" of the Class of 1866, they became enablers of the global communication system. As the system's gatekeepers, they defined its structure and geography and, in consequence, partook in shaping the logic of globalization. Over the years, only a select number of actors, such as William Siemens and John W. Mackay, managed to enter their circle. The sole success of all those attempting to "conquer" the Atlantic by cable helped to create a zeitgeist focused on engineering world projects and a belief in mankind's mastery (mankind defined as white, male, and Western) of nature and the West's supremacy over the rest of the world. It was a time when "myths [were] every day becoming realities" and the "apparent extravagancies of Utopians" turned into "realized dreams."[9]

LAYING THE GREAT ATLANTIC CABLE

In 1854, Cyrus W. Field was not the first to entertain the idea of a telegraph cable across the Atlantic, but the 1850s were exploding with ideas of a telegraph cable between Europe and North America. As early as 1850, even before the first commercial submarine cable of the Brett brothers became a lasting success, the British *Spectatory*, and similarly the French *Journal du Calais*, suggested the "most audacious speculation," namely to "extend [telegraphic] communication to America."[10] In Europe, the 1850s started off with an economic boom, and interest in telegraphy was not coincidental. British exports never grew more rapidly than at the beginning of the 1850s, and iron exports from Belgium more than doubled. In Prussia, the number of joint-stock companies jumped from 67 to 172 between 1851 and 1857. Politically, governments that had been shaken by the revolutions of 1848 gained time for recovery. Additionally, "new rituals of self-congratulation," the Great International Exhibitions of 1851 and 1853 in London and New York, punctuated the era, each of them "a princely monument to wealth and technical progress."[11] The first successful experiences with telegraphic connections across land and sea and the extent of the political, economic, social, and cultural ties that Europe had with North America soon turned the crossing of the Atlantic into a key topos. Proposals came from the scientific communities as well as the general public. From 1859 onward, the American Colonel T. P. Shaffner pushed for his scheme of a northern route, via the north of Scotland, the Faroe Islands, south Iceland, the southern tip of Greenland, and Labrador.[12] He was taking up an earlier idea that the British geographer James Wyld had proposed to the Danish government in 1852.[13]

At about the same time, two further enterprises suggested a southern route across the Atlantic: The South Atlantic Telegraph venture proposed a cable between the south of Spain and the coast of Brazil, making stops at various islands in between. The other project preferred a route from Portugal via the Azores to the southern states of North America. Both found little favor.[14] Lastly, in 1854 the American telegraph provider, Western Union, broached the idea of establishing telegraphic communication between Europe and America by means of land wires through Siberia and Alaska. In 1861, Western Union finished the first transcontinental telegraph line across the United States connecting Washington, DC, with San Francisco. This marked a grand success in the spirit of American westward expansion, and it allowed many of the roughly 350,000 Americans who migrated beyond the Mississippi between 1840 and 1867 to communicate more directly and quickly with the eastern United States.[15] The undertaking also signified the rivalry between Cyrus W. Field and Hiram Sibley, head of Western Union, as well as two alternative visions of America's future, one orienting itself toward Europe and the "Old World" and the other toward the promise of the West.[16] Work started on the Russian–American Telegraph, also called the Collins Overland Line, in 1865. It was intended as a straight line from St. Louis, Missouri, to St. Petersburg, Russia, and was supported by Samuel B. Ruggles of the New York Chamber of Commerce and Samuel F. B. Morse, one of the inventors of an electric telegraph.[17] As an integral part of the formation of the "American Empire," the project exemplified U.S. American frontier spirit and its westward course and helped put forward a convincing case for the Alaska Purchase in 1867. Furthermore, the Collins Overland Line was conceptualized as part of an even larger network scheme that not only would link the United States with the western part of North America and Russia, but also would include two additional lines to connect the United States with the coastal cities of China and with Central and South America. Collins's vision was "to link three continents with one continuous telegraph line."[18] Nevertheless, in February 1867, about half a year after the Atlantic success, Western Union abandoned the project. At the time, Western Union did not think that there would be enough traffic going from the United States to Europe to make up for any further investment in a line that would be slower and more likely to face interruption on its route to Europe than the transatlantic cable.[19]

Although Field was neither the first nor the only one to embark upon the project of wiring the world, his group was the first to execute the plan successfully. Knowing "nothing about telegraphy," Field still recognized the great potential of an Atlantic cable.[20] Following his interview with Gisborne, Field made inquiries with two leading American experts, Matthew F. Maury and his later Gramercy Park neighbor, Samuel F. B. Morse.[21] Maury informed Field about his oceanographic findings later published in *The Physical Geography of the Sea*, namely the existence of what he called the Telegraphic or Atlantic Plateau: a strip of almost level seabed between Cape Race in Newfoundland and Cape Clear in Ireland, which is nowhere more than 10,000 feet deep and "protected from the abrading action of [the Atlantic's] currents and the violence of its waves by cushions of still water."[22] Not only all Atlantic cables of the time but also today's fiber-optic cables follow this route along Maury's mythical plateau, "which seem[ed] to have been placed there especially for the purpose of holding the wires of a submarine telegraph, and of keeping them out of harms' way."[23] Morse showed himself as interested in and supportive of the project as Maury, advising Field on the business and laws of electrical telegraphy. Already in 1843, Morse had foreseen a time when an Atlantic telegraph project would be realized. When the offer came to work as honorary electrician for Field's undertaking, he accepted it gladly.[24]

On March 10, 1854, the "cable cabinet," which included Field and four other American gentlemen of fortune from Field's Gramercy Park neighborhood, founded the New York, Newfoundland and London Telegraph Company to establish telegraphic communication between Europe and America.[25] With its incorporation, the company received a fifty-year monopoly on landing rights in Newfoundland. David Dudley Field, Cyrus W. Field's older brother and legal advisor to the company, was far-sighted enough to anticipate that the question of landing rights would be crucial in securing the company's role in the future. Another company, the American Telegraph Company, was launched in 1855 to operate the terrestrial lines along the American east coast. It later merged with Western Union. In 1856, Cyrus W. Field and Samuel F. B. Morse left for Great Britain, where they found support among some of the most eminent telegraph engineers and scientists of the time, such as John W. Brett, William Thomson, and Charles T. Bright, as well as American expatriates in London.[26] Jointly with the telegraph entrepreneur John W. Brett and the engineer Charles T. Bright, Field established the Atlantic Telegraph Company in September 1856.[27] Its purpose was to secure British money for the undertaking, since the New York, Newfoundland and London Telegraph Company had been incorporated in the United States.[28]

Similar to his approach in the United States, Field tapped into already existing networks of cable entrepreneurs and circles of financiers in London, predominantly by introduction through John W. Brett.[29] Together with his brother Jacob, John Brett had gained fame in 1850/51 when they laid the first commercial submarine telegraph cable across the Strait of Dover; now Brett became Field's badly needed British advocate and sponsor. From Brett, whom Field had met through Gisborne, Field gained access to people essential for his undertaking. This included not only those financially or politically interested in a cable, such as Foreign Secretary Lord Clarendon or Secretary to the Treasury James Wilson, but also those able to manufacture, lay, and operate a cable of such a length. At the time, London was the only place where all these things could be had. One key contact was Richard A. Glass, owner of the cable manufacturing firm Glass, Elliot & Co., which had already supplied the Brett brothers' Dover-Calais cable of 1851. Moreover, Brett's Magnetic Telegraph Company and its respective shareholders "presented a presold market and the Magnetic Company's offices in London, Glasgow, Manchester, and Liverpool provided outlets for the sale of stock." In fact, the majority of the capital for the Atlantic Telegraph Company came from investors from the Magnetic, and one of its directors, John Pender, a merchant from Manchester, would later on play a decisive role in the entire cable business.[30] Finally, Brett also accompanied Field on travels to Manchester and Liverpool to address their Chambers of Commerce.[31] Other key contacts for Field were American expatriates such as Curtis M. Lampson, George Peabody, and Junius Spencer Morgan, as well as the American Chamber of Commerce of Liverpool. The Anglo-American fur trader Curtis Lampson had moved to London in 1830 as agent for John Jacob Astor. Later, he established his own business of C.M. Lampson & Co. From 1857 on, he served as director of the Atlantic Telegraph Company.[32] George Peabody and Junius S. Morgan, two American investment bankers in London, were among the larger shareholders and, during the 1860s, secured the continuation of the undertaking. George Peabody additionally served as one of the Atlantic Telegraph Company's directors. With their banking house Peabody, Morgan & Co., which in 1864 was renamed J.S. Morgan & Co., they played a large role in mobilizing European capital for American economic development.[33]

While in Great Britain the key to money and cable was an individual, John W. Brett, in the United States it was a place: Gramercy Park. In 1851, as a statement of his wealth and success, Field moved his family to Gramercy Park. This well-to-do neighborhood in downtown Manhattan was established in the 1840s, after the aforementioned Samuel B. Ruggles of the New York Chamber of Commerce had bought up twenty-two acres of swamp and farmland for real estate development. Ruggles had the land drained and set up a London-like square to appeal to the wealthiest residents of the city. They would obtain, by buying up one of the sixty-six lots, exclusive access to their private park through a golden key.[34]

At Gramercy Park, Field lived in immediate proximity to some of New York's most prominent citizens, such as the politician Samuel J. Tilden, the industrialist, inventor, and philanthropist Peter Cooper, the publisher James Harper, and the writer and diarist George Templeton Strong.[35] Some of his most important allies, such as Peter Cooper and Cooper's son-in-law, Abram Hewitt, and Cyrus W. Field's brother David Dudley Field, also lived at Gramercy Park.[36] In the United States, Field engaged his two most immediate networks in the enterprise: his family and his neighborhood. One of Field's most important allies was his brother, David, who was by that time already a renowned lawyer, law reformer, and codifier of law. David advised his brother on legal questions and during the negotiations with the Newfoundland government concerning landing rights. Matthew Field, an engineer and bridge builder, took on the work on the land lines through Newfoundland, and Henry Martyn Field, a travel writer and editor, authored the undertaking's history in 1866. He helped to spread the cable's fame and make his brother Cyrus unforgettable. During this early stage, personal relations, friendship, and family ties in the form of social capital played a central role in the setup

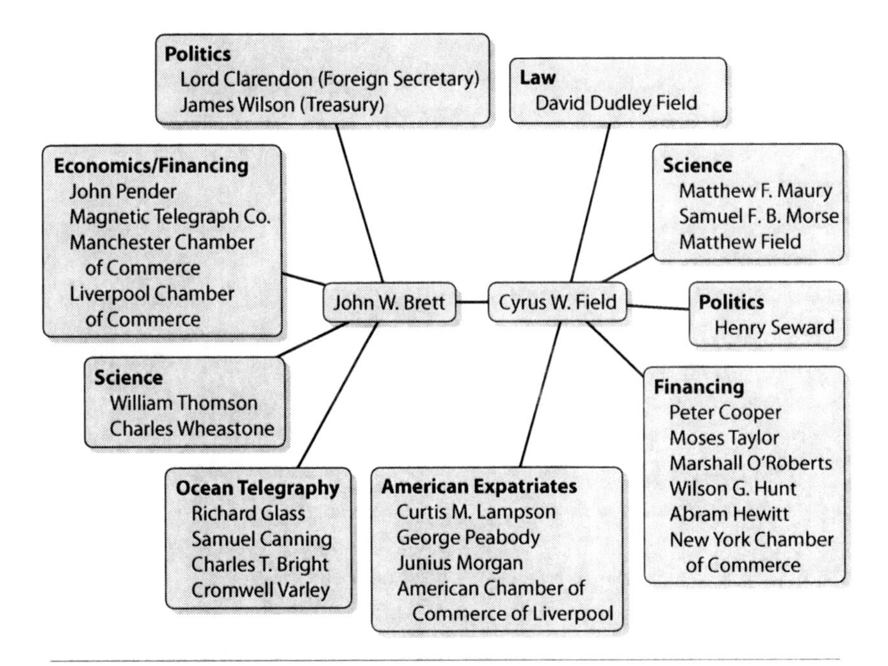

1.1 Simone M. Müller, "Social Networks and the Great Atlantic Cable."

of the schemes.[37] Of particular importance was the network's reach into all the different professional fields with influence on laying a submarine cable. Through their Anglo-American linkages, Field and Brett had connections into the fields of politics, economics and financing, science and engineering, and law.

While in 1856 Cyrus W. Field was on a promotion tour in Europe, his brother Matthew oversaw work on the land lines in Newfoundland. Together with the British engineer Samuel Canning of the cable manufacturer Glass, Elliot & Co., they worked their way through the wilderness of Newfoundland in a manner that caused the enterprise's biographer to comment that "[t]here is nothing in the world easier than to build a line of railroad, or of telegraph, *on paper*."[38] The 400-mile route went through desolate and predominantly uninhabited land with thick forests, rocky gorges, and swamps. In 1856, when the St. Lawrence cable sealed the telegraphic connection between New York and St. John's, Newfoundland, the entrepreneurs celebrated it like a battle won, as nature had had to yield to "man's unconquerable will."[39] The land connection established that the ocean telegraph was next. In 1857, the first attempt to cross the Atlantic via telegraph failed a mere 300 miles out in the ocean.[40] After two further failures, the Atlantic Telegraph Company successfully established a telegraphic connection between Ireland and Newfoundland in 1858. With a telegram of ninety-eight words, sent from Queen Victoria to the American President James Buchanan, the line was opened and the world seemed to step into a new era in global communication.[41] Unfortunately, the 1858 Atlantic cable never worked properly and failed entirely on September 1, after only 27 days.[42] As great as the first excitement over the newly established Atlantic cable had been during this summer, so profound was the disappointment over its failure.[43] After the cable's breakdown, the American Civil War dominated the scene in the United States, not only separating former cable alliances along the lines of the national North–South conflict, but also halting the project. In the meantime, people in Great Britain were more concerned with the Suez Canal, Darwin's *Origin of Species*, the London World Exhibition of 1862, and the opening of the first underground subway.[44] Although Field "compassed land and sea incessantly" for new cable possibilities, the scheme was not on the agenda of any serious investor or the government for the duration of the American Civil War. When in 1862 he toured the United States and Great Britain attempting to win new investors, he was barely successful.[45]

ON BOARD THE *GREAT EASTERN*

The story of the Class of 1866 takes off after the American Civil War and is closely connected to "the largest vessel in the world," the *Great Eastern*—a "white elephant" steam ship of gigantic size that left more than one of its owners bereft of wealth and sanity.[46] In the history of the wiring of the world, the *Great Eastern* serves as symbol and metaphor. Its enormity is symbolic, not only of the grandeur of the undertaking but also of the scale of engineering projects in the nineteenth century. These oriented themselves on the earth's surfaces and measured their success through size and magnitude, thereby underlining the projects' assumed global importance, on top of notions of Western supremacy and mankind's conquest of nature. Other examples of the time are the Suez Canal (1854/56–1869), the Gotthard Tunnel through the Alps (1863/69–1882), transcontinental telegraph projects such as Western Union's first transcontinental telegraph line (1861), the U.S. American Overland Route, a stagecoach and wagon road to the West Coast (1863–1869), the Canadian Pacific Railway (1881–1885), and the recurring plans for a tunnel underneath the English Channel.[47]

The ship is metaphoric in the sense that it represents a kind of outer space, in which forces came to play to forge the Class of 1866. The actors' "adventure" at mid-Atlantic served as an initiatory rite through which those partaking in it established themselves as full members of the globe's governing elite. Through their engineering success, the Irish and Scottish among them, such as Captain Robert C. Halpin, challenged the traditional markers of social belonging and moved into a growing Victorian upper-middle class.[48] The British members of the Class of 1866, as similarly the Allsopps and Guinnesses in brewing, Cunliffe-Listers in textiles, or Armstrongs in engineering and armaments, represent the changes in the social makeup of the governing elites since the 1750s and the ascent of bourgeois England in the nineteenth century connected to the increasing pace of industrialization, with its railroads, steamships, and canals. In the 1880s, the collective wealth amassed by the middle classes for the first time exceeded that of the upper class. Simultaneously, the conventional distinction between the aristocracies of birth and money and the oligarchies of manufacture and land were increasingly obliterated.[49] For centuries the status of those in power had been determined by a tiny "landed, hereditary, wealthy and leisured" elite; they now allowed access to a group of industrialists, engineers, and financiers.[50]

As chroniclers of the undertaking pointed out, the American Civil War, although it halted the enterprise of establishing an Atlantic cable, also served as a catalyst: it "stimulated capitalists to renew the attempt," adhering to their desire to avoid further misunderstandings between Britain and the United States.[51] In 1864, a new cable was ordered, financed predominantly by contractors' shares. In addition, the cable manufacturing company Glass, Elliot & Co. accepted Atlantic shares as part of their payment.[52] In June 1865, upon completion of the cable and the refitting of the *Great Eastern*, the contractors embarked on their third attempt to lay an Atlantic telegraph cable. Besides the enormous length of the Atlantic cable, the ship took on board about seven to eight thousand tons of coal, in addition to the approximately five hundred men on board.[53] They sailed from London to Valentia, Ireland, and then started laying the cable across the Atlantic to Heart's Content, Newfoundland. On board were entrepreneurs, financiers, engineers, journalists, naval officers, and mariners, as well as political figures and social starlets, who took passage for that part of the journey. Upon its departure, the ship resembled, in the words of French novelist Jules Verne, "a floating city."[54] The sight of the *Great Eastern*, with "10 bullocks, 1 milk cow, 114 sheep, 20 pigs, 29 geese, 14 turkeys and 500 other fowl all housed on the upper deck," also intrigued the onboard correspondent of the London *Times*, William H. Russell, who associated her with Noah's Ark or, as Henry Martyn Field wrote, "some large farm yard of England."[55] The names attributed to the great ship, ranging from a biblical connotation to that of a floating center of urbanity, all underline her assumed grandeur, her monstrosity, and the project's implied importance and its large-scale effects.

The *Great Eastern* was an incomparable spectacle, perfectly representing the zeitgeist of its engineering age. Upon her launch in 1858, the *Great Eastern* was 20,000 tons larger than any other ship and defended that title until she was scrapped in 1888. Originally designed to be a passenger steamer, the *Great Eastern* had been fitted to accommodate 800 first-class, 1,500 second-class, and 2,500 third-class passengers, or 4,800 passengers in all, or—if employed in the transport of troops—upward of 10,000 men, in addition to a crew of 400. Before being remodeled as a cable ship, she originally contained several compartments; five of them near the center formed five complete hotels for passengers, each comprising upper and lower salons, bedrooms, bar, and offices.[56] However, the steamer also had a "fatal attraction for disaster": it "killed its designer, drowned her first captain and ended as a floating circus."[57] With her size and dramatic history, the ship not only underlined the Atlantic telegraph's claim of importance, but also represented an essential part of its mythology. The steamer is just as well known as the cable itself, and the *Great Eastern* will always be remembered as the great ship that laid the Great Atlantic Cable. Both fed into each other, and neither would have been a success without the other.

Similar to that of the Great Atlantic Cable, the story of the *Great Eastern* is one of failures, a saga on the brinks of "becoming an epic disaster."[58] Its original owner, the Eastern Steam Navigation Co., had by 1857 amassed £90,000 in debt. There was talk of putting the ship up for auction or dismantling it before it had ever been at sea.[59] In 1860, after the formation of a new company, the Great Ship Company, the *Great Eastern* was set afloat as a passenger steamer, but never successfully. In 1863 she was laid up and offered for auction. Daniel Gooch, one of the Atlantic cable contractors who had also been on the board of the Great Ship Company, bought the *Great Eastern* together with Thomas Brassey, a railroad engineer, and a Mr. William Barber.[60] They fitted her as a cable ship and leased her to the Telegraph Construction and Maintenance Company (Telcon for short), the contractor of the 1865 Atlantic cable, of which both Gooch and Brassey were co-directors.[61] This purchase was one of the deciding factors in determining the future success of the Class of 1866 in the cable-laying business. It brought the operations of laying and operating cables into one hand. From the beginning, opponents had criticized the *Great Eastern* as a monstrosity, warning that it would take forever to fill her with people or cargo, that she would swamp markets, and that no market could be big enough to afford a return freight.[62] Now, it was the steamer's size that recommended her for the cable-laying task as she was the only ship in the world big enough to hold the transatlantic telegraph cable. Finally, fate would prove the renowned British engineer, Isambard Kingdom Brunel, right when he told Field: "There is the ship to lay the Atlantic cable!"[63]

The dual constellation of ship and cable was completed by a third component, a particular group of board members, the Class of 1866. Together, these three make up the basic ingredients of the various sagas of the wiring of the Atlantic. There is little information on the crew that worked in the tanks coiling the cable or that navigated the ship. According to company papers, many of them were Irish, hiring on at Valentia, the landing point of the cable on the European side. It seems that some used it as an opportunity for a cheap crossing to the United States or Canada. Their connection with the submarine cable business was short-lived and remained a one-time experience.[64] Answers are easier to come by for the actors' social, political, or cultural background when dealing with the protagonists of the cable project—the entrepreneurs, engineers, and electricians: they were the whiz kids of their time as well as representatives of an emerging, well-educated middle class pushing to the forefront of societal, political, and economic processes in the nineteenth century.

When embarking on the Atlantic cable undertaking, most protagonists already knew each other. In the 1850s, the field of engineering in Britain was a small world, and in particular, the electricians and engineers shared a certain career pattern. They belonged to the schools of Stephenson, Brunel, and Faraday and lived by a strong belief "in the marvelous achievements yet to be wrought by human invention."[65] Some had gathered experiences in the railroad business before they turned to telegraphy. Samuel Canning, who was assistant to Charles T. Bright in 1857 and Telcon's engineer-in-chief from 1865 on, and Daniel Gooch, a railway engineer, director of the 1865-established Anglo-American Telegraph Company, Telcon, and director of the Great Eastern Steamship Company, had both worked for Great Western Railways.[66] Others, such as Cromwell F. Varley (chief electrician of the Anglo-American Telegraph Company in 1865), Willoughby Smith (chief electrician of the same company in 1866), and Richard Glass (head of the cable manufacturing company Glass, Elliot & Co. and one of the Atlantic Telegraph Company's directors) had been involved with submarine telegraphy from its beginnings in the 1840s and 1850s.[67] Theirs was an even smaller community, essentially circling around the five telegraph companies that existed at the time, the two most important of which were the Electric Telegraph Company (after 1855 and a merger with the newly established International Telegraph Company, officially called the Electric and International Telegraph Company) and the British and Irish Magnetic Telegraph Company.[68] The field was completed by an even smaller number of submarine cable manufacturing companies: Glass, Elliot & Co. (since 1854, previously Kuper & Company), R.S. Newall & Co. (1840), and the Gutta Percha Company. In 1864, Glass, Elliot & Co. and Gutta Percha merged to form the Telegraph Construction and Maintenance Company (Telcon).[69]

In addition, the telegraph engineers were closely connected on a personal level. The network of William Preece, Latimer Clark, and Charles T. Bright in the 1850s and 1860s exemplifies this clearly. Together with his brother Edwin, Latimer Clark was one of the pioneers of submarine telegraphy in Great Britain. By 1858, when both Latimer Clark and Charles T. Bright came to work on the first Atlantic cable, their professional connection was presumably one of long standing, since both worked in a small field with essentially two influential submarine cable companies. Prior to 1858, Clark had worked as an engineer for the Electric and International Telegraph Company. Bright had worked in the same position first for the Electric Telegraph Company (predecessor to the Electric and International Telegraph Company) and from 1852 on for John Brett's Magnetic Telegraph Company. Their relationship was strengthened in 1860 when they entered into partnership as consulting engineers, drawing from the fame of the short-lived success of the first Atlantic cable. In the following years, among various other cable projects, they worked as consulting engineers for the Anglo-American Telegraph Company and its 1865 and 1866 Atlantic cable attempts. Bright and Clark's most enduring achievement arose from their joint proposal in 1861 that the British Association for the Advancement of Science should specify a system of electrical units.[70]

While Bright and Clark represent the interwoven professional relationships, William Preece exemplifies the field's connection to the government. Later on, Preece became one of the most important officials at the General Post Office. Preece was the son of a stockbroker and politician from Wales, who later moved his family to London, where William was educated at King's College School and King's College. In 1852, the Preece family met Latimer Clark. Together with his brother Edwin, Latimer Clark worked as an engineer for the Electric and International Telegraph Company, where both secured a job for Preece on the engineering staff. Aside from Clark, who was later an engineer for the Atlantic Telegraph Company and in 1860 served on the committee appointed by the government to enquire about submarine telegraphy, also on the payment list was the chief electrician of the Atlantic undertaking, Cromwell F. Varley.[71] Later, Preece supervised the telegraphs of the London and South Western Railway and, between 1858 and 1862, also the cables of the Channel Island Telegraph Company. In this position he was supervisor of James Graves, who became superintendent at the Atlantic cable station in Valentia.[72] The connection between William Preece and Latimer Clark grew even closer when Clark married Preece's sister in 1854. In later years, Bright, Clark, Preece, and Varley all served as presidents of the Society of Telegraph Engineers.[73]

Those involved with the Atlantic cable project not only knew each other but in most cases also shared certain characteristics in their social and regional background. None of them came from Britain's gentry, but all shared a middle-class upbringing. They were the sons of stockbrokers, chemists, coastguards, schoolmasters, or inventors. All were fairly well educated. They not only knew how to read and write but also had a more than basic understanding of arithmetic before they were taken on as apprentices to an engineer. Another common trait among many of them was their place of origin. John Pender, Daniel Gooch, Thomas Brassey, William Thomson, and James Anderson were of Scottish or northern English heritage. While this marked them as outsiders to the gentlemen's clubs of London, their entrepreneurial spirit fell in line with that of many other Scotsmen at the time who belonged in disproportionally high numbers to the merchants, soldiers, and missionaries making the British Empire. This strong impetus from Scottish actors drew from Scottish enlightenment and education as much as from the industrial structure "up north."[74] After the kingdoms of Scotland and England had been merged by the Act of Union in 1707 by removing Scotland's independent parliament, Scotland managed to keep some distinctive institutions, such as education, the Church of Scotland, and law. Drawing from the premises of Adam Smith or David Hume, this education system gave its students "a distinctive cast of mind, and an education more progressive" than that available at English institutions of learning.[75] Many of the engineers and scientists who nourished the electric revolution came from Scotland. Furthermore, the Industrial Revolution had left its traces on Scotland, turning Glasgow and Edinburgh into centers of finance and industry.[76]

John Pender, "a Scotchman," as the *Chicago Daily Tribune* found it important to point out in his obituary, first established himself as a textile merchant in Glasgow, before he moved his business to Manchester, at the time the hub of the world's textile industry, in the north of England.[77] Like Pender, many of the original investors were merchants from Liverpool, Manchester, or Glasgow, and early political support for the undertaking predominantly came from Scotland and Britain's north, through

Richard Cobden, John Bright, John Montague-Hay and William Montague-Hay, Marquess of Tweedale, Scotland.[78] This geographical commonality demonstrates two things: the Atlantic cable undertaking was attractive to people of finance and industry situated in the north of the United Kingdom, and it was particularly attractive to entrepreneurs who were not (yet) part of the London establishment. Their position as outsiders, in addition to their capital interests in transatlantic trade, enhanced their willingness to take a great risk such as attempting a telegraph cable across the Atlantic. The Great Atlantic Cable project was, from its British side, a development project of global scale undertaken by northern industrialism and not London's gentlemanly capitalism.[79]

The only outsiders to this established group of British telegraph entrepreneurs, financiers, and engineers consisted of the small group of Americans: Cyrus W. Field and his brothers, Samuel F. B. Morse, Peter Cooper, and Abram Hewitt. The Americans were different not only in their national background but also in their social and professional one. They emerged from a national context in which a broad bourgeois consensus blurred social distinctions and seemingly allowed for greater social mobility. Certainly, Cyrus W. Field, Abram Hewitt, and in particular Peter Cooper, who had been born into very poor circumstances, considered themselves part of a rising middle class and archetypes of the self-made man. Nevertheless, at the time of the cable undertaking, they represented the American elite. In effect, the two national groups played distinct roles. Globalization of the nineteenth century did not progress by all men being able to do all things, but by particular divisions of labor and enterprise.

Finally, the group on board the ship was complemented by members of the Royal Navy, such as Captain Robert Charles Halpin and the British navigator Henry A. Moriarty, who represented the governments' supportive measures. Captain James Anderson, commander of the *Great Eastern*, had formerly been employed by the Cunard Co. steam ship line.[80] All in all, the Class of 1866 was a relatively small group of no more than forty people, and so they remained.

The 1865 expedition was of great media interest. Prior to embarkment, the company was "besieged by applications . . . for permission to accompany the expedition" as newspaper correspondents from Great Britain, France, and the United States attempted to get on board.[81] Because it seemed impossible to satisfy everyone, all correspondents were excluded, except for one official correspondent the company itself provided, William H. Russell. This decision showed a genius for publicity. The man employed "to report events faithfully from day to day" was a famous correspondent of the London *Times*. Previously, he had reported from India and the Crimean War (1853–1856), and through his reports, he initiated a tradition of factual, eye-witness reporting that has remained important to this day.[82] To illustrate the voyage, several artists accompanied Russell, of whom several later wrote accounts for magazines, such as *Blackwood, Cornhill,* or *Macmillan's*. Russell's own account of the expedition is one of the most important primary sources on the laying of the Atlantic cable.[83] In addition to these news reports, various individuals, such as Daniel Gooch, kept and published diaries, and a number of biographies and autobiographies were produced in the expedition's aftermath.[84] Considering the amount of contemporary firsthand accounts published after the expedition, these cable undertakings are probably among the best-documented events in the nineteenth century. This was, however, no coincidence, but a conscious attempt at myth making.

A commonality of all reports is their emphasis on the special atmosphere on board, which created a "small world of its own."[85] On board the *Great Eastern*, a cable-laying ship, in the mid-Atlantic, they were in a peculiar situation: their ship was a space without a particular place, neither here nor there; yet it still represented the particular space where the story of the laying of the Atlantic cable took place. On July 23, 1865, the ship left Valentia, Ireland, and embarked on her trans-Atlantic journey. For the weeks to come, the men on board were thrown together in mid-ocean with minimal to no connection with the spaces of their real world, such as family, friends, and the more general societies of Europe and North America. They were cast off and outside of their traditional time, on their way between two worlds while creating a small world of their own. Yet breaking with traditional time contained yet another layer when one found oneself on board a cable-laying ship. In contrast to every other ship at that time, on board the *Great Eastern* they were not entirely cut off from the world and time they had just left. But for the first time, men were experiencing the simultaneity of times and so breaking with their traditional time. By the very cable they were laying out, the news flow from Europe continued being sent from the cable station at Valentia. It kept them well informed about what was going on in the "outside" world. They arranged to get Greenwich Time given to them every morning through the cable in addition to a daily telegram giving them the general news of the day. In his diary, Gooch remarked that it was "wonderful to get, while in the Atlantic, the news of the morning from *The Times*." Moreover, as they moved further west, they began to receive the news "at breakfast, the same as people living in the next street to the *Times* office."[86]

As a result, they created a space of their own outside of their world, yet it was in congruence with the outer world. Not only did they follow Greenwich Time in their recordings, but they also used their connectivity for the continuation of news flows.[87] In 1865 a ship's first own newspaper was born. In accord with his duties, William H. Russell kept the readers of the *Times* informed of the enterprise's progress on a daily basis. Twice a day, the telegraph at Foilhummerum Bay, Valentia, "spread to all parts of the earth a brief account of the doings of the Great Ship."[88] In return, news from ashore was sent to the ship, which helped the onboard publication of the *Atlantic Telegraph*—a quest taken on by Henry O'Neil, an accompanying painter.[89] During the 1866 expedition, when Willoughby Smith pursued the publication of the *Great Eastern Telegraph* (in addition to the *Test Room Chronicle*, which was published and illustrated by Robert Dudley), 164 messages were sent and received between ship and shore, containing 6,437 words or 30,059 letters.[90] Most of those dots and dashes contained little of news value, yet the repetition of the phrase "Can you read?", asking for confirmation that the signal had been successfully received, secured the unbroken thread with the Old World.[91] In mid-July 1866, about midway between Valentia and Newfoundland, people on board the *Great Eastern* learned about the meeting of the Reform League in Hyde Park, at which they pressed for manhood suffrage and the ballot. The meeting ended in riots. News not only from Europe but from the entire world reached the small cable community via a combination of mail steamer and telegraph cable. They learned about the resignation of the Brazilian cabinet during the Paraguayan War (1864–1870) and about serious Fenian insurrections in Canada. They heard that the King of Prussia, Wilhelm I, had written a long letter to the English Queen informing her about the Battle of Königgrätz and that the opening dinner of the Cobden Club in London, an exclusive gentlemen's club, had been a great success.[92] In fact, the ship's newspapers ensured "that the passengers and crew in the mid-Atlantic [were] actually better informed as to the world they [had] left than half the dwellers in [their] country towns at home."[93]

The newspapers' purpose on board was not only to inform but also to entertain. Painter Henry O'Neil went to great lengths "to cause . . . some fun at lunch."[94] His aim was to fight off some of the monotony for all those who were not directly involved in the paying out of the cable. As Daniel Gooch, director of the Atlantic Telegraph Company as well as the Great Eastern Steamship Company, noted in his diary in early August 1865, it was the mere sight of a ship passing by that created "quite a sensation, as we have not had anything to look at over the mighty expanse of water."[95] Their "amusements of the day," as O'Neil reported in the *Atlantic Telegraph*, often consisted "through the kindness and liberality of the Admiralty" of nothing but "from daylight till dusk – Looking out for the 'Sphinx,'" one of the accompanying ships.[96] O'Neil wrote poems and small, often ironic, commentaries. He also set up a fake auction, where "the property of various gentlemen leaving then present quarters" was up for sale. Among those were aspects of the *Great Eastern* itself—"cards to view apply to Mr. Gooch on board"—or the "Good Will of the Atlantic Telegraph Co.," which was "invisible property in Mr. Field's possession."[97] The highlight of the auction was "some Mile of Telegraphic cable taken from a depth of 2,000 fathom," which, it had been calculated, would "by cutting it into slices of 1/4 inch in thickness sufficient [enough 'coins'] be realized to pay off the entire debt of Great Britain."[98] The auction was planned for August 12, 1865, probably as a diversion from the problem of the break of the Atlantic cable, which had occurred ten days previously. The days since the break had been spent in various attempts to grapple the lost cable before they had decided on August 11, "shattered in hopes as well as in ropes," to return home.[99]

The 1865 expedition, disastrous as it had been financially, served to increase the courage of the promoters and their belief in the probability of an early and complete success. They issued a paper stating the benefits of the expedition and demonstrating their perseverance. Among other points, the list gave credit to the facts that the *Great Eastern* had shown itself to be "the very type of vessel" for the cable work, and although the picking-up gear had proven insufficient, no serious fault could be found with the paying out machinery. Furthermore, they had demonstrated the feasibility of grappling in mid-Atlantic, proving the possibility of recovering a cable at such great depths.[100] In late 1865, in order to overcome financial difficulties and with the object of raising fresh capital, the Atlantic Telegraph Company was amalgamated with the newly incorporated

Anglo-American Telegraph Company.[101] The entrepreneurs' goal for 1866 was not only to lay a cable across the Atlantic but also to recover the 1865 cable to have two functioning telegraph cables.[102]

Despite the breakage of the 1865 cable, the atmosphere and spirit on board the *Great Eastern* in the summer of 1866 was good. Aside from the ship's newspapers, two plays brought new highlights to onboard entertainment. Functioning almost as plays within the play of the Atlantic cable enterprise, they mirrored the group's situation and served their self-reflection. Both *Contentina* and *Being a Cableistic* relate the story of the 1865 and 1866 expeditions.[103] In *Contentina or THE ROPE!! The GRAPNELL!!!! And the YANKEE DOODLE!!!!* written by J. C. Deane and G. V. Poore and illustrated by Robert Dudley, Cyrus seeks his love Contentina who is held captive by her father Neptune—an allegory of the broken 1865 cable that lay unfinished on the bottom of the Atlantic for a year. Yet, as the character Glass in *Being a Cableistic*, which drew on Richard Glass and was acted by Robert Dudley, told Neptune, "by the way, I've called today – We're going to try another lay."[104] Both plays share a happy ending and paint the picture of a group of young and adventurous men defying Neptune himself for the benefit of mankind. For many, and particularly for those on board, the Atlantic cable represented yet one more example of men's triumph over the "opposition of nature" in the form of Neptune's despotism.[105] The media discourse in newspapers and magazines followed the success of the cable in using similar martial vocabulary to draw resemblances to former generals and conquerors.[106] In the play, the character of Neptune represented not only nature but all that was backward, not "modern," and in opposition to the Atlantic cable project. In Neptune's complaints on how the cable "ruined" his parks and "spoiled" his walks, a general discourse on the disadvantages of telegraphy had gone submarine.[107] The agents of the Atlantic cable positioned themselves not only as benefactors of mankind but also as bringers of modernity.

Aside from their liminal situation between two worlds, it was the very act of establishing telegraphic connection between those two worlds that furthered the group-building processes. Although the Atlantic cable enterprise received enormous attention, public as well as political support had continuously dwindled after their first failures in 1857 and 1858. Smaller and larger investors, the newspapers, and the general public had lost faith in the feasibility of the undertaking, and it had become increasingly difficult to find financial backing. After all, in its beginning, submarine telegraphy represented a high-risk and high-cost venture. Due to the two early failures, the investors had literally "sunk their £1000 shares in the ocean."[108] As Henry Moriarty, navigator during all four cable attempts and "leased" to the enterprise from the Royal Navy, explained in retrospect in a letter to the *Standard*, "their scheme ranked in public opinion, only one degree in the scale of absurdity below that of raising a ladder to the moon."[109] The group of adventurers who had come together on board the *Great Eastern* shared a vision of telegraphic communication between the Old World and the New and from then on to "a girdle round about the earth" against all odds and failures. Yet it would be erroneous to view them as "[s]tarving adventurers" who were "ready to embark in any Quixotic attempt" because they had nothing to lose.[110] Quite the contrary; they had poured large parts of their private fortunes into the undertaking, and, seen in relative terms, the enterprise was to a great extent self-financed. These investors were well-established entrepreneurs, such as Richard Glass, Cyrus W. Field, John Pender, Daniel Gooch, and Peter Cooper, who all had their own economic interests in the project.[111] Against all the failings and discouragements they faced, a mixture of faith and fanaticism seemed to have kept them going.

On July 27, 1866, only fourteen days after they had embarked on their expedition in Valentia, the *Great Eastern* and its cable crew reached Heart's Content, Newfoundland, with a functioning telegraph cable. They finally connected Europe and North America through instantaneous communication.[112] On August 9, the telegraphic fleet set out again to grapple and complete the 1865 cable, which they successfully did. By mid-September, two cables were open to the public for trans-Atlantic telegraphy. Although public enthusiasm did not reach the exorbitant extent of 1858, it was nevertheless great. Festivals and parades were held, and numerous official cable dinners were given, as for example one given by the Liverpool Chamber of Commerce upon the return of the *Great Eastern*. Newspaper accounts from the summer of 1866 overflowed with commentaries about what the cable would do for world economy, world politics, or global peace. *Trewman's Exeter Flying Post* even claimed that "if this Atlantic Telegraph only works it will in a couple of years accomplish the mightiest revolution in history!"[113] The joy about the success of the two Atlantic cables was probably greatest among the cable agents themselves. As Daniel Gooch reported, "wild scenes of excitement" took place at Heart's Content and Valentia. The "old cable hands" held the cable up and danced round it, "cheering at the top of their voices." Demonstrating the contemporary image of the telegraphs as technological extensions of the nerves of mankind, "one man actually put it in his mouth and sucked it." Indeed, it must have been "a strange sight," yet nothing could illustrate more vividly the notion of the Old World and the New connected in thought.[114]

"OLD ATLANTIC FRIENDS": THE CLASS OF 1866 AND MEMORIALIZATION

For most of the cable pioneers, the success of the Atlantic telegraph marked the decisive turning point in their lives and careers. All of a sudden, they were no longer considered mad to believe in their success. Beforehand, the enterprise had elicited utter disbelief in its feasibility, not only from those pursuing rival projects, such as the American telegraph engineer T. P. Shaffner with his North Atlantic route, but also from within their group of supporters. By the end of those eight strenuous years from 1858 to 1866, many of the investors had withdrawn and directors resigned. Even Abram Hewitt, assistant and son-in-law to the entrepreneur and Great Atlantic Cable promoter, Peter Cooper, was having doubts. He later admitted that he had regarded "the investment of so much money in a doubtful enterprise as a piece of folly."[115] In the aftermath of 1866, their work was celebrated as "a glory to [their] age and nation," and they themselves were honored "amongst the benefactors of their country."[116] The highest honors were bestowed upon them from various governments and institutions of the world.

In connection with the 1865/66 expedition, four of its members, the cable manufacturer Richard Glass, the engineer Samuel Canning, scientist William Thomson, and Captain James Anderson, who had steered the *Great Eastern*, received the honor of a knighthood. Staff Commander Henry Moriarty, received the most honorable Order of Bath in the class of a companion (C.B.), the lowest class in the British order of chivalry founded 1725 by King George I. Curtis Lampson (1806–1885), deputy chairman of the Atlantic Telegraph Company, and Daniel Gooch had conferred upon them the dignity of a baronetcy and became among the few from the middle classes to access the ranks of peers and the hereditary status of English nobility. Lampson, who stemmed from an old New England family and had become a naturalized British citizen in 1849, was the first former American citizen to receive such an honor.[117] The Chambers of Commerce of Liverpool and New York gave out gold medals to commemorate the cable laying.[118] Finally, Charles T. Bright had already received his knighthood after the first short-lived cable attempt of 1858, making him the youngest knight in British history. Others among them, such as John Pender, would be honored for their cable work later in their life. It was not only their home country, Great Britain, that recognized their success with state honors, but also other countries such as Portugal, the United States, and France. Samuel Canning, for example, had the order of *St. Jago d'Espada* conferred upon him by the king of Portugal, and Charles T. Bright was offered the French *Légion d'honneur*.[119] Through these acts, the industrial nations of the Western world installed them as heroes of a technical modernity.

Submarine telegraphy and the success of the Great Atlantic Cable defined and marked the lives of these cable pioneers. From 1866 onward, it seemed as if submarine telegraphy was their sole activity and concern, or as Captain James Anderson phrased it, it came to be their fate "to be connected with [ocean] cables ever since they [had] been successfully laid."[120] Even the ship, the *Great Eastern*, "proved herself [only] well equipped for one distinctive task—that of laying the transoceanic telegraph cables."[121] The success of 1866 had put the actors of the Class of 1866 in a most advantageous situation. Financial, governmental, and public support for further ocean cable projects was readily available, and they were the only ones at the time who had the resources and the network to do it. The future careers of the "cable kings" John Pender and James Anderson and their buildup of a globe-spanning communication network in the 1870s through the Eastern and Associated Companies was based on the network established by the Great Atlantic Cable project.

The careers of some of the engineering pioneers also greatly benefited from their newly acquired fame. Independent engineering advisors or consulting firms flourished in the 1860s and 1870s; as early as 1860, for example, Charles T. Bright and Latimer Clark entered into partnership. In 1865, William Thomson and Cromwell Varley formed a patent partnership together with the electrical engineer Fleeming Jenkin, which eventually earned all three men large royalties from cable companies.[122] Soon thereafter, Samuel Canning, having previously left the service of

Telcon, entered into a consulting partnership with Robert Sabine.[123] All of these consulting firms were highly influential in the expansion of a global ocean cable network and the design of cable laying. They worked as consultants for cables in the Mediterranean, the Atlantic, the Persian Gulf, and the West Indies, as well as for those to Australia, Southeast Asia, and South Africa. In this way, they took on employment from a number of different governments. For the West Indies project, for example, Charles T. Bright simultaneously held contracts with the English, Spanish, French, and Danish governments.[124]

Outside of the cable business, many of these men also took on positions of influence. From the mid-1860s on, Charles T. Bright, Daniel Gooch, and John Pender were all elected as members of the British Parliament, where they exerted their influence on decisions relevant to the cable business such as the Telegraph Purchase Act of 1868. This nationalized all British landlines and thus freed money for reinvestment.[125] Charles T. Bright, candidate for Greenwich, a city where most ocean cables were manufactured, ran his election campaign on the Atlantic ticket, stressing his status as an Atlantic cable engineer.[126] Members of the Class of 1866 also exerted tremendous influence on international decisions. Cyrus W. Field, for example, represented the New York Chamber of Commerce during a conference on the usage of the Suez Canal and subsequently received a personal invitation from Ferdinand de Lesseps, builder of the canal, upon its opening.[127] In 1875, John Pender allegedly talked the British government into buying Khedive's Suez Canal shares worth £4,000,000.[128] The success of the Great Atlantic Cable pushed many a career of the enterprise's protagonists. Not only as professionals but also as politicians, they now employed positions of importance from which to influence the course of politics, science, and commerce in the British Empire and the United States.

Yet the laying of the Atlantic cable had not only changed their lives as individuals, but also coalesced them into a group, the Class of 1866. Those engineers, electricians, and entrepreneurs connected with the Great Atlantic Cable were among the first to pursue submarine cable business on a large scale, but they also represented a kind of cable royalty, which gave them further access to the highest circles of the world's ruling class. The idea of a "class of cable engineers" was an open concept, growing in congruence with the expansion of submarine telegraphy and electrical science and finally signifying a profession.

The cable actors, however, remained a closed entity. The Class of 1866 represents this notion of a closed community. The men were united in their quest to succeed, and experiences of common failure bound them even closer.

Their loyalty stretched not only to the project but also to each other. Although not all of them became friends, something akin to a culture of friendship developed, even leading them to consider the group as a "Telegraph family— . . . brothers—Friends."[129] Aside from close business relations, most of the members of the Class of 1866 were connected by mutual sympathy and formed long-lasting friendships; often, it was a combination of both. For instance, John Pender and the American Abram Hewitt exchanged Christmas greetings as well as stock market tips across the Atlantic. They discussed cable issues, and Hewitt served for the decades to come almost as an unofficial advisor to John Pender on the American situation concerning government support and public opinion for further cable projects. From their correspondence, it appears that Pender would not undertake anything that demanded American involvement without first consulting Hewitt.[130] Personally, the two also got along splendidly. In the summer of 1883, both went on a cruise around the Mediterranean together with their families, which was partially for pleasure and partially devoted to cable business.[131] The network of the Class of 1866 was bound not only by business relations but also by personal friendship. The social element played a tremendous role in the enduring bonds of the Class of 1866.

Although the close network of the Class of 1866 proved to be a valuable source for business and philanthropic projects, it was not always tight knit. The community was challenged over the basic question of "who was to have the most credit" for their work even before they had safely returned to Great Britain in the fall of 1866.[132] Antagonism on board predominantly played itself out between Samuel Canning, engineer-in-chief, and James Anderson, captain of the *Great Eastern*, and their associates. Their controversy between brain and brawn predated a greater dispute concerning the role of science and business within the future of submarine telegraphy that was predominantly fought between the engineer William Siemens and John Pender in the 1870s. A similar dispute took place concerning the official honoring of the cable agents by the British state. John Pender never forgot how he was ignored in 1866, while others, who had in his eyes contributed less to their joint success, were knighted. In an angry letter to Lord Wilton from October 1866, he justified his own position in the cable enterprise, concluding that he had "great cause to complain of being overlooked in the late public notices & in the distribution of Royal favour."[133] Although the members of the Class of 1866 considered themselves as "family" or "friends," strains were soon put on the cohesion of the group. These resulted from a fundamental quarrel about the standing of the engineer within international business.

The late 1880s marked something of an endpoint to the Class of 1866 and their monopoly on the wiring of the Atlantic. By then, many of its members had retired from the cable business or died. The cable manufacturer Richard Glass died in 1873, as did the oceanographer Matthew F. Maury, who had discovered the Atlantic Plateau. They were preceded in death by Samuel F. B. Morse, who died in 1872. The American investor Peter Cooper died in 1883, Daniel Gooch in October 1889, and Charles T. Bright a year earlier. The Anglo-American director of the Atlantic Telegraph Company, Curtis M. Lampson, died in 1885, and Cromwell Varley died in 1883. George W. Campbell, director of Telcon, died in the early 1890s, and Willoughby Smith, former chief electrician of Telcon, retired from his management post in 1888.[134] Simultaneously, the 1880s marked the end of the Anglo-American Telegraph Company's monopoly on the North Atlantic market, as it had to yield to a rival and give in to a duopoly; independent consulting engineers also lost influence, as the contracting companies had acquired a staff with greater experience and efficiency. This limited the work of the "engineering hero" and eventually rendered him unnecessary.[135] Only a small number of the old Atlantic heroes, such as John Pender, James Anderson, William Thomson, and the American Cyrus W. Field, had remained; their success remained unbroken until into the twentieth century.

An important aspect to the success of the story of the wiring of the Atlantic and its actors lies in their attempts and success at mythologization. In the literature, the Great Atlantic Cable of 1858–1866 is often depicted as the turning point for global communication. Among other terms, it is labeled "[t]he Wire that changed the world" or, in an analogy to modern communication, "the Victorian Internet."[136] Its historiography truly reads like a "heroic story of the transatlantic cable."[137] There is no doubt as to the importance of the first Atlantic telegraph and its status as a milestone for globalized communication, but today's memorialization of 1866 and the North Atlantic as historical focal point is also due to the efforts of the Class of 1866. They did a great deal to keep alive their idea of the history of submarine telegraphy, the story of the Atlantic telegraphs, and their own roles within that narrative. They authored books and memoirs, held commemorative "cable banquets" and "telegraph soirees," and monumentalized memory in the form of cable ships, such as the *John Pender*.[138]

In the aftermath of the euphoria of 1866, in particular those who owed their rise in social status to the success of the Atlantic cable felt an urge to report how they, as "self-made men," had attained "their high position in the world."[139] They gave talks and lectures and wrote papers on submarine telegraphy and their own contribution to it.[140] Some clearly aimed to be scientifically enlightening; others seem to have been driven by mere vanity. Kindly bestowing upon the students of the new Swindon Mechanics' Institute in "exceedingly simple . . . utterances . . ., like all really great men . . ., how he came to contribute so much light to this poor dark world," Daniel Gooch, for example, elicited this rather ironic description of his talk in the *Pall Mall Gazette*.[141] The members of the Class of 1866 soon engaged in the ritual of telling and retelling, celebrating, and commemorating, deliberately concealing stories of failure. Little does the world know, for instance, that Cyrus W. Field's "cable cabinet," which so successfully wired the Atlantic in 1866, failed miserably three years later with a scheme for a cable across the Pacific.[142] These processes of commemoration, which were perpetuated throughout the nineteenth century, were as much part of a ritual of renewing their personal ties to each other as they were a tool to shape public perception and manifest their status of stardom.

The process of retelling and sharing the experience of laying the Great Atlantic Cable was central to the entire expedition and confirmed its status as a media event for which star journalist William H. Russell and artist Robert Dudley had been employed.[143] Their narrative was sent to all the principal journals even before the *Great Eastern* arrived in London, "so that the public were at once placed in possession of every fact connected with the proceedings, almost simultaneously."[144] Yet, aside from being right "on the scene" to relate everything "as it happened," Russell also asserted the claim to be the "historian of the enterprise," as had the American *New York Herald* reporter John Mullaly in 1858 when narrating about the first Atlantic cable expeditions of 1857 and 1858.[145] The Great Atlantic Cable made history, while its agents documented history through Russell's embedded journalism as well as a series of published firsthand accounts, such as those of George Saward, secretary of the Anglo-American Telegraph Company; Henry Martyn Field, brother of Cyrus W. Field; Willoughby Smith, telegraph engineer on the *Great Eastern*; and Charles Bright, son of Charles T. Bright.[146] Near contemporaries, however, realized that there was yet another history of the telegraph waiting to be written. Already in 1891, the Atlantic electrician Willoughby Smith insightfully stated that "[t]he correct history of submarine telegraphy ha[d] never yet been written."[147]

Apart from writing their own history, another instrument for shaping memory were the frequent cable banquets, mostly organized by the American Cyrus W. Field or the cable entrepreneur John Pender. These were meant to commemorate important as well as less important dates in the young history of submarine telegraphy. Over the years, neither the guest list nor the banquets' general setup, with direct telegraphic connections to all parts of the earth for the guests to use free of charge or the tradition of reading out congratulatory letters or cablegrams, changed. The guest list usually contained the Class of 1866 and others involved in the cable business in addition to people of distinction and high social standing. The company present at John Pender's private house to celebrate the laying of the ocean cable to India included the Prince of Wales, the Duke of Cambridge, and Ferdinand de Lesseps alongside James Anderson and Robert Halpin, then captain of the *Great Eastern*.[148] In fact, as G. W. Smalley remarked during one of these banquets in 1873, the assemblies resembled "telegraphic family part[ies]," which were intended as a ritual of self-congratulation; neither the public, nor the press, "so far as it represents the public [was] very much concerned" during these events.[149] The get-togethers seemed to be more important than the celebrated event itself, as the latter was rather randomly picked. In 1868, a banquet was held to celebrate Cyrus W. Field's stay in London; in 1872, coinciding with the Alabama Claims crisis, a series of claims put forward by the U.S. government against Great Britain for damages caused by its assistance to the Confederates during the American Civil War, the partygoers came together for a traditional American Thanksgiving. In 1873, Field called for a "commemoration of the signature of the agreement on the 10th of March 1854, for the establishment of a Telegraph across the Atlantic."[150] Similarly, in 1879, Field invited guests to celebrate the twenty-fifth anniversary of the first company ever formed to lay an ocean cable, and in 1885, the cause for celebration was the twenty-seventh anniversary of the first Atlantic cable of 1858. In 1898, the idea of a cable memorial was debated in the Anglo-American press, and a committee was set up to consider the form of the memorial.[151] In books, banquets, and monuments, the Class of 1866 constantly returned to their formative enterprise of the Great Atlantic Cable. They created their own master narrative and perpetuated their image as the "sole" telegraph experts.

The wiring of the Atlantic between 1854 and 1866 is an oft-told story. The purpose of this chapter is not to relate once again the sequence of events but to point to the emergence of a network of agents from the undertaking, the Class of 1866, which would come to dominate the system of global communication. Business and scientific networks, such as John Pender's and James Anderson's global telecommunications network of the Eastern and Associated Companies or the Society of Telegraph Engineers, fed off the original cable group and its network as social capital. Personal networks in the forms of family or business relations, friendships, and neighborhoods played a central role in the success story of the Great Atlantic Cable. Without the advocacy of John Brett and his far-reaching network into the fields of politics, finance, science, and submarine telegraphy or the support of the American expatriates in London, Cyrus W. Field's mission in London would have failed, and the cable would have suffered the same fate as the rival enterprises of a North Atlantic or a South Atlantic route, which never got beyond the planning stage. Field's domicile in the wealthy New York neighborhood of Gramercy Park, with its connections to America's leading families, also played a central role. As the Great Atlantic Cable demonstrated, the direct and personal connection between people was a prerequisite and was elemental for the success of these large-scale projects nourishing "globalization."

The actors of globalization were the rising middle classes of Europe, who enjoyed the advantages of an educated upbringing and the benefits of an era of capitalism and industrialization, which furthered the emergence of trade and investment on a global scale. The great majority of the Class of 1866 did not belong to the English gentry or London's gentlemen's clubs. Originating from the United States, Scotland, or Britain's industrial north, they were initially "outsiders" to the system of London's gentlemanly capitalism. The success of 1858 and, in particular, that of 1866, however, lifted the Class of 1866 into the globe's governing elite, and people like John Pender, Charles T. Bright, Samuel Canning,

James Anderson, Cyrus W. Field, Daniel Gooch, and William Thomson emerged as the "global telegraph and cable barons of the age."[152]

The success of the Class of 1866, however, lay not only in the wiring of the Atlantic but also in their great success in mythologization. From the first enterprise onward, the laying of the Great Atlantic Cable was staged as a media event, which in its aftermath was continuously commemorated through banquets, speeches, and personal memoirs. At the time, the Class of 1866 was so successful in shaping the history of ocean telegraphy that most of its rivals felt obliged to establish a symbolic personal connection with the Great Atlantic Cable. It seemed as if each contestant in the business had to undergo a similar initiation rite before he could be taken seriously in the market. The Siemens brothers, for example, were convinced that only a major ocean cable would earn them the merits necessary for substantial business.[153] Other cable entrepreneurs, such as the American James Scrymser, took the Great Atlantic Cable as the moment from which their own history with submarine telegraphy began.[154] The Great Atlantic Cable enterprise was a moment of manifold beginnings. It should be remembered not only as the technology's breakthrough but also as the event from which spun the structuring of the globe as a communicational sphere.

2

THE BATTLE FOR CABLE SUPREMACY

Indeed, now that the great enterprise is completed, there can be no doubt that in a few years the entire globe will be spanned by the telegraph wires, and the news of the planet will be given every morning in the London papers. The Atlantic makes the only great break in the continuity of the land of the globe. The three old Continents are a single mass. The wires may be carried from Singapore over island after island until they reach Australia. In the New World they may be laid easily from Labrador to Patagonia. By bridging the Atlantic these two great systems are brought into connection.

"Epitome of this Morning's News," *Pall Mall Gazette*, July 27, 1866.

IN 1866, most contemporaries were distinctly aware that the Atlantic cable only marked a beginning. It laid the foundations for a modern Pangaea, re-created by submarine cables. Apart from forming a closed network of cable actors, the Class of 1866, the Atlantic cable's great significance was its claim to global reach. Its importance within a history of global communication and its distinction from other world projects lay not in its singularity but in its implied multiplicity. In the words of the *Pall Mall Gazette*, the idea behind ocean telegraphy was not to have one cable, but many. The then omnipresent picture of Shakespeare's sprite Puck and his girdle "round about the earth" as an allegory for submarine telegraphy was only complete as one "linking up the whole world."[1] In this story of rapid network expansion, the 1870s were a crucial turning point. The 1850s had seen a series of long-distance cable failures, with the two prominent cases being the Atlantic cable and the Red Sea cable. In 1861, of 11,364 miles of cable that had been laid, only 3,000 were working.[2] The 1870s, in contrast, marked a period of rapid network expansion with cables from Europe (Porthcurno) through the Mediterranean (Malta–Alexandria) to India (Bombay–Aden–Suez, 1870), from Indonesia to Australia (1872), and from Portugal to Brazil and Argentina (Carcavelos–Madeira–Cape Verde Islands, 1874), as well as to China (Singapore–Cochin China–Hong Kong, 1871), Japan (Vladivostok–Nagasaki–Shanghai–Hong Kong, 1871), and South Africa (Aden–Zanzibar–Mozambique–Durban, 1879).[3] In addition to these roughly 100,000 miles of undersea cables, increasing to 115,000 miles by the end of the 1880s, some 650,000 miles of telegraph wires had been laid over land by the same time.[4]

The Class of 1866 was responsible for most of the submarine telegraph network expansion. The Atlantic had only been their first coup. Now they were at the forefront of a global submarine telegraph explosion, spearheaded by John Pender, Lord William Montague-Hay, and Daniel Gooch.[5] Despite this global expansion, the Atlantic remained their home base. Its maritime space not only encompassed the major axis between London and New York, but extended, economically as well as culturally, as far as Buenos Aires, Africa, and continental Europe.[6] It was the most lucrative (telegraph) market of the time, and company after company attempted to enter the Atlantic telegraph field and challenge the established actors' monopoly. While the rest of the world's oceans became more or less divided into spheres of influence, the Atlantic remained a hotly contested realm.

In the 1870s, the Class of 1866 already had an established pattern for dealing with rivals. First, they launched a hefty price war, which usually brought competitors to the edge of bankruptcy by lowering the tariff to nonremunerative rates. Second, they forced each new rival to join the Atlantic pool, a strict working agreement with the original Atlantic cable companies. It was a "system of competition ending in an amalgamation."[7] The Siemens brothers were the one main player who tried to challenge the Class of 1866's model of global telegraphy. Although often marginalized in the history of Atlantic telegraphy, these actors, particularly William Siemens, constituted *the* antagonist to John Pender, the evolving leading figure of the Class of 1866. Indeed, the main battle for telegraphic supremacy in nineteenth-century cables was fought between, on one side, John Pender, the Atlantic pool (a joint working agreement of cable companies employed on the North Atlantic), and the Globe Telegraph and Trust Company, and on the other side, William

Siemens, Siemens Brothers, and the Direct United States Cable Company. Their "telegraph war" was not only an economic contest between two business rivals but also a dispute over the systemic paradigms of global communication.

Submarine telegraphy was foremost a commercial undertaking. Until the 1890s, economic interests were one, if not *the* most forceful, of the motors behind the expanding cable network. Simultaneously, commerce and trade almost instantly felt the benefits of messages' immediacy. Yet financial and investment elites were not the only people to harbor an economic vision of a global communication system. The shareholder lists of the Direct United States Cable Company document how submarine telegraphy spread beyond the Class of 1866 to include smaller investors, many of them female, from outside of the financial and trading centers. By the late nineteenth century, a sense of global connectivity and the importance of a global perspective had spread to the masses.

FORGING AN EMPIRE OF CABLES

The successful wiring of the Atlantic placed the Class of 1866 in an advantageous position. They had proven the feasibility of long-distance ocean cables and triggered the rapid expansion of this new technology. Thereby, they had established themselves as men of "enterprise, energy, [and] tenacity" who could easily be entrusted with undertakings of a similar grandeur. In the post-1866 era, one name stands out: John Pender.[8] The textile merchant from Scotland was one of the directors of the Atlantic Telegraph Company. In 1865, he was responsible for the merger of the Gutta Percha Co. with Glass, Elliot & Co. to form the Telegraph Construction and Maintenance Company (Telcon). After the failure of 1865, Pender offered a personal guarantee of £250,000 to finance the manufacture of a new cable. With the success of the 1866 transatlantic cable, he set out to wire the world. In the 1870s, he launched a series of cable companies. By the end of the decade, he presided over an incomparable global telegraph empire, the Eastern and Associated Companies. This conglomerate not only continually formed new companies to complete the "girdle round about the earth," but also eliminated competitors to such an extent that all that was left was Pender's company, which in contemporary imagination looked like a giant kraken-like sea monster.

After the Atlantic, Pender's next coup was a submarine cable to India. Before the 1840s and regular steamship travel to India, it took five to eight months for a letter from Britain to arrive in India; a response could usually not be expected within two years. With the opening of steamship mail, this changed to six weeks in each direction.[9] British politicians and merchants had discussed a telegraph line to India already in 1853. In 1857 a group of officials and telegraph entrepreneurs laid the Red Sea and the Persian Gulf cables.[10] The connection met the government's concern after the Indian mutiny of 1857, when a widespread anticolonial rebellion threatened British rule. Slowness of communication with London had presented serious obstacles to British suppression of the uprising.[11] Pender's India cable was finished in 1870. It was manufactured by Telcon and paid out by the *Great Eastern*—two further Atlantic veterans. It was laid in several sections through the Mediterranean, touching at Gibraltar, Malta, Alexandria, Suez, and Aden. To limit the financial risks, one company was established for each of these sections. In 1868, the Anglo-Mediterranean Telegraph Company was established to lay a cable from Malta to Alexandria. In 1869, Pender founded the British India Submarine Telegraph Company for laying a cable from Bombay to Suez, followed by the Falmouth, Gibraltar and Malta Telegraph Company to finish the cable connection via Gibraltar and Portugal to England. In 1872, these three companies were merged together with the Marseilles, Algiers and Malta Telegraph Company to form the Eastern Telegraph Company. John Pender served as its chairman.

For a telegraphic extension further east, three additional companies were established: the British-Australian Company, the British-Indian Extension Company, and the China Submarine Telegraph Company connecting Europe with Australasia and China. In 1873, these were amalgamated as the Eastern Extension, Australasian and China Telegraph Company.[12] Both the Eastern Telegraph and the Eastern Extension Telegraph Company formed the centerpieces of the Eastern and Associated Companies and their global system. This further comprised the Black Sea Company (1874) and the Eastern and South African Telegraph Company (1879). The arrival of electric telegraphy as "big business," in addition to the Telegraph Purchase Bill of 1868, through which the British government nationalized all landlines, helped this rapid establishment of submarine telegraph companies in the 1870s.[13] The Telegraph Purchase Bill enabled new telegraphic ventures to secure "a good deal of the money let loose" by the winding up of the landline companies. It freed about £8,000,000 for reinvestment by those "who looked favourably on electric telegraphs as a subject of save [sic] and sure remuneration."[14] By World War I, the Electra House Group, as the telegraph conglomerate of the Eastern and Associated Companies was also called, had become one of the world's most powerful multinational corporations.[15] Near the turn of the century, it owned over 50,000 miles of submarine cable, or about one-third of the total cable mileage of the world. It represented a joint nominal capita of over ten million pounds sterling and carried about two million messages per annum.[16] With the Eastern and Associated Companies, John Pender and his fellow telegraph owners held a near monopoly of lines between Britain and North, Central, and South America, and total control of the Britain-India-Australasia route.[17]

John Pender is usually seen as the rightful head of the Eastern and Associated Companies' conglomerate. This great man thesis, however, underestimates the contributions of his fellow campaigners, in particular James Anderson, as well as the influence of a certain business structure established in the post-1866 era. Following the 1866 triumph, Scotsman and captain of the *Great Eastern*, James Anderson, left the ship to be bestowed with a knighthood. Subsequently, he gave up the position of a mariner to promote himself as a cable entrepreneur and right-hand man to John Pender. Already in 1869, he was back on board the *Great Eastern*, this time as general superintendent of the new French Atlantic Company entrusted with laying the French Atlantic cable.[18] From his position among the ship's staff, he had moved up to become a cable entrepreneur, and from directing the cable ship to directing the cable undertaking.

Throughout his life, James Anderson held several important positions in the cable business. He was managing director of the Eastern Company, the Eastern and South African Telegraph Company, and the West African Telegraph Company. He was chairman of the Brazilian Submarine and the Direct Spanish Telegraph Companies and director of the Anglo-American, the African Direct, the Eastern Extension, and the West India and Panama Telegraph Companies and the Globe Telegraph and Trust Company. Lastly, he was trustee of the Submarine Cables Trust.[19] In British news, Anderson was the most visible of the cable actors. He served not only as, in effect, general managing director of the global system but also as its "public relations manager," although this was not his proper job description. In his letters to the press, he promoted, expounded, or defended cable policies. He had to explain tariffs, justify routes, or obfuscate cable breakages.[20] James Anderson also shielded John Pender and his cable conglomerate during hostile acquisition of other cable companies. He was the frontman for the "dirty work" of business takeovers, while Pender remained in the background. It was James Anderson, and not John Pender, who served as director for the Direct Spanish Telegraph Company and the Brazilian Submarine Company, when they were forced into a working agreement with the Eastern and Associated Companies in 1874.[21] Anderson represented a group composed "of those whom energy, merit and intelligence had raised to their position and kept there" and who strongly influenced the processes of globalization.[22]

Sole focus on Pender also neglects how the Eastern and Associated Companies' cable network functioned in its early period and how the agents it recruited operated: they held multiple directorships in several cable companies and cross-directorships in cable construction, laying, and operating companies. At the time, scarcely any of these cable companies were assisted by government monopolies, subsidies, or guarantees. Rather, this extension of global communication in the 1870s was due to the enterprise of a few enthusiastic individuals.[23] Most companies' boards of directors were overwhelmingly composed of veterans of the first Atlantic projects. Akin to the Atlantic Telegraph Company in the 1850s, which had recruited most of its shareholders from John Brett's Magnetic Telegraph Company, later companies also drew from the original Class of 1866 and others involved in early submarine projects. Aside from John Pender, names such as Daniel Gooch, James Anderson, William Montague-Hay, and Baron Emile d'Erlanger reoccur in several of the boards of directors. Despite the geographical distance to London, the American Cyrus W. Field remained an important actor within this cable clan. He served as a director of the Anglo-Mediterranean Telegraph Company and the Globe Trust and Telegraph Company, which would later play a key role in the Pender-Siemens battle over Atlantic cable supremacy. The boards of directors reveal a small group that wove its telegraphic net round about the earth. Recruiting mainly from their peer group, they created a closed market and so inhibited new competitors. In the years to come, open positions were often filled by family members. Pender's three sons, James Pender, John Denison-Pender, and Henry Denison-Pender, all entered the cable business.[24] After Pender's death in 1896, James Pender and John Denison-Pender become the new central

TABLE 2.1 "Members of the Class of 1866 on the Board of the Electra House Group in the 1870s"

Atlantic Tel. Co.	Anglo-American Tel. Co.	Telcon	Great Eastern Steam-ship Co.	French Atlantic Tel. Co.	Anglo-Mediter-ranean Tel. Co.	British India Tel. Co.	Marseilles, Algiers & Malta Tel. Co.	Falmouth, Gibraltar & Malta Tel. Co.	British Australian Tel. Co.	British Indian Extension Tel. Co.	China Submarine Tel. Co.	Eastern Tel. Co.	Globe Trust
John Pender	John Pender	John Pender			John Pender	John Pender	John Pender	John Pender	John Pender	John Pender	John Pender	John Pender	John Pender
Cyrus W. Field	Cyrus W. Field				Cyrus W. Field								Cyrus W. Field
				(James Anderson)	James Anderson				James Anderson			James Anderson	James Anderson
Thomas Brassey	Thomas Brassey	Thomas Brassey	Thomas Brassey		Thomas Brassey								
Charles T. Bright													
Samuel Gurney		Samuel Gurney			Samuel Gurney	Samuel Gurney							
Stuart Wortley	Stuart Wortley												

	William Montague-Hay	William Montague-Hay		William Montague-Hay	William Montague-Hay		William Montague-Hay	William Montague-Hay	William Montague-Hay			William Montague-Hay	William Montague-Hay
		Daniel Gooch	Daniel Gooch										Daniel Gooch
		George Elliot											
		Richard Glass											
		Sherard Osborn											Sherard Osborn
				Julius Reuter									Julius Reuter
				Emile d’Erlanger	Emile d’Erlanger	Emile d’Erlanger	Emile d’Erlanger	Emile d’Erlanger			Emile d’Erlanger	Emile d’Erlanger	

figures in the Eastern and Associated Companies network. Men of rank and wealth usually filled the few remaining positions on the boards of directors.

Apart from multiple directorships, the phenomenon of cross-directorships extended the influence of Pender, Anderson, Gooch, and company beyond one single cable company to the whole entrepreneurial side of the submarine cable business. The cable clan's control ranged from the manufacturing and operating processes to the laying of ocean cables. This was fundamental to establishing and securing their global monopoly and allowed a handful of people to control an entire group of cable companies and thus almost all global communication. The Atlantic Telegraph Company and the Anglo-American Telegraph Company controlled the operation of cables. Through Telcon, the Class of 1866 controlled the manufacturing processes. Thus, they not only manufactured their own cable, but also controlled who else could order cables. Lastly, through the Great Eastern Steamship Company, which had leased the *Great Eastern* to Telcon, the laying operation was also under the control of the Class of 1866. The key players were Daniel Gooch and the famous railroad contractor Thomas Brassey, who served as directors on the board of all three companies. Pender himself had inaugurated the formation of Telcon. However, he left control as chairman soon after to Daniel Gooch. In 1888, his son, James Pender, was introduced into the cable business as director of Telcon.

Such total market control strongly depended on the actors' settling within the British Empire. Despite the breakthroughs in technology and a professionalization of the business, submarine telegraphy remained a costly expenditure. Initial costs for a cable consisted of the manufacturing and laying costs, in addition to the costs of setting up the cable stations with housing, staff, and equipment. Costs further depended on charges for landing rights and the necessity of ocean soundings. The greatest part of the costs came from the two major raw materials needed to manufacture the cables: copper and gutta percha. The British monopoly on manufacturing submarine cables based itself directly on the country's imperial outreach, as the substance was found only in the Dutch, British, and French colonies of the Far East.[25] Plantations in Southeast Asia supplied gutta percha, the insulating material for cables, and patents by London financiers controlled its use. Until the 1890s, the manufacture of submarine cables was almost entirely confined to factories on the banks

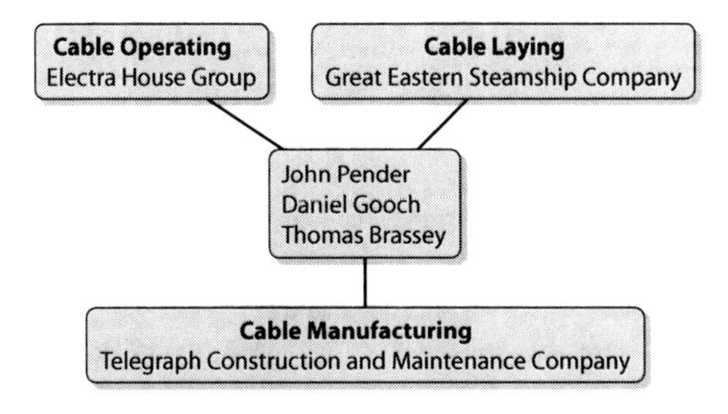

2.1 Simone M. Müller, "Submarine Cable Business—Vertical Networks."

of the Thames. In 1931, Telcon still manufactured the great majority of the world's submarine cables. Many countries had to import the cables, the equipment, and often the technicians and operators as well.[26]

Throughout the nineteenth century, control of both the manufacturing and operating processes remained elemental to the success of cable entrepreneurs. Almost total control over the entire cable industry and all production processes was key to the exorbitant success of the Class of 1866. This continued well into the early twentieth century. In 1907, the Australian cable reformer Henniker Heaton, under way on his mission for a "cable" penny post, a tariff system where telegrams could be sent for one penny, complained about the "Cable Kings [who] hold in their hands the powers of life and death, so to speak . . . [by] controlling the cables of the world." He referred to the boards of directors of the Eastern, the Eastern Extension, the Eastern and South African, and the West African Telegraph Companies, which were in effect made up of the same six people.[27] Even smaller companies that broke into the telegraphic market and established telegraph systems in Southeast Asia, southern Europe, or South America held connections to the Class of 1866.[28] While fierce competition on the Japanese and Chinese telegraph market with the Great Northern Company was soon resolved through dividing up spheres of influence, the southern European and Latin American systems were soon incorporated into the Eastern's network through a tight working agreement, playing itself out through financial and personnel entanglements.[29] James Anderson served as chairman for the Brazilian Submarine and the Direct Spanish Telegraph Companies.[30] The daily business of both companies was run by two Atlantic people, both of whom had been assistant electricians during the 1858 Atlantic attempt: Charles Gerhardi now served as manager of the Direct Spanish Telegraph Company, and Richard Collett was the secretary of the Brazilian Submarine Company.[31] Yet the Class of 1866 would not dominate the cable market completely for long nor could they suppress any rival, as the clash for supremacy between Siemens Brothers and the Class of 1866 unfolded in the 1870s.

WAGING TELEGRAPH WAR ON THE ATLANTIC

In the aftermath of 1866, the newspapers reported that "a hundred busy brains [were] planning a dozen new Atlantic companies" as new entrepreneurs dreamed of an "El Dorado to be found in transatlantic telegraphy."[32] The success of 1866 had produced a "strong confidence" that the public "placed in the science of submarine telegraphy" and an "eager manner in which scheme after scheme [was] accepted."[33] The U.S. Congress was bustling with adjudging on ocean telegraph bills, many of which never came to anything, and Britain's Public Record Office houses proof of numerous Atlantic cable companies, such as the Direct Atlantic Cable Company and the New Atlantic Cable Company, that never got beyond the planning stages.[34] The Direct Atlantic Telegraph Company in particular conveys how Atlantic cable fever had infected a broad spectrum of investors. Clerks and accountants set up the company with total capital of £100. The statement of incorporation reveals that they had little idea of how to proceed. It remained unclear whether the company would establish "a line or lines" of telegraphic communication; in addition, the question of landing places remained unresolved. The contractors were merely planning to connect "some place in the United Kingdom and some place in the United States or the Continent of America."[35] Their entrepreneurial spirit was certainly adventurous, but not wholly naïve. They witnessed the boom of ocean cables and realized how easy it seemed to embark on such a profit-promising enterprise.

In this Atlantic cable frenzy, only the French Atlantic Cable Company laid a submarine telegraph connection. In 1868 Baron Emile d'Erlanger and Julius Reuter set out to put France in direct telegraphic communication with

North America and launched the French Atlantic Telegraph Company (La Société du Câble Transatlantique Français). Within just eighty days, mainly British, French, and German shareholders invested the necessary capital.[36] As the *Daily News* argued, "[f]or no other reason but to establish a submarine telegraph company could a million of money have been found in such an incredible short space of time."[37] This new provider of Atlantic communication was independent for an extremely brief period. Manufacture and paying out of the cable was undertaken by Telcon, which engaged the *Great Eastern* and employed the same staff as for the prior Atlantic attempts.[38] There was also great continuity in terms of financing. The Anglo-American bank J.S. Morgan & Co. loaned much of the money, £500,000 of £715,000.[39] Both American bankers, Junius S. Morgan and George Peabody, the two principals behind J.S. Morgan & Co., were closely associated with Cyrus W. Field and the Great Atlantic Cable.[40] Certainly, these men would not invest in a scheme that competed detrimentally with their previous investments. Rather, we might ask if the French Atlantic Company was ever conceptualized as true competition. Already five months after opening the cable for public traffic, the French Atlantic Company entered into a joint-purse agreement, the "Atlantic pool," with the Anglo-American Telegraph Company. This joint-purse agreement represented one big financial combination: it was made up of the net earnings of each company and the revenue earnings accruing to each company therefrom. These were fixed in proportion to each company's respective contributions to the pool.[41] In 1873, it was fully amalgamated into the Anglo's system.[42]

Thus far, transatlantic telegraphy is the story of the Atlantic pool, a financial and working agreement of Atlantic telegraph companies led by the Anglo-American Telegraph Company. Within the pool, the business situation on the North Atlantic moved from a monopoly to a duopoly in 1888 and became an oligopoly by the turn of the century.[43] Although this is generally correct, it leaves out one important link in the chain: manufacturing companies such as Siemens Brothers. In 1873, the Siemens Brothers company entered the Atlantic market with ambitions to break the Anglo-American's monopoly. It was not the first telegraph undertaking of this transnational family business, as the Siemens brothers (the family clan of Werner, William, Carl, and Georg Siemens) had already laid and manufactured shorter cables.[44] Between 1868 and 1870, they laid the Indo-European telegraph line, an overland connection between London and Calcutta crossing British, Prussian, Russian, and Persian territory. This also brought about their first encounter with John Pender and his Eastern system, which attempted to amalgamate the Indo-European telegraph line into its network. As early as 1868, James Anderson and William Siemens had a lengthy disagreement displayed in the London *Times* on the (dis)advantages of land and submarine lines to India, every one of them highlighting their own project.[45] An Atlantic connection, however, represented the Siemens brothers' most important coup because they believed that this would establish their fame and fortune within the cable system. In 1872, Carl Siemens expressed the conviction that until Siemens Brothers had laid a major submarine cable, they would not be taken seriously in the business.[46] Moreover, the brothers intended to challenge not only the established system of cable making but also cable operation itself.

The Siemens clan commenced with the telegraph business in the 1840s when Werner Siemens, the oldest of the brothers, established in cooperation with Johann Georg Halske the *Telegraphenbauanstalt* (telegraph manufactory) in Berlin. Carl Wilhelm Siemens, often known under his Anglicized name William, immigrated to Great Britain in 1844 seeking employment as a civil engineer. From 1850 on, he worked as an agent for Siemens & Halske. In October 1858, his work was officially recognized with the founding of Siemens, Halske & Co., London.[47] The same year, William Siemens negotiated a contract as an advisory engineer with the cable maker Newall & Co. It was in this function that William Siemens was employed on board the *Agamemnon* laying the 1858 Atlantic cable.[48] William Siemens also acquired a reputation as an independent authority on submarine cables. In 1861 he served as expert to the Joint Committee to Inquire into the Construction of Submarine Telegraph Cables.[49] In 1863, Siemens, Halske & Co. dissolved their connection to Newall & Co., moved to Woolwich, and set themselves up as independent manufacturers and contractors of submarine cables. Georg Halske withdrew from the Berlin business in 1865, and the firm was reorganized as Siemens Brothers. Halske's withdrawal put all power in the hands of the brothers and strengthened the appearance of a multinational family enterprise with Werner Siemens in Berlin, Carl Siemens in Russia, and William Siemens in London.[50]

The first cable of importance laid by the Siemens Brothers was the Cartagena (Spain)-Oran (Algeria) cable for the French government. It was an unfortunate beginning, as the cable broke ten miles before it reached Cartagena. After the Indo-European telegraph line, mainly an achievement of the Berlin firm, William Siemens secured a contract in 1873 for the Platino-Brasileira line (1873–1874), which was an extension of the telegraph link between England and Brazil. It was to lead southward along the coast of Brazil and Uruguay. Although this cable was a technical success, it proved to be a "singularly ill-fated venture" in other regards.[51] The vessels chartered to transport the cable across the Atlantic sank twice. For years it stuck with Siemens Brothers that they had "lost two ships," although they could not truly be held responsible for the weather on the South Atlantic.[52] Still, these incidents put them at a disadvantage against the Class of 1866, who with their "magnificent triumph" across the Atlantic magnified that they could get the job done.[53]

According to Werner Siemens, German financiers had approached him in 1871 about a *direct* telegraph cable between the United States and Germany.[54] So far, the Atlantic telegraph cables only landed on Newfoundland territory. Messages first had to be transferred to the landlines travelling south from Heart's Content and St. Pierre, Newfoundland, before they reached New York. When in 1873 the Direct United States Cable Company (DUSC) was established, it took up this desideratum. The idea behind such a direct cable was not only to speed up communication, but also to create space for competition and cheaper transmission rates.[55] Over the course of 1872, the brothers occupied themselves with the task of laying an Atlantic cable while they explored their options and solicited potential partners. At the time, financial support for submarine schemes was still readily available, and Carl Siemens even considered the option of laying two Atlantic cables as money was to be had in abundance.[56] But money was not all that mattered, in particular after John Pender launched the Globe Trust and Telegraph Company in November 1872. The Globe, as this investment company came to be called, represented a new type of financial intermediary. The first such company was founded in 1868. Its structure resembled that of equity funds of today, as it was a device for taking out the extreme financial risk of cable laying by spreading it over a number of companies.[57] A Globe share comprised units of shares in all of the Globe's subsidiaries. It reduced the risks of investment to a minimum by extending the risk over many lines and might as well, in the words of Cyrus W. Field, "have been called the Submarine Telegraph Insurance Company."[58] With the Globe Trust, submarine cables investment became an option for inexperienced, small, and more risk-averse investors. It catered "to the emerging middle classes with savings and no knowledge of where to invest."[59] This security for the shareholders came at a high price for the entrepreneurs: Pender's Globe Trust was "militantly monopolistic."[60] Although Cyrus W. Field denounced any monopolistic ambitions, he considered "unity" among the cable companies to be their "strength and power."[61] According to Daniel Gooch, another director of the Globe, it was a wise principle "that these large cable companies . . . should . . . combine to form a property which shall be a safe and sound investment for people, who are not acquainted . . . with this class of property."[62] In effect, the Globe's ultimate aim was to put all its rivals out of business.

Set on avoiding a "partnership" with either the cable pool of the existing Atlantic telegraph companies or the Globe Trust, the Siemens Brothers' final strategy was well considered. First, they manufactured the cable themselves in their workshop in Woolwich. Second, they launched their own cable ship, making them independent of the *Great Eastern*—until then the only steamer big enough to carry an Atlantic cable. At the turn of the year 1872/73, William Siemens commenced negotiations with the boat builders Mitchel & Co. about an explicitly fitted cable laying ship, the *Faraday*.[63] The name of the cable ship, referring to the scientist Michael Faraday, revealed a deeper divide between Siemens and Pender over the role of science in business in the controversy yet to come. Faraday discovered the scientific principles that allowed electricity to be used for practical purposes, and without his achievements, all electrical inventions, including the telegraph, would have been impossible. Siemens's use of his name represented his approach toward submarine telegraphy from the science and not the business perspective. Siemens emphasized technological progress and competition over economic gains, whereas Pender preferred a monopolistic market structure.

In March 1873, DUSC was launched with a market capitalization of £1,300,000. The company was registered in London. William Siemens held the position of consulting director, and the German engineer, George von Chauvin, one of the rising employees of Siemens & Halske, became manager and electrician. Siemens Brothers alone held £50,000 worth of shares in return for being entrusted with manufacturing the cable at the price of £1,100,000.[64] The Siemens brothers also broke fresh ground in the financial structure of the new company. The case of the French Atlantic Cable Company and its financial backing through J.S. Morgan & Co. had demonstrated that Pender and his growing cable empire held a stronghold within London, then the world's financial center. Also the Globe Trust was dominated by John Pender. Because, according to Werner Siemens, "the English financial market was closed off to [them] by the overpowering rivalry" of the other Atlantic cable companies, most of the company's capital came from the European continent.[65] By July 1873, 187 shareholders had purchased the 65,000 shares, each worth £20. Yet out of these shareholders, it was a small group of nineteen who owned almost 48,000 shares worth £957,820 or roughly 75 percent of the total capital. Among these were the leading and emerging banks of Europe *outside* of London, such as Société de Crédit Mobilier in Paris (7,048 shares), the Banque Centrale Anversoise in Antwerp (6,167 shares), the Deutsche Bank in Berlin (2,657 shares), and the Banque de Paris et des Pays-Bas in Paris (2,203 shares), as well as the Anglo-Austrian Bank in London (1,323 shares).[66] Other major investors included a select number of wealthy individuals, predominantly merchants or "networking businessmen," as economic historian William P. Kennedy termed this social group that came to be so important for the nineteenth-century investment market.[67] Among them were Alexander Frederick Kleinwort, a German expatriate in London (4,406 shares); Louis Lemmé from Antwerp (1,542 shares); Georges Brugmann from Brussels (1,982 shares); and, of course, members of the Siemens family (3,432 shares all together).[68]

Although the company was incorporated in London, in 1873 three-quarters of the company's capital was held outside of Great Britain, but not outside of Europe. The major centers of trade and finance in Europe, such as Antwerp, Brussels, and Paris, albeit secondary in their position to London, were the main recruiting grounds for big investors. Most of those residing in London, such as Alexander Kleinwort, Charles Günther, and Johan Conrad im Thurn, were German merchant émigrés.[69] These men were in all likelihood part of the same social circles as William Siemens in London, and their common German heritage probably played a role in their decision to invest in the Siemens project. Similarly to the Great Atlantic Cable and the Americans in Britain, or more generally to British investment schemes in the dominions, tapping into the network of expatriates was crucial. Within these networks, trust was generated and information circulated.[70] Finally, judging from the company's financial structure, it almost seems as if continental Europe was striking out against London, which endeavored to claim all of the world's communication. These efforts emerged in the 1870s not due to nationalist animosities but because of economic disadvantages in other commercial centers, such as Paris, Amsterdam, Antwerp, and, later, Berlin. In addition to the transatlantic transmission rates, the merchants on the continent also had to pay for the landlines leading from the Atlantic cables' landing places. This increased costs excessively and disadvantaged them compared to their British competitors in transatlantic trade. Such considerations explain the relatively high numbers of Belgian shareholders. In the mid-nineteenth century, Great Britain and Belgium remained the two most highly industrialized countries per capita. Although Belgium had a relatively small economy, it was highly important. In 1873, it produced about half as much iron as its much larger neighbor France.[71]

The *guerre à l'outrance,* war to the uttermost, as William Siemens termed the upcoming fight against the Atlantic telegraph monopoly, began well before the Atlantic cable was even landed. It encompassed all means of intrigue, slander, and possibly sabotage, and both parties primarily used the British press to influence and agitate the public and the stock markets.[72] Also paying out the cable proved cumbersome. Several attempts were necessary before it was safely laid across the Atlantic. Additionally, Siemens Brothers faced a court injunction prohibiting the cable landing on the coast of Newfoundland, which was initiated by the Anglo-American Telegraph Company.[73] Finally, on September 15, 1875, the DUSC cable was opened for public traffic. Already on the first day, brokers raced the Atlantic cables. The new DUSC cable beat the old Atlantic ones by over an hour. Its shares rose by £11, and within the course of a few months, DUSC covered 30 percent of the transatlantic telegraph traffic.[74]

According to the Siemens brothers' letters, the "Anglos," as they called the Atlantic telegraph clan around Pender, Gooch, Anderson, and company, undertook everything possible to defend their monopoly.[75] In December 1875, Carl reported to Werner that the fight was becoming fierce as "these dogs were spreading any sorts of rumors." Carl was referring to a "ridiculous article" bashing their cable, which had appeared in the *Daily Telegraph.* He concluded that from now on "the big guns should be hauled out against this robber band."[76] The Siemens brothers were so suspicious that when their cable broke three times in less than a year, they believed that it had to be "malevolent" and "purposeful destruction."[77] Consequently, they had their cable tested by the two most eminent authorities in submarine telegraphic science, Professor William Thomson and F. J. Bramwell. These two confirmed that "these fractures [were] the result of violence willfully applied," a fact that was duly circulated in various newspapers.[78] It is difficult to determine whether the cable broke due to sabotage or other reasons. At that time, it was fairly common for submarine cables to break, as they had to withstand strong undercurrents, rocky ocean beds, and, particularly in more shallow water, ships' anchors. Yet sabotage cannot be ruled out.

After a legal suit and possibly sabotage, the Anglos attempted two relatively simple business strategies: price war and hostile takeover by purchasing shares. Soon after the opening of the DUSC line, the Anglo-American Telegraph Company initiated the price war, which soon brought both competitors under three shilling per word. Yet it had little effect in forcing the new contestant out of business.[79] Hence, John Pender and James Anderson began to quietly amass a large number of DUSC shares. When, in early spring 1877, the Siemens family and their supporters realized that Pender was secretly buying up shares, they attempted anything possible "to undo the gains made."[80] The matter would come to a head at the DUSC's shareholder meetings in spring 1877.

Sources from both sides of the argument, such as Emma Pender, wife of John Pender, as well as the Siemens brothers' correspondence, report how the threat of rioting and violence was in the air at the DUSC's shareholder meetings in the spring of 1877.[81] One DUSC shareholder in particular, Henry Labouchere, journalist, editor, and politician, unnerved Emma Pender as he attempted to threaten and blackmail John Pender. During shareholder meetings in February 1877, when Pender was received with both "cheers and hisses," Labouchere was Pender's harshest and most outspoken critic.[82] He sought any means to thwart Pender's plans of a merger with the Anglo-American Telegraph Company and declared the shareholders' vote Pender had promoted (and won) on the issue as illegal and irreconcilable with the DUSC's articles of association. His opposition to Pender became particularly vivid during the court case following the unresolved voting issue: as a last resort, some DUSC shareholders and directors initiated a lawsuit over the results of this vote, which would have given Pender authority to proceed with his takeover. Labouchere again showed that he was a man of quick temper:

> Labouchere was in court & . . . the man simply raved. He swore he would be revenged. He would publish every "black" that Pender had ever committed & if he could not get enough of true ones to destroy him by he would publish lies: "I have the power to make half the world believe and the rest will follow whether they believe it or not." . . . Well the meeting of the Globe & Direct took place & Labouchere seemed actually to control every word & action of the Direct [DUSC] board.[83]

In March 1877, the court case was decided in Pender's favor, declaring the vote as valid. Still, the big showdown was yet to come. DUSC's final shareholder meeting in April 1877 lasted "over five hours," and newspaper reports described it as a combat until the very end within a brutal "system of warfare."[84] According to William Siemens, it had been a "fight over life and death."[85] His brother Carl reported to Werner that William had "raged and let hell loose over Pender's clique."[86] Emma Pender too remarked in her letter to William des Voeux, her son-in-law, that there were, in the words of a lady, "plenty of disagreeables" involved in the takeover. There were further "disturbing private interests" and "biters & assailers in every corner." According to her reports, William Siemens, in particular, "ha[d] occupied a disgraceful position." As a result, he would "probably have to disgorge"; and she continued, managers and directors of the DUSC "have fought hard & fight still."[87]

In the end, all their fighting came to nothing. By mid-1877, the Anglos had gained control over the company and forced it to surrender. A special resolution was published in July 1877 stating that the company would "be wound up voluntarily" and "liquidated."[88] In the following weeks, a new company was established with the exact same name and even some of the old directors. Its management was integrated within that of the Anglo-American Telegraph Company, and a joint-purse agreement, sharing gains and losses according to a fixed proportion, was adopted in order to "harmonize . . . the . . . companies' operations and dedicate . . . both firms to obstruct any and all rivals from entering the world's most valuable communication market."[89] Such a step had become necessary, as the original DUSC had stated in its articles of association that "no arrangement should be come to by which the company should participate in the profits of the existing Atlantic Telegraph company."[90]

Within this "war," Pender and Siemens were fighting on several fronts. It was a clash of not only two rivaling companies but also two different business models with warring concepts of the importance of research and development. Reconsidering the telegraphic war on the Atlantic from the investment perspective, it was not so much the Anglo-American Telegraph Company that opposed the DUSC, but Pender's Globe Trust and Telegraph Company. Four years before those stormy DUSC board meetings in 1877, Werner Siemens confessed to his brother how much he "hated this monster consortium" (the Globe) as "progress [was] suffering" from its monopolistic approach.[91] After the defeat of the DUSC, William Siemens used his position as president of the internationally operating Society of Telegraph Engineers as a platform for a political statement. In his inaugural address of January 1878, he actually meant to discuss the advances in the field of submarine telegraphy as accomplished by various scientists in different countries. Yet, he used the opportunity to launch a major attack against Pender and the Globe. He declared that the Globe constituted a major threat to the "ingenuity and enterprise" of the engineer. According to Siemens, the "free exercise of these faculties" was menaced "not by legislative action, but a powerful financial combination," the Globe. This combination intended "to merge the interests of all oceanic and international lines and the construction of new lines into one interest" and attempted to suffocate the "irrepressible spirit of British enterprise."[92] It seems to be an odd coincidence that the Eastern Telegraph Company held its ordinary general meeting the very next day, during which Pender duly backtracked: he was perfectly honest about his position on the place of scientists and that he regarded William Siemens as such. Scientists were, in his view, all very well in their place, but their place was "in the laboratory, or at any rate not in the directory of big business." As to the financial combination Siemens had been referring to, "it must be one which only exist[ed] in Dr. Siemens' own imagination."[93]

None of the competitive struggles between different submarine telegraph companies around the globe had ever been or would ever again be fought so fiercely. Certainly, much of the dispute arose from personal animosities. Yet, the clash between Pender and Siemens was not only over business matters but also over the future course of submarine telegraphy:

competition versus monopoly. William Siemens, as an engineer, represented a scientific approach to ocean cables; he was eager to improve the technology through large investments in research and development.[94] He held competition to be the motor for these scientific improvements, as rivals would strive for ever better and faster ocean cables. Competition would also lead to tariff reductions. John Pender, in turn, represented the practical man and a service-oriented business model, with a rather "conservative" approach to research and development. He saw no benefit in "ruinous competition" or costly developments that might, or might not, lead to improvements for providing faster and more reliable service in ocean cabling. Pender's victory marked by the DUSC takeover had serious implications for the future of submarine technology, which by the 1890s had fallen into a "culture of lassitude" where it was content to use tried and tested technologies.[95]

The dispute between Pender and Siemens over the place of scientists had implications for the future of not only submarine telegraphy but also industrial and scientific Britain. In the eyes of Siemens, it came down to a fight between progress and stagnation and was thus a challenge to the very premises of their age. The mid-nineteenth century had predominantly been the age of smoke and steam—of the steam engine, the railway, and the telegraph—in sum, of British industrial development. Thereafter, three new kinds of industry (chemical, electrical and optical, and refrigeration) developed that were based much more strongly on advanced scientific knowledge, were centered on the research laboratory, and were located outside of Great Britain.[96] Scholars usually mark the 1870s as the beginning of Britain's industrial decline relative to Germany and the United States and connect it to this very "failure" of companies to invest in research and development.[97] Although the scale of research and development in Great Britain was larger than usually proclaimed, its history after the 1870s illustrates the extent to which technical development diversified:[98] in the mid-nineteenth century, much of the world's technical history had been essentially British, but by the end of the nineteenth century, British and world technological history diverged considerably.[99]

In the end, this division between the scientist and the practical man had a most profound impact on the Class of 1866 itself. This group of Atlantic cable actors originally consisted of entrepreneurs and engineers, financiers and electricians, practical men as well as scientists. To reach their goal, an even division between both sides was fundamental. Now, in the entrepreneurs' view, science became more and more a costly and unnecessary impediment to great profits and the scientist, engineer, and electrician an appendage to the business corporation. The implementation of Pender's monopolistic business model targeted at capital accumulation did not necessarily mark the end of telegraph engineering science. Still, Pender's intervention more or less ended any potential reign of the engineer-entrepreneurs. In 1856, the Atlantic Telegraph Company had been launched by the engineers John Brett and Charles T. Bright in cooperation with the businessman Cyrus W. Field. In the 1870s, the ranks of directors barely included any active scientist or engineer. Almost simultaneously, the work of the independent consulting engineers started to become more and more limited.[100] Finally, for William Siemens and Siemens Brothers, defeat had not been as absolute as one might think. They had lost the DUSC but won immeasurable "prestige" as cable manufacturers and cable layers; they established themselves as a viable alternative to Telcon.[101] From this time onward, almost every competitor attempting to defy the Eastern and Associated Companies' global system, such as Jay Gould's American Telegraph and Cable Company or John Mackay and Gordon Bennett's Commercial Cable Company, placed orders with Siemens Brothers to produce and lay their cable. This was particularly the case on the North Atlantic; Siemens Brothers laid seven out of sixteen Atlantic cables finished during the course of the nineteenth century.[102] Telcon and Siemens Brothers only merged their operations in 1935, when the history of long distance ocean cables slowly came to an end.[103]

THE NEW WOMAN AND THE COMMON MAN ON THE TELEGRAPH STOCK MARKET

While competition raged in the 1870s between the system's "great men" and their competing business models, the telegraph companies also changed drastically at the level of smaller shareholders. By the 1870s and 1880s, telegraph companies attracted more than just large investors with major capital. The DUSC shareholder lists reveal not just its large investors but also the average people who owned only one or two shares. As an investment, submarine telegraphy was attractive not only to merchants, gentlemen, and stock brokers but also to clerks, ministers, schoolmasters, and, in particular, women—people who might never have actually used the cable service around the globe. Still, these people understood it as a key technology of their time and a safe place to invest their savings.

The industrial economy of the nineteenth century was fundamentally different from the mercantile economy of the seventeenth and eighteenth centuries in both its geography and internal structures. Up to the early nineteenth century, enterprises had characteristically been financed privately through family assets and expanded by reinvesting profits. This often meant that most of the company's capital was tied up and that companies relied heavily on loans for routine operating expenses. With the increasing size and cost of industrial undertakings, such as railways, metals production, or submarine telegraphs, it became more and more difficult to mobilize enough money, particularly in countries just entering industrialization and lacking large accumulations of private capital. Yet even in countries such as Great Britain and France, where such reservoirs of capital were readily available, new ways of mobilizing capital and channeling savings into enterprises were sought. In addition to the investment banks, such as Crédit Mobilier, joint-stock companies dominated the new economy of the time. In Great Britain, they were established and regulated through the Companies Acts of 1856 and 1862 and financed "by the investment of the faceless thousands." The Limited Liability Act of 1855 further took out some of the risks of investments; turning the company into a "person," it protected investors from losing all should the venture fail.[104]

Like many of the other grand industrial and infrastructural schemes of the time, the submarine cable boom of the 1870s owed much to the reorganization of capital as well as the "vulgarization of the money market" of the early nineteenth century.[105] It would have been impossible for John Pender to establish the series of companies in the 1870s merely with the money of a small group of investors, let alone his own private capital. This was even more the case because he received little financial support from the various governments concerned. After the cable failures of 1858, submarine telegraphy developed to be a primarily privately financed enterprise. Of the total mileage of over 165,000 nautical miles of cables paid out by 1898, private enterprises had provided nearly 90 percent. Different governments from all around the world supplied the remaining 10 percent.[106] In 1888, when £40,000,000 had been invested in the global diffusion of communication, only £4,000,000 had come from the British government.[107]

For a long time, ocean cables were too capricious a technology to be the basis for safe and steady business. Several authors concluded that submarine telegraphy "was nothing, if not adventurous."[108] Especially in the early decades, the companies were, similar to the railway companies during the boom of the 1840s, frequently spoken of as "gambling speculations."[109] Among the "casual sources of detriment" to the cables themselves were ships' anchors, rocks, sharks, sawfish and swordfish, teredoes (a marine worm), and "other 'common objects' of the deep sea in different latitudes."[110] In times of peace, the greatest threat to ocean cables remained, despite all ocean soundings, the unknown and unseen of the deep sea, in addition to the common fisherman.[111] The initially common occurrence of cable breakages had considerable effects on the cable companies' worth on the stock market. When a cable was damaged, it could take weeks or even months before the connection was repaired. If the company did not own a second cable along the same route or could not offer alternative transmission routes, its business came to a complete standstill. During this period, the value of shares usually plummeted.

On September 29, 1875, for example, the *Birmingham Daily Post* confirmed "rumours" that the DUSC Atlantic cable was broken, while the company still attempted to keep up the appearance that all was in working order. They even sent their messages, as Emma Pender reported, "in a covert & dishonourable manner" through their rival's Atlantic cables.[112] It was only on November 3, 1875, five weeks later, that the company announced that the broken ends of the cable had been buoyed and that "communication [would] shortly be restored."[113] When hardly a month later the DUSC cable broke again, rival Anglo-American shares "rapidly advanced, closing at a rise of 8s [shillings] for the day," while DUSC shares lost considerable value.[114] For the shareholders, interruptions usually meant that they could be "deprived of their current dividends so long as the interruption continued."[115] On the contrary, great gains could also be made, and in the long run, cable companies were a profitable investment, although entrepreneurs attempted to disguise just how profitable their cable companies were. By 1898, the Eastern and Associated Companies' shares had "only" gained 50 percent in contrast to their value two decades earlier.[116] Companies adopted various measures to obscure high profits and keep further rivals from entering the market. Early on, the companies started to pay out lower dividends and invest the remainder into a reserve fund to cover costs of cable maintenance. In the early twentieth century, most of the companies even covered the enormous costs of new cables from the reserve fund, thereby avoiding high interest on loans.[117]

But to whom were the companies paying out their dividends? For the DUSC, complete shareholder lists are accessible from 1873 until the 1920s, when the British General Post Office bought the company. These Annual List(s) of Members and Summary of Capital and Shares, which the Companies Act required British companies to maintain from 1844 on, give detailed information on name, street address, occupation, number of shares, as well as date of purchase and sale. These details reveal social and cultural aspects of the financial networks backing the emergence of a global communication system.[118]

One of the first and most obvious findings is a stark increase in DUSC shareholder numbers mirroring the general trend of the stock market's growing importance in the Western world in general and in Great Britain in particular: over the course of the nineteenth century, middle- and upper-class England became a "nation of shareholders," and by 1900, roughly two-fifths of the nation's wealth was invested in company shares.[119] Although the amount of DUSC shares, 65,000 at £20 initial nominal value, remained relatively unchanged, the number of shareholders increased steadily from 187 original shareholders in July 1873 to 922 shareholders in April 1877. When the company was relaunched in November 1877, the number dropped to 436 shareholders but rose again to 1,372 in March 1882 and 1,795 in 1887. Thereafter, shareholder numbers remained relatively steady. In the 1909 annual shareholders' lists, there were 1,780 DUSC shareholders.

Such an increase in investors in the 1870s and 1880s signaled a shift from "big money" to "small investment," that is, those with 100 shares and less. The rise in popular investment had started in Europe with the "railway mania" of the 1840s. Additionally, the emergence of financial journalism, in the form of investors' manuals and newspapers with stock price lists, charts, and instructions on problem solving, turned investors into "literate citizens" and helped the common men to "master a language that . . . before was too obscure."[120] In the concurrent reshaping of the financial market, both railways and telegraphs played an important role. Similar to the railways, the telegraphs were both a means of diffusion of stock market information and thus central to the rise of popular

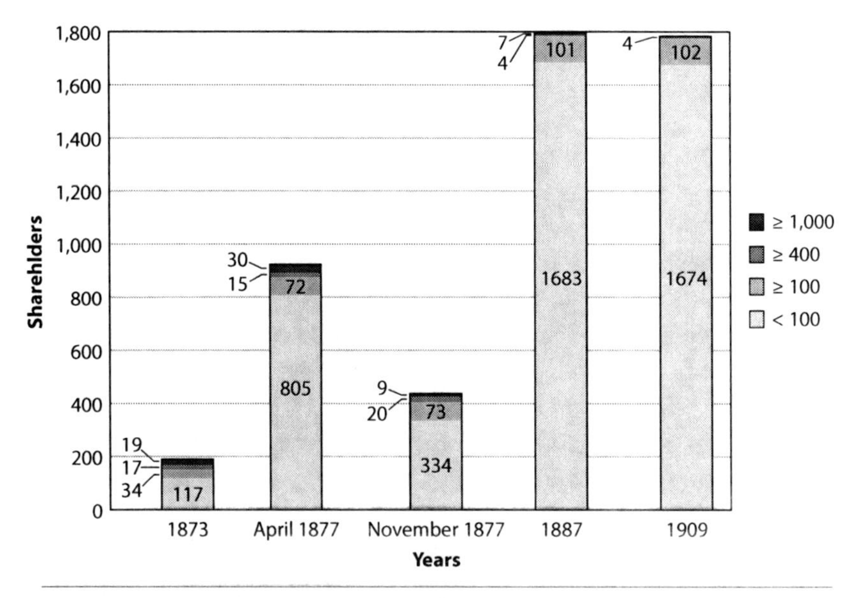

2.2 Simone M. Müller, "DUSC Shareholders Over the Years." Data source: Direct United States Cable Company Ltd., Annual List of Members and Summary of Capital and Shares of the Direct United States Cable Company Limited, 1873, 1877, 1887, 1909, Public Record Office, National Archives Kew.

investment as well as objects of investment.[121] In 1909, only 106 DUSC investors, representing 5 percent of all shareholders, owned 100 shares or more. Only five of these investors owned more than 400 shares. Solely, the Globe Trust and Telegraph Company, in itself a joint-stock company representing a larger number of small shareholders, held 9,945 shares. Comtesse Isabelle Gontran de la Baume-Pluvinel from Paris, with 420 shares, was the largest individual shareholder. This is in contrast to 1873, when 71 shareholders, representing 37 percent of all shareholders, owned 100 shares or more; 19 of these original 1873 shareholders had bought 1,000 shares or more. They represented almost 48,000 shares worth £957,820, or about 75 percent of the company's total capital.

The shift in DUSC shareholders toward larger numbers of smaller shareholders depended on several factors. It correlated with an increasing public recognition of submarine telegraphy as a technology-in-use, as well as its acceptance as a relatively safe means of investment. The best indicator for this development is the media coverage cable-laying enterprises received. After 1866, the extent of media coverage dwindled continually. Each cable-laying enterprise "gave greater experience to those involved," which meant that for the cable actors as well as for the general public, "the pioneering aspect of it all soon settled into tedium."[122] Also, the setup of the Globe Trust in November 1872 helped to install submarine telegraphy as, in the words of Globe director Daniel Gooch, "a save [sic] and sound investment."[123] The great jump in the number of shareholders after 1877 relates to the joint-purse and working agreement, as well as the end of the tariff war between the DUSC and the Anglo-American Telegraph Company. Although the Atlantic pool's monopoly prohibited fair competition and potential lowering of tariffs, from an investment perspective, it placed the company on safer footing. The Atlantic pool agreement foresaw that in case of cable breakages, messages would be rerouted through the other members' cables, limiting the traffic loss.[124] Another aspect that made DUSC shares attractive was the company's high technological standards for its cables. It was the first to inaugurate fast working on Atlantic cables. The company used a new design invented by Siemens Brothers allowing for faster transmission speed. In addition, in 1878, its cable was duplexed. Now the cable could send sixteen words per minute each way over a length of 2,420 nautical miles.[125] Prior to the DUSC cable, receipt of a reply to a cablegram between New York or Boston and Europe within thirty to forty minutes had been considered a remarkable transmission time. Already in the 1870s, DUSC people brought this transmission time down to ten minutes. In 1893, *The Sun*, a New York newspaper, reported that the results of the Oxford and Cambridge boat race were received in the United States via DUSC cable, within thirteen seconds of the finish of the race.[126]

Alongside the number of shareholders, their regional constituency also changed. Over the years, DUSC capital became increasingly British in terms of its shareholders, as the stock market moved, tightly connected to the spread of the telegraph network, from the great commercial and industrial centers into Britain's midsize towns. At the same time, the number of non-British investors significantly decreased. In April 1877, 79 entries had an address from outside the United Kingdom (UK). These foreigners made up roughly 8.6 percent of the total number of shareholders and represented 22,748 shares (about £454,960 in nominal value) or 35.2 percent of the total capital. Strongholds of non-UK shareholders were the centers of commerce such as Antwerp

(18 shareholders at £142,500 nominal value = 11.0 percent of the total capital), Brussels (6 shareholders at £17,560 nominal value = 1.4 percent), and Paris (30 shareholders at £147,100 nominal value = 11.4 percent). These numbers could have been even higher, because some non-UK stockholders still might have used the London stock market. The annual lists also reveal the company's particular connection to Germany at the time. Eight shareholders came from Berlin and one each from Bremen, Breslau, Cologne, Frankfurt, and Königsberg. They represented a nominal value of £94,660, or 7.3 percent of the company's total capital. Additional places of foreign investment were Florence, Lyon, and Vienna. In the early years of the company, the foreign shareholders' addresses correlated directly with locations of industrialization and trade, places where people placed a high importance on up-to-date international information. About half of the foreign investors gave "merchant" as their occupation, in addition to some "bankers," or "agents du change."

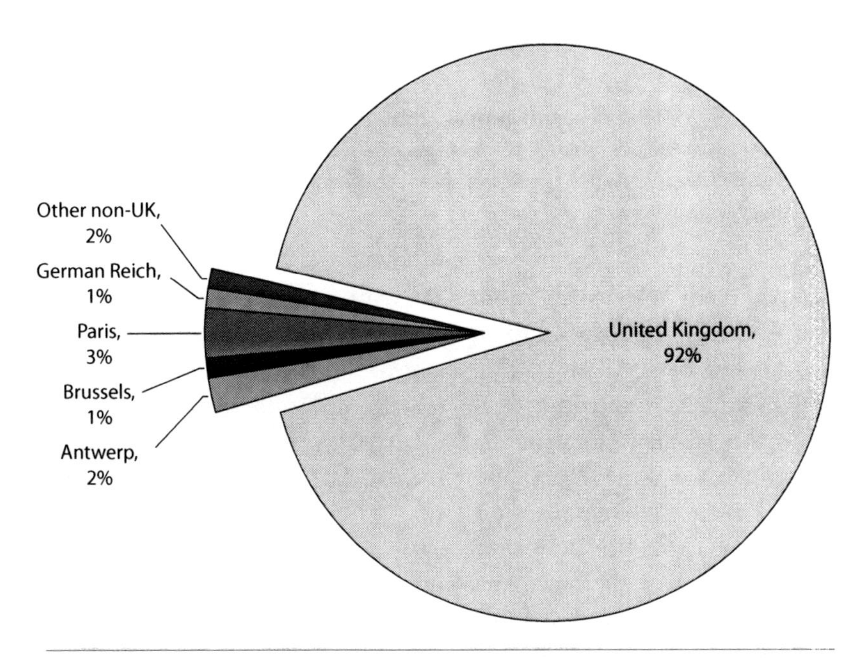

2.3 Simone M. Müller, "Foreign Shareholders, 1877." Data source: Direct United States Cable Company Ltd., Annual List of Members and Summary of Capital and Shares of the Direct United States Cable Company Limited, 1877, Public Record Office, National Archives Kew.

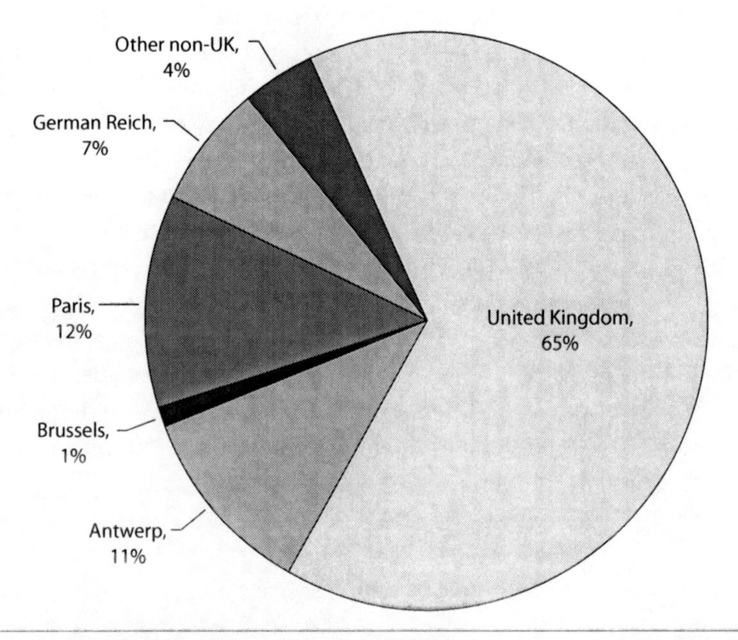

2.4 Simone M. Müller, "Foreign Capital, 1877." Data source: Direct United States Cable Company Ltd., Annual List of Members and Summary of Capital and Shares of the Direct United States Cable Company Limited, 1877, Public Record Office, National Archives Kew.

Among them were such prominent names as the Russian banker Joseph Günzburg, then residing in Paris; Charles Meara of the Antwerpian trading house C. Meara & Co.; and Gaston Dreyfus, one of the best-known bankers of the time. Of the remaining "gentlemen," many, such as the Antwerpian merchants Louis Lemmé, proprietor of the trading house Louis Lemmé & Co., and Gustav Grisar of G. & E. Grisar, also worked in the trading profession.

Later on, with the terrestrial telegraph system, the stock market also reached from the commercial centers into the British countryside. Stockholders no longer came solely from the cities of London, Glasgow, Liverpool, or Dublin, but from Chiswick, a popular country retreat for Londoners with nothing but an agrarian and fishing economy; Ewell, a similarly small place outside of London; Huddersfield, a small cotton-mill town between Manchester and Leeds; or Wincanton, Somerset, a small town between London and the southwest of England. Stockholders increasingly originated from small or midsize towns that were of little importance from an industrial or financial point of view. Such geographic diversification showed that the sense of a global connectedness, or rather its importance, had moved from the centers of finance and trade into the countryside. It now equally reached St. Albans in the north of London, Oswestry in Wales, or Scarborough on the North Sea coast of North Yorkshire. At the same time, as its stock "moved" out into the countryside, the company became increasingly British. In 1909, only 26 shareholders had an address from outside the United Kingdom. They made up 1.5 percent of the total number of investors representing shares worth £33,980 nominal value, or 2.7 percent of the company's capital. In a sense, the company was being financially "nationalized."[127] Moreover, the Globe's hostile takeover of the DUSC in 1877 furthered this process. From November 1877 on, the Globe Trust and Telegraph Company was always by far one of the largest shareholders.[128] It remains unclear whether this nationalization was because the buyers became increasingly national or because small shareholders invested in domestic companies that they knew.

The increasing nationalization of small investors meant that the social composition of investors changed drastically. Investors' occupations became increasingly diverse in social structure. Submarine telegraphy was no longer an investment for those who directly benefited from the use of submarine cables, such as merchants, bankers, shipowners, or stockbrokers, such as Louis Lemmé, Gustave Dreyfus, or Werner Siemens. It also appealed to other professional groups, such as clerks, schoolmasters, ministers, and students who merely saw it as an investment. With high probability, John Bidgood, a gardener from Exeter (six shares), Thomas Forster, an art student from Wills (four shares), and John Anderson, a bootmaker from Bridgenorth (eight shares), were not involved in any transatlantic trading or financial transactions and might never have sent a cablegram to North America or some other place in the world. Yet they might have recognized the cables as a key technology of their time providing a connectivity to faraway places around the globe—a connectivity that they certainly consumed and experienced in the form of coffee or tea, news reports, and world exhibitions or simply by the imperial expansion of their home country, most noticeably seen in 1877 when Queen Victoria had herself proclaimed Empress of India. Investment in ocean cables was particularly attractive for those living right next to a cable station or working at one of the cable stations of the world. Several stockholder entries have addresses in Heart's Content, Newfoundland; Brest, France; and Weston-super-Mare, England, three of the seven main Atlantic cable landing places.[129] From April 1877 on, James Graves, superintendent of the Atlantic cable station in Valentia, Ireland, is listed. He owned the considerable number of 155 shares and exemplifies how much the cable operators identified with their profession. Certainly, the staff also had the advantage of inside information over other investors.[130]

Among those small investors, women showed a particular interest in ocean cables as a means of investment. This might come as a surprise, as women are usually reported as having played no great part in the history of ocean cables and the globalization of communication. Their greatest presence lies in their absence. Contemporaries and scholars today usually consider women en passant while attributing rather passive roles: they are portrayed as the patient, faithful, and devoted wife to the ingenious entrepreneur; she bears all hardships willingly and silently, although the ocean cable business is an increasingly "unwelcome intrusion" into her social and family life.[131] These women, as for example Mary Stone Field, deal with their husbands' months of absence as well as with the fact that they are risking their family's financial security with stoic indulgence. Throughout the entire twelve years of the Atlantic cable project, the Field family's fortune "had been linked precariously with the success of this strange wire at the bottom of the ocean." Mary Stone Field had born it all without even one complaint.[132] Even at the cable stations, women were bound to the household sphere. In contrast to the landlines, at the Atlantic cable stations, female operators were not employed before the twentieth century.[133] Women were seen as beautiful adjuncts to the male undertaking, and generally, the narrative places them at the margins of the development of global communication and processes of globalization. These seem to be "a fundamentally masculine activity."[134]

Yet, women played a considerable role within the story of the globe's telegraph network as financiers. Their capital generally made up a substantial part of financial resources backing the industrial economy of the nineteenth century.[135] Contrasting the separate spheres thesis, contemporary share registers and investment advisers' records show that women were active in the market. They were not necessarily solely forced into the restricted roles of wives, mothers, or helpmeets and thus excluded from active participation in economic and social life.[136] Emma Pender's diaries and letters, for instance, reveal an economically independent and shrewd business woman who bought land and shares and borrowed and lent money.[137] An article from 1886, commenting on the visitors to the Atlantic cable stations, remarked that "elderly ladies" who came to visit not only "display[ed] an evident degree of common sense" concerning the working of the telegraph, but were also "frequently . . . pecuniarily interested."[138]

The DUSC shareholder lists reveal that the numbers of female investors increased considerably until, by the beginning of the twentieth century, women made up almost half of the company's shareholders. For July 1873, there are only seven entries marked as female. These seven made up 3.7 percent of all shareholders and represent 1,852 shares, or about 2.8 percent of the total capital. By 1887, this number had significantly increased: of 1,795 shareholders, 444 entries, roughly 25 percent, are distinctly marked as belonging to women.[139] They represented 10,748 stocks with a nominal value of £20, which was about 16 percent of the company's capital. This rise of female investment between 1873 and 1887 is closely connected with the Women's Property Act of 1882. The English Married Women's Property Act of 1870 had already recognized a woman's right to maintain property separate from her husband's control. Yet in 1882 these rights were considerably expanded.[140] These changes directly translated into the cable company's shareholders list.[141] Until the 1880s, married women's shares had to be listed under their husband's name. Now, they appeared as fully independent entities. Only an insignificantly small number of women among the DUSC's members are listed together with their husbands or a male guardian. The analysis of shares from February 1909 shows that the increase steadily continued. By the beginning of the twentieth century, 784 shareholders were female, or 44 percent of the total.[142] Yet they only represented 17,264 shares at £20, worth £345,280 in total, which translated to about 26 percent of the company's total capital.

Female shareholders' relatively small capital investment made them the typical small investor. In 1887, only sixty-eight women owned more than thirty shares; more than half of them owned ten shares or less. By the 1880s, however, the time of big investment in telegraphy was generally over. Still, over the years, there are some exceptions, usually wealthy widows, such as Elise Louise Adlegonde Cromlery from Belgium or

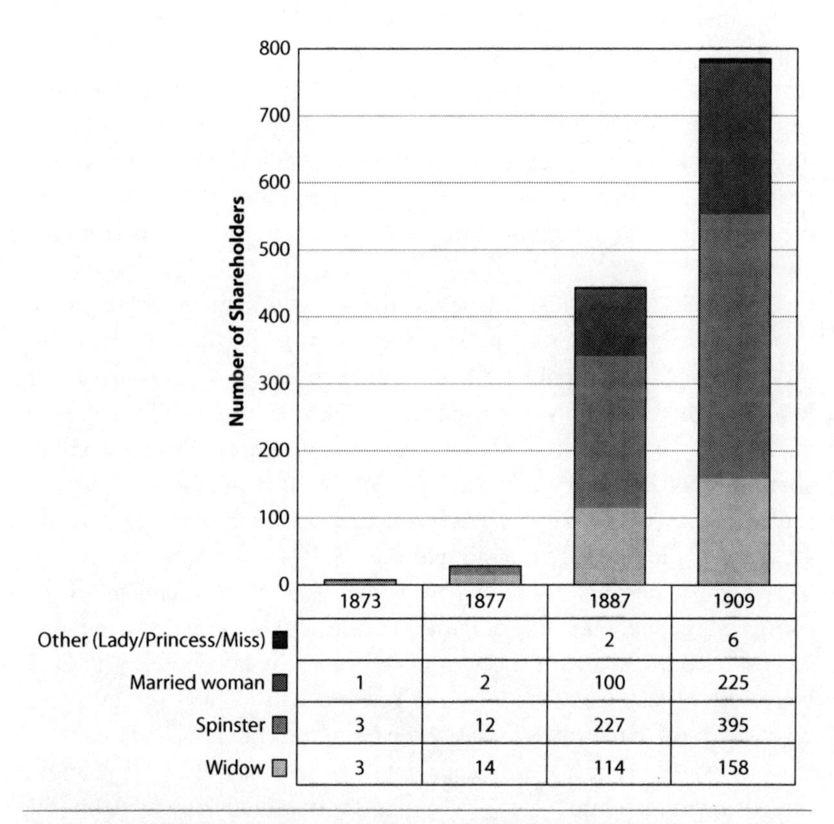

	1873	1877	1887	1909
Other (Lady/Princess/Miss)			2	6
Married woman	1	2	100	225
Spinster	3	12	227	395
Widow	3	14	114	158

2.5 Simone M. Müller, "DUSC Female Investors, 1873–1909." Data source: Direct United States Cable Company Ltd., Annual List of Members and Summary of Capital and Shares of the Direct United States Cable Company Limited, 1873, 1877, 1887, 1909, Public Record Office, National Archives Kew.

Isidora Collier de la Martiere from Paris, who held exceptionally large numbers of shares. In 1909, only 164 women held 30 shares or more, and only 24 held 100 shares or more. Yet Comtesse Isabelle Gontran de la Baume-Pluvinel, who came from an old noble family, represented the largest individual shareholder, man or woman.

Women flocked to investment in telegraphy for a variety of reasons. The rise in female investment coincided with the era of the New Woman at the end of the nineteenth century when women challenged traditional limits of a male-dominated society. Stock brokerage represented a means for new female freedom as reached in the Women's Property Acts.[143]

Theoretically, women could not only invest but also participate in the various shareholder meetings and vote for the board of directors. Practically, they were never employed as directors or even reported to have attended a shareholder meeting. Stock investments had a special appeal for middle-class women, who were generally denied access to the professions and excluded from entrepreneurial activities but who still needed to make money. This was particularly the case when they were unmarried, which explains the high number of spinsters among female cable investors.[144] Throughout the years, the group classified as "spinsters" generally made up the majority among women investors. In 1909, for example, of 784 female shareholders, 395 or roughly 50 percent were spinsters as compared to 29 percent who were married and 20 percent who were widows.

Typical female investments, such as government bonds, banks, railways, utilities, or debentures, were safe and low risk. The ocean cables, with their likelihood to break, fierce competition, and ultimately great highs and lows on the stock market, should not have ranked high among female investors.[145] Still, female investment in ocean cables was as high as 25 percent in 1887 and 44 percent in 1909. Women might have been attracted to submarine telegraphy by the relatively high revenues, the fact that the cables were not as risky as thus far perceived, and the accessibility of the companies' product. From the very beginning, ocean cables had been a very public project, highly visible in the media. Women were usually kept out of traditional circles and places of information, such as clubs or fraternal lodges, where relevant business and stock market information was being traded.[146] In contrast to such male secrecy about stock market information, the progress and the failures of the cables could be easily followed in the daily papers. In 1876, a charge was even filed against DUSC when it had allegedly not immediately reported on its cable's breakage.[147] Information relevant to stock market investment could be gathered at home so that women could act relatively independently without having to leave "their sphere." There are hardly any primary accounts of women discussing their financial strategies to allow a conclusive statement on the specific motives for women to invest in submarine telegraphy. Nevertheless, the shareholder lists underline the importance of female capital for establishing global communication and facilitating globalization processes.

Cable entrepreneurs and investors were very diverse and ranged from the "great men" to the small investor. In contrast to the much publicized story of John Pender as the sole anchorman of ocean cabling, a closer look at the boards of directors reveals how a small group, the network of the Class of 1866, controlled almost all global communication via cross-directorships and multiple directorships. Throughout the nineteenth century, their monopoly was repeatedly challenged, most successfully by Siemens Brothers in the 1870s. At the end of a fierce cable war with Pender's business consortium, Siemens Brothers managed to establish itself as a viable alternative on the cable manufacturing market. The antagonists' battle for Atlantic cable supremacy not only shifted power relations on the market, but also revealed a deeper divide over how these cable actors envisioned global communication. Yet, these visions of the globe's "electric union" concerned not only its economic structuration but also the system's moral implications. Ultimately, the cable network produced both a unified market of goods and a unified market of morality; along with the transportation of stock market information, contemporaries believed the cable network would help the spread of values and ideas such as universal peace or Europe's civilizing mission.

3 | THE IMAGINED GLOBE

Father rejoiced like a boy. Mother was wild with delight. Brothers, sisters, all were overjoyed. Bells were rung, guns fired, children let out of school shouted, "The Cable is laid!" "The Cable is laid!" The Village was in a tumult of joy.

David Dudley Field to Cyrus W. Field, *Telegram*, August 9, 1858.

THUS RAN the congratulatory telegram David Dudley Field sent to his brother Cyrus, vividly capturing the excitement sparked by the successful laying of the 1858 Atlantic telegraph. Not only Field's hometown cheered the cable's advent; celebrations were held "ocean-wide" as both sides of the Atlantic commemorated "with electric enthusiasm . . . the nuptials of the Old World and the New."[1] During the summers of 1858 and 1866, banquets, speeches, sermons, and parades erupted all over America, Great Britain, Ireland, and Newfoundland. Even at places not directly linked by the cable, orators dwelled on the merits that the Atlantic telegraph seemed to promise for the world's union, its "civilization," and "universal" peace.[2] The system of ocean telegraphy de facto came to represent a Eurocentric exclusiveness in the form of an economic undertaking of enormous scale, intended to foster global trade as well as to serve its agents' financial interests. In its public perception, the arrival of the Atlantic cable set not only a small village in the middle of Massachusetts, but almost the entire Euro-American world, into a state of enthusiastic frenzy. The agents' breakthrough with a technology that only some would be able to use was indissolubly connected to a vision, a promise almost, of a brighter future for all. The new technology's avowal of a peaceful and civilized modernity entailed an imagined global unity that reached far beyond the telegraphs' actual means of point-to-point communication between the world's centers of urbanity.[3] Ultimately, the cable network produced not only a unified market of goods but also a unified market of morality; it would not only transport stock market information but also help the spread of values and ideas.[4]

Contemporaries discussed the imagined global impact of the telegraph, referring to ideas of an electric union, universal peace, and the telegraphs' civilizing mission. The rhetoric of universal peace served as placeholder. It exemplifies how they structured their global images as well as visions of participants in, and beneficiaries of, the advantages of global communication, according to Western-centric thinking. The mid-nineteenth-century notion of an electric union and the pacifist concepts connected with it are some of the first expressions of a global imaginary in the history of modern globalization. For the cable actors, these discourses formed their economic and political realm of action—they were duly implemented into their sales rhetoric—and influenced their cultural way of thinking. A variety of explanatory ideologies nourished contemporaries' understanding of universal peace between the 1850s and the First World War, drawing from the political-economic philosophy of Manchester Liberalism, the idea of a *Societas Christiana*, or *l'esprit d'internationalité*. All of these ideologies were expressions of an elitist worldview that unthinkingly excluded the vast majority of the planet. Finally, telegraphy also functioned as an instrument for Euro-America's civilizing mission—a concept that is inextricably linked to the notion of the telegraphic progress as well as the engineer as the "great civilizer." All of these ideas were inherently connected to the expansion of submarine telegraphy. They are verifications of an underlying philosophy of technology that expressed itself in the utopian and Eurocentric ideas of a world society of "kindred nations" in a world that was progressively "civilizing" itself according to the European model. During the Great Atlantic Cable undertaking, these ideas of universal peace and telegraphy's civilizing mission were as prevalent and important to the undertaking as the furtherance of world trade and economic prosperity.

TECHNOLOGY, UNIVERSAL PEACE, AND THE WORLD'S ELECTRIC UNION

> Glory be to God in the Highest, on earth peace, goodwill towards men. The Queen desires to congratulate the President upon the successful completion of this grand international work.[5]

On August 16, 1858, Queen Victoria and the American President James Buchanan exchanged congratulatory telegrams to officially open the Great Atlantic Cable. With her wish for "peace" and "goodwill," Queen Victoria gave prominent expression to a broader public discourse on the Atlantic telegraphs and universal peace. The expression "peace and goodwill" is a quotation from the Bible that contemporaries easily understood. The phrase, taken from the Christmas story in Luke 2:14, celebrates the advent of Christ as the Prince of Peace.[6] Its usage in the submarine cable context demonstrated that, aside from an economic reality, people in Europe and in North America connected an imagined globe with the cables' proliferation as well as the advent of a new era of a Christian civilization. Indeed, as the British member of parliament John Bright emphasized in a speech at an Atlantic cable banquet in 1864, there seemed to have been not one man in 1858 who had not felt "that a new world and a new time were opened to him."[7]

The Great Atlantic Cable distinguished itself from other engineering undertakings of the time. One of these distinctions was the ubiquitousness of the cable's public perception all across the Western world. According to cable chronicler Henry Martyn Field, in a "history of popular enthusiasms . . . large space [must be given] to the Atlantic Telegraph."[8] The response of a generally Euro-American public to the laying of the first Atlantic cables resulted in vast amounts of newspaper accounts, pamphlets, poems, songs, leaflets, and cable souvenirs, such as the jeweler Tiffany's engraved piece of cable. Similar to Queen Victoria's telegram, many of these accounts centered upon the cables' inherent moral and civilizing promises and a "drawing together of all parts of the globe in one single world."[9] The sources mirror a still undiluted belief in progress inherently connected to the era's technological developments. As Charles Bright, son of the famous cable engineer Charles T. Bright, concluded in his 1903 cable book, "[a]nticipation and reaction to the cable became a celebration of the union of all the families of man under the dominion of one science and one art, made visible in steam locomotives and electric wires."[10]

Many newspaper articles contained the topos of "instantaneity."[11] The cable, "a tie nearly as subtle as that of love," had connected the two hemispheres by means of "instantaneous intercourse . . . as though no Atlantic rolled between [them]."[12] Indeed, the "marvelous cable" turned the Atlantic into a mere "whispering gallery" and enabled "each world, . . . almost instantaneously, to feel the heart-pulses of the other beat."[13] Now that

England and America had been brought into speaking distance, the cable would "be the first to tell England and America what each [was] thinking of the other."[14] Many a contemporary hoped that instantaneous communication was the key to a world in peace. This idea was based on a theory of communication that saw the act of communicating as the transmission and reception of ideas. It presumed that transnational and cross-cultural contacts enabled people to better understand the complexities of another community and that communication enhanced empathy for differences. The more frequently this happened, the better.[15]

The concept of peace through telegraphic, and thus instantaneous, communication, however, did not originate with the Atlantic cable of 1858, nor did it vanish with its failure only a few weeks later. Rather, it represented the pinnacle of a discourse that accompanied means of modern communication from their very beginnings. Already with the advent of optical telegraphs around 1800, enthusiasts in the United States "hailed the potential of the new medium."[16] Haunted by the "specter of disunion," early Republicans, such as Treasury Secretary Albert Gallatin, urged that there was no other solution "to the 'inconveniences' and even 'dangers' posed by the enormous size of the United States" but to establish speedy communication throughout the entire country.[17] In the electric telegraphs, which during the 1830s quickly displaced optical telegraphy, peace advocates saw their greatest strength in their inherent speed by which distance could be crossed in almost no time. Through telegraphic instantaneity, communication theoretically controlled actions even over great distances. Misunderstandings could be corrected at the moment they appeared, instead of giving problematic messages days to sink in before the explanatory letter arrived. According to Napoleon III, in the age of the telegraph, "any misunderstanding . . . might be readily rectified."[18] The magazine *Punch* saw submarine telegraphy especially well-suited to meet expectations of the creation of peace.[19] These cables established communication *between* but not, as most land telegraphs did, *within* nations or communities. They created connections between regions that had for centuries been separated by seemingly unbridgeable geographical impediments. Already in 1855, Samuel F. B. Morse predicted that one of the effects of the telegraph would be "to bind man to his fellow-man in such bonds of amity as to put an end to war."[20]

The idea of the globe's electric union also manifested itself in iconography. One engraving in particular, showing two winged female figures, came to represent the cables' mission of peace. It demonstrates the modularity and translation of the idea of an electric union into various contexts around the globe. The cartoon "Effect of the Submarine Telegraph: Or Peace and Good-Will Between England and France," which in contrast to its messages still needed to be transported between the Americas and Europe by ship, originated in the context of the Brett brothers' submarine cable across the English Channel of 1851. In the picture, two "mermaids" of Greek stature with angelic features follow the cable's path along the ocean ground. They deliver the olive branch, representing peace, from England to France. Just as the cable had taken its way from the British Isles to continental France, the mermaids were represented as the peaceful outreach of the English to their long-time enemy. Broken weapons, remains of sunken ship wrecks, and human skulls, representing centuries of belligerent relations between those two nations, are scattered on the ocean bed. Alongside the cartoon, *Punch* published the

3.1 Unknown Artist, "Glory to God in the Highest, and on Earth Peace, Good Will Toward Men!," in *Harper's Weekly. A Journal of Civilization*, September 4, 1858, p. 16.

"mermaid's song" dwelling on the marvel of an "enchanted wire" running "from shore to shore." Through a "conduit of a magic fire" kindly spirits are wishing "To England, Peace! – to France, Good-will!"[21]

The frequent use of the mermaids within pacifist concepts soon turned them into a household word. In particular the biblical reference to "peace and goodwill" developed into a well-known topos, which traveled telegraphic space. In September 1858, the American magazine *Harper's Weekly* printed the very same picture in their special edition on the Atlantic cable expedition. The engraver only added small details to the original picture of 1851, such as the cable ship *Great Eastern* and star-shaped ornaments for the mermaids, thus making them look as if they were dressed in a Star-Spangled Banner. In effect, he simply exchanged France for America.[22] Similar to the article in *Punch* from 1851, *Harper's Weekly* expressed its hope that a telegraphic line would ameliorate existing Anglo-American tensions. In this pictorial transfer of the mermaids' song across the Atlantic, they substituted the Anglo-American context for the Anglo-French one, without making any distinction between the countries' varying histories of transnational relations. Communication via ocean cable appeared as a panacea for international tensions and conflicts, a panacea that was—using the metaphor of an ever expanding net of wires—globally applicable and ever spreading. In the mid-nineteenth century, the image of the mermaids and their promise of universal peace traveled along the telegraph lines, accompanying the globalization of communication.

It is difficult to determine whether the discourse on universal peace was primarily manipulated by the cable agents' promotional rhetoric, was due to sheer exaggeration, or stemmed from the sincere belief that ocean telegraphy could truly accomplish a global electric union. Nevertheless, the cable agents' and the publics' universal peace aspirations are not to be taken as utterly naïve or to be dismissed too lightly. Their hopes were not entirely without foundation, as the timing or, rather, the sequence in which people learned of events could be more important to resolving conflict than the actual chronology. Prior to the introduction of ocean telegraphs, these could vary considerably. On Christmas Eve 1814, for example, a peace treaty ended the War of 1812 between Great Britain and the United States. However, news of that treaty did not reach Washington before February 1815. In the meantime, on January 8, 1815, U.S. troops led by Andrew Jackson defeated the British at the Battle of

New Orleans. According to historian David Nickles, faster communication, such as a transatlantic cable, would have "averted that engagement, saved many lives, and, in all probability, prevented Jackson from later becoming president of the United States."[23]

In addition to ocean cables, other enterprises of civil engineering were also seen as centerpieces of a transboundary distribution of universal peace. In the decades after 1860, when many natural, century-old barriers, such as mountain ranges and oceans, were overcome to facilitate transnational railway and steamship connections, contemporaries hoped that this would develop new neighborhoods of kindred nations and relationships of mutual respect and friendship.[24] In 1861, upon the opening of the South Eastern Railway Company's new rapid steamship between France and England, an article in *Punch* concluded that it would undoubtedly improve Anglo-French relations. According to the author of "Neighbours Getting Over Their Distance to One Another," it was the kettle "in the vapour of which young James Watt prophetically saw the first steamer," which would turn out "to be the most powerful pacificator the world has ever known." Emphatically, the writer suggested that the Peace Society should adopt the "kettle as their crest."[25]

As the engraving "L'Arrivée à Paris" from 1884 and the 1869 gold medal commemorating the opening of the Suez Canal demonstrate, all of these grand engineering projects seemed to be embedded into similar pacifist discourses, irrespective of whether it was the Great Atlantic Cable, the Suez Canal (1869), the tunneling of St. Gotthard in the Alps (1881), or the scheme of an English Channel tunnel. All of these projects overcame geographical obstacles that had for centuries been considered to be permanent. These technological breakthroughs allowed, as Reverend Cortlandt van Rensselaer, an American Presbyterian, pointed out, that "man walk[ed] beyond the bounds of his domain" and reached out into the world with a peaceful hand.[26] All these projects would be spearheaded by Eirene, the goddess of peace, and "prépare la paix du monde."[27] Ferdinand de Lesseps, the builder of the Suez Canal, expressed the very same notion. To him, the great construction projects of the nineteenth century represented "enterprises of universal interest" that had an identical purpose: "to bring peoples closer together and thereby to bring about an era in which men, by knowing one another, will finally stop fighting."[28]

Yet to whose distance was de Lesseps, Field, or one of the many other creators of global development projects at the time referring? These pacifist concepts need to be read through the lens of Euro-American power relations. In the case of the Atlantic cable, this certainly means the American Civil War and the emerging concept of Anglo-American special relations.[29] According to Queen Victoria, the Atlantic cables established not only telegraphic communication but also "an additional bond of union" between the United States and Great Britain.[30] Moreover, they were meant, as the American President Buchanan emphasized, to serve the promotion of "perpetual peace and friendship between the *kindred* nations."[31] Due to a common, if not shared, history of colonialism, and through mutual language and heritage, it became increasingly common in the nineteenth century to refer to both nations as familial. Texts implicitly drew from the racial concept of Anglo-Saxonism, which developed at the time. Beginning with the early 1800s, Englishmen and Americans started to compare the Anglo-Saxons to each other. Their conclusion was that the "innate characteristics of the race," and not environment or accident, had led to their rise and success.[32]

The experience of how close both nations came to war during the 1860s helps to explain some of the later concepts of peace. After the War of 1812, relations between Great Britain and America developed peacefully, and by the outbreak of the American Civil War in April 1861, they were more dependable than they had been since American independence.[33] Still, it is remarkable that city dwellers living along the Atlantic shoreline were debating early terrestrial telegraphic lines in the United States in the 1840s as an early warning system in case of foreign, i.e., British, attack.[34] The Civil War put new strains on the vulnerable Anglo-American relations. Events such as the Trent Affair of 1861, when Americans took two Confederate diplomats hostage from the British mail steamer *Trent*, or the Alabama Claims in the war's aftermath, which were claims lodged by the United States against Great Britain for damages caused to the United States by British support for the Confederacy in the Civil War, contributed to mistrust and general estrangement. Both incidents express the North's fear and the actual possibility that Great Britain might take sides in the Civil War.[35] Indeed, Anglo-American relations remained marred by mutual distrust, suspicion, and antipathy until the end of the nineteenth century.[36] Wishes for "peace and goodwill" expressed both nations' hope that cables could serve to facilitate diplomacy, the emerging concept of "kindred nations," and Anglo-American special relations.

Nevertheless, universal peace and "peace and goodwill" would not concern every nation or community on earth, only the few Euro-American industrial nations. Universalism was "incorporated into the nation-states and their societies."[37] The term was employed within a Eurocentric rhetoric of imperial and industrial power relations and only valid as such. With regard to the submarine cables laid in the Irish Sea, an English journalist even warned how fatal a message it would be for Ireland, "if [the British] interpreted the first dispatch of this new agent of intercourse [the cable] as a decree of perpetual subjection to our country"; quite the contrary.[38] Still, considering how entirely the world was dominated by European powers in the Age of Imperialism, there is a certain logic to contemporaries' Eurocentric assumption that "la paix Européenne" would suffice to spread peace "universally."[39]

Modern means of communication and transport in the nineteenth century, ranging from the Atlantic cable to the Gotthard Tunnel, played a crucial role in contemporaries' imagination of a unifying world. They saw the abolition of distance by bringing neighbors into speaking distance with each other as the enabling structure for peace. In this train of thought, the Atlantic cable garnered special attention. Its speed brought a new sense of instantaneity and, with it, a new sequence in the chronology of events irrespective of distance. With its globe-spanning implications, it symbolized the means to make this peace universal. However, the notion of a global electric union only served as placeholder for various regionally generated ideas of morals, values, and progress.

MANCHESTER LIBERALISM AND *SOCIETAS CHRISTIANA*

Beyond a mid-Victorian belief in progress employed within a Eurocentric system of imperialism, there were two additional sources of influence within peace rhetoric: Manchester Liberalism and its sense of a correlation between the spread of markets and peace, and the notion of a unified Christian globe. The sources reveal an often blurred divide between the British and the American actors concerning their "global" imaginary of a unifying world. Although religion, for example, had little influence on James Anderson and John Pender, it played a tremendous role for the Americans Cyrus W. Field and Samuel F. B. Morse. The congratulatory telegrams of the mayors of London and New York in 1866 best illustrate this divide. On August 4, the Lord Mayor of London telegraphed that he hoped that their "commerce [would] flourish" and that "peace and prosperity [may] unite [them]."[40] His wording is in accordance with the logic of Manchester Liberalism. Only hours later, he received the reply from the mayor of New York. The American in turn highlighted, "the Providence of God" that had directed the "energy and genius of men" in this work. The cable may be "instrumental in securing the happiness of all nations and the rights of all people."[41] This juxtaposition of commerce and the providence of God plays a tremendous role in explaining the pacifist discourse of the Great Atlantic Cable. It also influenced the setup and organization of the global media system and the cable agents' actions within and beyond this system. Finally, it marks a fine Anglo-American distinction between an emphasis on commerce versus religion.

In mid-nineteenth-century Europe, liberalism was probably the most popular model of explaining international relations and was soon incorporated into the British peace movement.[42] The British cable concepts of peace were heavily influenced by a concept that had originated with the English revolution in the seventeenth century. It considered peace as the basis for utilitarianism or, according to William Penn, as a means for the "protection of property."[43] Britons of the time did not view striving for personal economic prosperity and riches through trade and expansion as contradictory to aspirations for universal peace. Rather, contemporaries believed in a positive correlation between free trade, the establishment of international markets, and peace. They followed the argumentation of commercial liberalism. Primarily influenced by Adam Smith's *Wealth of Nations*, commercial liberalism in the nineteenth century had developed as a new international order that based itself on the freedom of trade and the rights of citizens to freely engage in private actions across the borders of states. From the eighteenth century on, influential thinkers such as James Mill, John Stuart Mill, Jean-Baptiste Say, and Richard Cobden had made the argument that close economic contacts contributed to peace by making war irrational and useless.[44] The development of ever-faster means of transport and communication were essential tools for integrating markets into ever-larger spheres of activity, while the peaceful impact of the mid-nineteenth-century trade revolution was expected to be on a "planetary scale."[45]

Both key protagonists of Manchester Liberalism, John Bright and Richard Cobden, supported the cable project. The two merchants from the Manchester area had been closely associated with each other ever since their engagement in the Anti-Corn Law League in 1839. The League challenged British protectionist policies that strictly regulated foreign imports in the agriculture sector and caused tremendously high food prices. Repeal of the Corn Laws became the symbol of their campaign for free trade. Its success in 1846 made Bright and Cobden national celebrities. To a large degree created by the Anti-Corn Law agitation, the economic-political movement of Manchester Liberalism based itself on the principles of laissez-faire, noninterventionism, and free trade and saw the spread of universal peace as a logical corollary to its theories. In the words of John Bright, a Quaker, Manchester liberals believed that free trade "would unite mankind in the bonds of peace."[46] The railroads, steamboats, cheap postage, and telegraphs were vital means to "keep the world from actual war."[47] Global integration, in their view, resulted in a particular kind of global interdependence that rendered war impossible. From the early 1850s on, Cobden and Bright also dominated the British peace movement. They were the heads of the British Peace Society's Manchester and Salford Auxiliary and participants in the various international peace congresses taking place in the aftermath of the Europe-wide, but largely unsuccessful, revolutions of 1848, in which tens of thousands were killed.[48] The "Manchester Peace Gods," in the words of the magazine *Punch*, equally aimed at the commercial as well as the pacifist integration of the world.[49] The Atlantic telegraph seemed to provide them with both.

Both Bright and Cobden were from the beginning extremely supportive of the Great Atlantic Cable project. John Bright not only backed the undertaking in the British House of Commons, but also remained supportive during the lean period of the early 1860s.[50] The relationship between Field and Bright was especially close. During the American Civil War and the Alabama Claims crisis, they exchanged information on their respective country's sentiments.[51] Even after the cable enterprise was over, Bright's relationship to the Class of 1866 remained close. The politician attended most of the commemorative cable banquets in the decades to come. Although he died before the 1866 cable was completed, Richard Cobden was similarly attached to the undertaking and had been one of the early visionaries of a telegraphic connection to America. According to Field, in the aftermath of the Great Exhibition of 1851, Cobden had already negotiated with the Prince Consort that the exhibition's profits should go into establishing telegraphic communication across the Atlantic. Later on, when the idea of an Atlantic cable materialized in the form of Field's cable company, he aided the project politically, supporting, for example, the notion that the British government should supply one half of the capital necessary for the undertaking.[52]

The geographical composition of the cable agents and shareholders also suggests a strong influence from the Manchester school's way of thinking and its conception of a peace movement through free trade. In addition to John Pender, a large group of early Atlantic cable investors were merchants from Manchester. The "cotton metropolis" of North West England had, in the nineteenth century, become synonymous with the model of industrial capitalism, which was undisturbed by state intervention and followed the principles of Manchester Liberalism.[53] Due to the commitment of Bright and Cobden, it had also become the center of the British peace movement. The last of the international peace congresses organized in the aftermath of 1848 took place with about 500 delegates in Manchester in 1853.[54] Furthermore, the Cotton Famine (1861–1864) in North West England during and partially due to the American Civil War and interruptions in cotton trade forcefully showed the impact of war on international trade. It demonstrated that peace was indeed advantageous to commerce.[55] In the public speeches or letters of many of the British cable agents, this idea is tangible. In 1886, for instance, James Anderson, general manager of the Eastern Telegraph Company's submarine system, delivered a speech before the British Chamber of Commerce. He attested to the strong influence of the Manchester School. According to Anderson, submarine cables were not laid for a time of war, but as a means for an international economy. They enhanced the development of a "growing federation of commerce in which all nations [were] free to join." Foreign trade became "an extended home trade" and produced an interconnectedness that made war unprofitable.[56] The notion of a supremacy of economic and commercial ideals played a tremendous part in the setup of the global system. For actors like John Pender, James Anderson, and William Hay, who dominated the global development of ocean cables, social or cultural ideals were only secondary and side effects of economic and commercial ideals.

Although the followers of Manchester Liberalism had turned away from Christian pacifism, this idea reigned strong among the Americans and their interpretation of the cable project. Probably the most common idea within American sources, which also originated in a totally different economic context of the American School's focus on protectionism and self-sufficiency, was the notion of unity between mankind and peace to all nations.[57] People from all over the world were asked to join in the German immigrant William Spitznasski's, "Festlied für den Atlantischen Telegraphen" ("Hymn to the Atlantic Telegraph"), which he had composed for the official celebration of the City of New York on September 1, 1858. One of the stanzas sung "to human mind's great praise" expressed particularly well the aspiration that "Peace be on earth to ev'ry nation!" This was based on the Christian understanding that all men are of one creation, or in the poetry of Spitznasski: "One harmony is all creation—One family the human race."[58] This idea of a conjuncture of science and divine providence was frequently repeated in many an American clergyman's sermon. According to Reverend Cortlandt van Rensselaer, the completion of an Atlantic telegraph cable represented a victory of morals as much as technology: it had an important "educating influence on the popular mind" and magnified "the triumph of mind over matter." Simultaneously, it assisted "in bringing God to view as the great and glorious Ruler of the Universe."[59] Clergymen connected this notion also to the Atlantic cable itself, which they translated into religious terms as the harbinger of Christ as the Lord of Peace. The cable agents rose to be God's instruments and were incorporated into the world's history of salvation. In the words of the Presbyterian clergyman William Adams, they were "disciples of that true 'star-eyed science' which walk[ed] hand in hand with the one true religion of [the] divine Lord."[60]

The religious context is as important for framing the discourse on universal peace as it is for contextualizing the cable agents in their scientific endeavor. Although today the conjuncture of science and religion usually "conjures up an immediate image of conflict and confrontation," this was not the case in the 1850s and 1860s.[61] Rather, the new technology was widely embraced by American spokesmen of religion, such as Reverend van Rensselaer or William Adams, who was a close friend of both Cyrus W. Field and Samuel F. B. Morse. Archbishop John J. Hughes had an engraving made into the cornerstone of the Anglican St. Patrick's Cathedral in New York, celebrating Cyrus W. Field and his scientific wonders.[62] Finally, as the Direct United States Cable Company (DUSC) stockholders' lists disclosed, a large number of British clergymen saw it as a lucrative means of investment.[63] The idea of a Christian union, or a

Societas Christiana, which encompassed all people in the body of Christ, represented the underlying explanatory model for a global electric union in the American sources. In this view, technological progress was a manifestation of God's providence, and the cable protagonists served as his agents in preparing the advent of Christ. From this understanding of technology, American spokesmen of religion and American cable entrepreneurs concluded that their work and their telegraphic network were also intended to spread the Christian religion.

As Morse revealed in 1857 in a letter to his wife, the support of the Christian community was very important to him personally and presumably also to some others.[64] Some of the cable agents, such as Cyrus W. Field, Samuel F. B. Morse, and Peter Cooper, fully embraced this idea of a Christian union and extended it to their cable work.[65] Although many of the British cable agents, such as Daniel Gooch, had a distinct religious background, the Americans in the group emphasized it. In his autobiography, director Peter Cooper, for example, portrayed himself as a "truly religious man" who held "Christianity and progress . . . to be closely related." In this he accorded with Field and Morse, as well as his son-in-law Abram Hewitt, who all believed that "material and scientific advances would eventually be followed by spiritual and cultural advances."[66]

In 1859, with the support of Abram Hewitt, Peter Cooper launched the Cooper Union for the Advancement of Science and Art, an institute that provided free higher education to men and women.[67] The institute would, time and again, play an important part as a meeting place in the social history of the American cable fellows as well as in their endeavors for the furtherance of Anglo-American relations. Samuel F. B. Morse and Cyrus W. Field employed their cable business travels around the world for missionary purposes. In 1870, they embarked on a cable trip to Europe and included a detour to Russia. They were part of an American delegation of the Evangelical Alliance that had set out in cooperation with the European Alliances "for the purpose of inducing his imperial majesty, the Czar of Russia, to stop the persecutions of the Protestant Letts and Estonians in the Baltic Provinces, and to grant religious liberty to all his subjects."[68] Under the slogan of "We are one body in Christi," the Evangelical Alliance promoted universal Christian unity and showed a strong inclination toward the promotion of religious liberty. The delegates considered their petition to the czar to be not only "of the utmost importance . . . for Russia, but prospectively also for the cause of Christian missions in Turkey," feeling that the "proclamation of religious liberty throughout that vast empire would be one of the greatest events of the century, equal in importance to the emancipation of the serfs by the present emperor."[69] In all likelihood, it had been their friend, clergyman William Adams, member of the Board of Foreign Missionaries, who had established the connection to the American Evangelical Alliance.[70]

Another of Cyrus W. Field's and Peter Cooper's "missionary" interests lay with the London branch of the U.S. Sanitary Commission, of which Field, alongside George Peabody and Junius Morgan, was a founding member.[71] The U.S. Sanitary Commission was a government agency that coordinated the voluntary work of women during the American Civil War, for example, taking care of the wounded or disabled. Its New York branch used Cooper Union as a meeting place.[72] Due to Field's influence, Richard Glass, of the British cable manufacturer Glass, Elliot & Co., donated "for the benefit of the Sanitary Fund 1,000 tons of coal to be delivered at his own expense."[73] Religion played an important part in the lives of Americans involved in the cable enterprise. They connected their religious ideals of liberty and a Christian union of the world with their cable work and used their cable travels as a means to diffuse their religious beliefs.

Nevertheless, the idea of the interrelationship between progress and Christianity had its loopholes. The pacifist concepts drew a thin yet clear line between Christians and "pagan peoples" and ultimately between the "civilized" and the "uncivilized." Peace and goodwill were not to be bestowed upon the entire globe, even less so upon all its inhabitants. On the contrary, the electric telegraph was heralded as a "treasury of Christendom" and, as such, restricted to the "Christian nations to whom that art has been vouch-saved." Telegraphy, and more distinctly the ocean cables, represented a means of distinction, separating "us" from "the other" and defining the latter as "pagans," "barbarians," or "savages."[74] Although the telegraphs created a global unity among the "Christian fellows," they were only passing through the other's territory, leaving them unconnected: they were not part of the network because only Christianity represented the "basis of Civilization."

According to contemporaries, the acceptance of the Christian faith was fundamental to advancement; those who had not yet accepted it "relapsed into their primitive stupidity" and could neither grasp the meaning of the telegraph nor benefit from it in any way.[75] In 1872, the Eastern Telegraph Company published a series of Christmas greetings from "'round the world" that had passed through its cables. Many of them entailed that same Christian sentiment of telegraphic exclusivity. From the "four quarters of the world, . . . glad words of Christian fellowship" were "flashed at one time and with one will by the new speech of Civilization": telegraphy. Still, although ocean telegraphy allowed men to speak "with one voice all 'round the globe," the "fraternal message" had to overcome various "impediments." These were "trackless voids of wild water, . . . desolate waste places of the earth, . . . unsubdued [sic] regions of solitude or barbarism, and [a] swarm of Pagan peoples." Before the "civilised earth" could speak "at the same time in the same words of united and brotherly feeling," it had to pass geographic obstacles such as the North Atlantic and human obstacles such as non-Christian peoples.[76] The company saw Christianity and progress as interrelated. Not only would technological progress, such as the ocean telegraphs, spread Christianity, but Christianity was the enabling structure for (technological) progress in the first place.

In the early years of global communications, contemporaries connected two different and almost conflicting imaginaries of a world in electric union with it. These marked the distinction between British and American protagonists. Both sets of concepts of peace, for instance, grew out of different national settings concerning political, economic, and social ideals: In Great Britain, the economic-political ideas of Manchester Liberalism nourished concepts of a pacifist universalism. In the United States, where the economic-political system focusing on state protectionism of the American School contrasted Manchester Liberalism, a religious justification of a Christian union was much more prevalent. In both cases, pacifist concepts comprised economic and political ideas as well as cultural concepts.

CRITICS, CHANGES, AND *L'ESPRIT D'INTERNATIONALITÉ*

Even in the 1850s and 1860s, discourses on universal peace were not unanimously positive. A general disillusionment over the telegraphs' ability to create peace as well as the obvious exaggeration of some statements spurred on critical remarks. Submarine telegraphy had not eased tensions and the danger of war during Napoleon III's coup d'état in 1852, prevented the Crimean War in 1853, or rendered the American Civil War impossible.[77] Rather, telegraphy had then found its first military tactical application because the telegraphs offered unprecedented access to front-line events. After all, William H. Russell, official Great Atlantic Cable correspondent, had become famous with reports from the Crimean War.[78] From the 1870s on, pacifist concepts of communication outlived themselves. Instead, contemporaries increasingly discussed the question of cables as a "powerful instrument of war."[79] However, until the Spanish-American War in 1898, ocean telegraphy's suitability for belligerent conflicts remained a theoretical question.[80] From the 1870s to the 1890s, concepts of internationalism served as an explanatory model for the moral implications of a global media system.[81] Through ideas of an international law code and the International Telegraph Union, the two legal reformers Louis Renault and David Dudley Field attempted to create the structure for a global electric union. They were driven by the idea of *l'esprit d'internationalité*, which would teach nations to follow certain common principles in their mutual relations and result in a peaceful international system.[82]

Early critics of a unified world via telegraphy targeted their objections at the very quality that promoters of peace highlighted: speed. According to these critics, the quality of communication was defined by its content and not by its speed. The *Pall Mall Gazette*, for instance, questioned how much the Atlantic telegraph "contribute[d] to that warm affection between different fractions of our race." It wondered whether the people of Great Britain and the United States would "really love each other more warmly because [they could] send a message to New York and receive an answer within five hours."[83] Aside from the obvious "exaggerated rhetoric" that the subject "almost invariably [seemed] to provoke," critical journalists highlighted the mere instrumentality of telegraphy.[84] It could "threaten as well as compliment—[could] bear tidings of woe as well as messages of peace."[85] In their argumentation, critics depersonalized the cable, which had beforehand been esteemed not only as the *messenger* of peace but as the *maker* of peace, and reduced it to a technical and mechanical application within human intercourse. It remained, after all, only "a musical instrument, on which operators may play any tune they choose."[86] In the decades to come, the tune more often than not consisted of brief and suggestive messages. This hindered rather than enhanced transnational understanding. In fact, "[b]y its impact on both the role of public opinion and the quality of decision-making, telegraphy tended to make the management of diplomatic crises more difficult and arguably increased the likelihood of war."[87] From the 1870s on, the mermaid's song and the notion of peace and goodwill became less frequent in cable rhetoric.

Spurred on by such disillusionment, two things happened to submarine telegraphy in cultural rhetoric. First, a shift occurred from the personified telegraph cable to the application of telegraphy and, as such, a clear emancipation of man from machinery. The technology should serve man's purposes and not follow any kind of intrinsic destiny. Human reason dominated the new technology, and as such, "[t]ruth and error, an honest and a malevolent purpose" could use it alike.[88] In addition, the cable agents themselves backed away in their sales rhetoric from any promises to generate peace. In a speech celebrating the landing of the French Atlantic cable at Duxbury, Massachusetts, in 1869, James Anderson admonished his American audience that no one knew "whether the electric cable will become the great implement in war or an instrument in the cause of peace." Rather, he only phrased it as a possibility: the cable "*may be* a great promoter and sustainer of the whole world, and of civilization, of good feeling and of good fellowship."[89] Second, a demystification of submarine telegraphy occurred, which went hand in hand with a certain "normalization" of the reception of future cable projects. In his diary of 1874, Daniel Gooch, again embarking on the *Great Eastern* on an Atlantic cable-laying enterprise, remarked on how decidedly the nature of cable laying had changed: "all now is a simple matter of quiet business and there is no fuss or reporters."[90]

The protagonists of the cable system promptly reacted to these changes in the public's and, more importantly, the government's or rather the military's perceptions. Their channel to secure a status of neutrality for the ocean cables and to protect their property became the International Telegraph Union (ITU) and its instruments of international law. The ITU is one of the oldest of the international organizations that emerged after 1850 as a response to the increasing mobility of goods, capital, and labor across national borders. It was established in 1865 by the leading telegraph authorities of the various European governments and was soon joined by many nations outside of Europe. The organization's main objective was the promotion of a uniform system of traffic exchange and universal tariffs. It did so through internationally binding telegraph conventions, which were the results of the regular telegraph conferences.[91] Already in 1871, at the ITU conference in Rome, Cyrus W. Field submitted a position paper to the delegates on the protection of ocean cables in the case of war. His aide-mémoire also contained a letter by Samuel F. B. Morse who urged the delegates to consider the telegraphs "a sacred thing, to be by common consent effectually protected both on the land and beneath the waters."[92] Both men followed a U.S. government initiative of 1869, which had then been dropped due to the outbreak of the Franco-Prussian War. The U.S. proposal suggested that governments should refrain from any kind of supervision of the cables' messages and that each state should enact laws for the security of submarine cables within their jurisdiction. The agreements found should then also be valid in times of war.[93]

Despite fierce opposition at the Rome conference, Field had good fortune. Still at the conference, Field telegraphed to Morse about their success: the conference had adopted the proposition "to recommend the different governments . . . to enter into a treaty to protect submarine wires in war as well as peace." It was further suggested that no international cable should be laid without the joint consent of the governments proposed to be connected.[94] At the time, the ITU only represented the European countries and their systems, but some of the members brought extra-European colonial territory into the agreement: Britain brought India; France brought Algeria, Tunisia, and Indochina; and Russia and Turkey brought their Asian possessions. Most of the important ocean cable companies at the time were also present and ready to commit themselves to the decisions taken.[95] Even in 1871, the agreement took on the appearance of being "globally" binding in the sense of a Eurocentric electric union, or in the words of Cyrus W. Field, the convention represented "twenty-one countries, six hundred millions of people, and twenty six different languages."[96] For the history of ocean telegraphy and the systemic premises of global communication, this episode at the ITU conference marked an important shift: the breakthrough of internationalism. Although Morse used religious rhetoric in his letter to Field, their approach morphed in the international society to support practical applications over the *Societas Christiana*; an electric union was not guaranteed by the cables as harbingers of Christ, but through the instruments of international law.[97]

The 1871 decision was not a legal agreement but only a recommendation to explore the matter. On leaving the conference in Rome, Field traveled all over Europe to urge the American ministers in each of the cities visited "to help on this treaty."[98] Yet for more than a decade, little came of this universally agreed upon recommendation. In 1878, Louis Renault, professor of international law at the University of Paris and member of the Institut de Droit International, established in 1873, convinced the institute to set up a commission of academic experts to deal with the issue.[99] The commission was composed of an international group of well-known experts on international law: Johann Caspar Bluntschli from Switzerland, Nicolas Saripolos from Greece, John Westlake from Great Britain, Louis Renault from France, and David Dudley Field, Cyrus W. Field's brother, from the United States.[100] The commission based its report on Renault's 1877 report *Etudes sur les rapports internationaux: La Poste et le Télégraphe* and recommended that governments embrace a common strategy on cables during war.[101] Renault argued that electric telegraphy could indeed accomplish much in the development of international relations. Nevertheless, to exploit its full potential, it had to be seconded by international law, which was, in the words of Dudley Field, "a power like gravitation that held the world together."[102] Indeed, in the same way that natural obstacles, such as the Atlantic, were overcome, legal obstacles also needed to disappear. The electric union could not be a limited sphere.[103] To reinforce this idea, Renault introduced another important idea of a world in union: *l'esprit d'internationalité*. This global imaginary was one of universal codification, regulating the international movement of goods, labor, capital, and communication, which would in the end produce a peaceful international system.[104] David Dudley Field in 1873 became the first president of the Association for the Reform and Codification of the Law of Nations, later renamed the International Law Association (ILA); from 1866 on, the ILA promoted the scheme of an international code. Through such a code, international differences would be solved by arbitration, which would render wars unnecessary.[105] Within such a system, Renault and Field argued the administrators of these two organizations had done more for civilization and international understanding than celebrated diplomats.[106] From the 1870s on, *l'esprit d'internationalité* represented the dominant force in the emergence of an international system promoted by organizations such as the ITU, ILA, and Institute of International Law. Both David Dudley Field and Louis

Renault saw the global media system as one field of international cooperation that needed universal codification but that entailed a vision for the future.

The institute's recommendation, as well as other appeals, fell on deaf ears. In 1881, the Eastern Cable Companies as well as the Vereinigte Deutsche and the Great Northern Telegraph Company brought a pamphlet before the London Board of Trade and the Foreign Office. In it, they meticulously listed all cable breakages and damages, their causes, and their costs. In a time of fierce price competition, the question of cable security was preeminent for the cable companies.[107] Yet, not until 1882, when France invited a group of international government representatives to consider the protection of submarine cables in times of peace and war, did governments take action. An international group of administrators, headed by David Dudley Field, met six times between October 1882 and July 1887. The forty-six delegates produced the draft for the International Telegraph Convention for the Protection of Submarine Cables of 1884.[108] In 1887, twenty-four nations, among those the major powers of the time, ratified the Convention. It achieved little, however, concerning the question of submarine cables during war or the moral implications behind an international law code. Already in 1882, the British delegate introduced the proposal that all agreements should only concern cables in times of peace, not war. France seconded the motion.[109] Consequently, the treaty only addressed the "interruption of cables in the ordinary way."[110] Article XV explicitly stated that all stipulations were made "for the time of peace only" and would in no way "restrict the action of belligerents during time of war."[111]

In his closing speech of the 1883 conference, Louis Cochery, French Minister of Posts and Telegraphs, revealed the crucial turning point: in the end, the issue of cables in times of war was reserved for diplomats, not legal or telegraph experts.[112] The conference's inability to draft a binding resolution showed the limits of international law, which in 1880 David Dudley Field had still heralded as the agent of "the Brotherhood of men" and "Peace and Good-Will."[113] Although international regulations increasingly became a matter for administrative experts, questions of war remained reserved for the diplomats.[114] However, diplomats showed little interest in the matter. It was James Anderson who pointed out the obvious: the idea that "submarine cables could or should be relied upon in time of war [was] nonsense." Any commander-in-chief who could not "arrange his plans when the cables were cut [was] not fit to be at the head of the concern."[115] Additionally, as the Dutch delegate had already pointed out in 1871, it was unconceivable that a state would adhere to cable neutrality as long as the messages' content was not fully controllable. In case of war, "any government would destroy a telegraph line to stop transmission of enemy communication."[116] Finally, great power relations also played a role. With her supremacy on the high seas and most cable companies registered as British enterprises, Great Britain would have gained the greatest benefit from such a neutrality treaty among the European powers. Presumably, it was such control that prevented other powers from signing.[117]

Throughout the nineteenth century, governments showed little interest in solving the issue of cables at war. Rather, "[c]ommensurate with its vast importance," no subject appeared to have been "more censurably neglected," according to the *Pall Mall Gazette*.[118] This changed with the Spanish-American War of 1898. To the principal imperial powers, the war brought attention not only to the legal rights of cable property but also to the fact that reliable submarine communications under exclusive control were "absolutely necessary."[119] During the war, the U.S. Navy cut several cable connections in the West Indies, the cables connecting Florida with Cuba, and the connection between Asia and the Philippines. They acted upon a policy developed by General A. W. Greely, Chief Signal Officer of the U.S. Army. It stated that a cable with two terminals in enemy country could be cut any time, cables between belligerent parties were subject to harsh censorship, and cables between an enemy state and a neutral state could also be cut, if only within the three-mile shoreline.[120] This policy also redefined the world's seas. Up until then, the international community saw the oceans as "the great international highway, belonging equally to all nations." Now, the political boundaries of a state also included such portions of the high sea that a nation could "by her commercial and naval vessels and her submarine cables, reach out and secure."[121]

In response to the unresolved international legal situation on cables and war, new attempts were made to solve the issue. In 1899, at the first Hague peace conference, the Danish delegate proposed to treat submarine cables as analogous to terrestrial cables and put them under national jurisdiction. The proposition failed due to British opposition.[122] In 1900, General Greely expanded upon his earlier ideas and had a *Handbook of Submarine Cables* prepared with an outline of practical rules of action to at least regulate the issue for the American military.[123] In 1902, the Association of International Law agreed on rules that a submarine cable uniting two neutral territories was inviolable and that cables may not be cut in neutral waters.[124] In 1907, the Hague Conventions of land warfare slightly modified these rules. The conventions primarily drafted by Louis Renault stated that submarine cables connecting with a neutral territory shall not be seized or destroyed except in the case of absolute necessity.[125] The same year, Renault received the Nobel Peace Prize.[126]

Although the Hague Convention represented a major achievement in protecting submarine cables internationally, all attempts amounted to nothing. Facing military conflict, internationalism in the form of a binding international law code to regulate the electric union failed soon after 1907. During the Italo-Turkish War of 1911 to 1912, the Italian navy cut all cable connections with Turkey through the Mediterranean within a couple of days.[127] On August 4, 1914, almost immediately upon the commencement of war between Great Britain and Germany, Britain had the civilian cable steamer *Alert* cut those five Atlantic cables that connected the German Reich with the rest of the world. Germany, in turn, unsuccessfully tried to destroy a British cable station in the Indian Ocean. Moreover, telegraphic messaging increased drastically during war time, and all companies found their cables used to full capacity. Cutting cables, the belligerent parties entirely disregarded which company operated the cables. In the case of the "German" Atlantic cables, these were operated by the American-based Commercial Cable Company. The Americans had entered into a joint-purse agreement with the Deutsch-Atlantische Telegraphengesellschaft with regard to the latter's two Atlantic cables laid via the Azores in 1900 and 1902.[128] The events of the First World War seemed to squash all ideas of an electric union and the benefits of instantaneous communication around the earth. From a universal panacea, the technology had finally developed into a powerful weapon of war. Looking retrospectively upon the First World War, James Bryce, academic, jurist, and Britain's ambassador to the United States, even concluded that "had it not been for the extraordinary development of means of communication Europe would not have burst into a world-wide conflict with almost explosive violence." Whereas in the "good old days" of stagecoach, horse, and sailboat, it would have taken months for a conflict to involve an entire continent, telegraphy brought all European nations simultaneously face to face with crisis.[129]

From the 1850s until 1914, contemporaries used the logic of three very different concepts to explain the idea of an electric union as accomplished via an ocean telegraph network. From the 1870s on, the dominant explanatory model was that of international law and a codified world in union. Louis Renault and David Dudley Field were the key protagonists. Both believed that the codification and regulation of the international movement of goods and people as well as knowledge and information would result in the creation of an international system that was inherently peaceful. In the end, the ideas connected to the implementation of a Manchester liberal system of international trade, the *Societas Christiana*, and *l'esprit d'internationalité* of a universal law code could not prevent the entire world from going to war in 1914. The implementation of a global communication system had not lived up to its initial promises of peace and goodwill.

THE ENGINEER AS THE "GREAT CIVILIZER"

Alongside the ideas of universal peace and an electric union, there was another cluster of concepts connected with the development of a global communication network—namely, the notions of *civilization, being civilized,* and lastly, *the diffusion of civilization.*[130] Telegraph technology was fundamental to a Western understanding of civilization, and ocean telegraphy should be "an instrument . . . to diffuse religion, civilization, liberty and law throughout the world," in the words of U.S. President Buchanan in 1858.[131] Similarly, the American telegraph engineer George Squier, of the U.S. Signal Corps, concluded in 1909 that "[t]he mails, the telegraph and the telephone [were] civilizing the world."[132] The cables equally signified Euro-American superiority as well as a Euro-American mission statement to diffuse its civilization. In 1879, for example, Rose Pender, daughter-in-law of John Pender, published her travel diary from a cable business trip to the African continent together with her husband. In *No Telegraph; or a Trip to Our Unconnected Colonies,* she argued that only a submarine telegraph connection and instantaneous communication with London could bring civilization to these "distant and wild places."[133] Within this context, the Atlantic cable symbolically crowned the Euro-American mastery over nature and its forces achieved thus far.

Even more so than the cables, the telegraph engineers, electricians, and cable operators symbolized the interrelatedness of technology, civilization, and their diffusion. Musing on the functions of the engineer, William Preece, engineer-in-chief at the British Post Office and brother-in-law to Atlantic engineer Latimer Clark, concluded that the engineer represented "the great civilizer" of their time.[134] As such, the telegraph agents became, willingly and unwillingly, an integral part of Euro-America's ideology of a civilizing mission.[135] They not only "conquered" nature, in the form of the world's oceans, and gathered knowledge about the unknown, such as the deep sea, but also took possession of their surroundings as spokesmen of a "civilized" Euro-America by naming indigenous children, villages, or mountains of countries they operated in. However, issues of civilizing were discussed not only along the more obvious "West–rest" divide but also within the transatlantic realm, bringing the debate over a civilization-wise distinction of Old World versus New World to the forefront.

Western Europe and North America's industrial and technical developments in the nineteenth century were fundamental to their understanding of civilization. They distinguished their nations from all preindustrial, meaning un- or semi-civilized, ones. Although to be civilized or civilization served as a sort of self-description, the terms only came alive in their antithesis to savagery or barbarity. Those employing the terms divided the world according to degrees of progress and development.[136] Fundamental to the concept was the idea of mastery; to be civilized meant "to be free from specific forms of tyranny: the tyranny of the elements over man, of disease over health, of instinct over reason, of ignorance over knowledge and of despotism over liberty."[137] As brought to life in the play *Contentina* staged on board the *Great Eastern* in 1866, the world's oceans were perceived as some of the greatest despots of the time, symbolized in the figure of Neptune. They separated peoples and markets and made intercontinental travel and trade an unsafe and tedious matter. The Atlantic, in particular, appeared as the greatest obstacle separating the Old World and the New World.[138] Before the age of transatlantic telegraphy, travel and communication across the Atlantic amounted to a matter of weeks, if not months. For that reason, transatlantic commerce was slow and insecure. With the transatlantic submarine cables, a Euro-American "humanity" had accomplished mastery over the Atlantic.

The work of the American naval officer Matthew F. Maury was fundamental to the taming of the Atlantic. The "Pathfinder of the Seas," whom Cyrus W. Field had consulted immediately with regard to an Atlantic cable, had done important work in the first half of the nineteenth century mapping this ocean.[139] His work on ocean currents and winds made transatlantic voyages and voyages between the Americas considerably shorter and safer.[140] His greatest contribution was the discovery of the Telegraph Plateau, which was one of the key prerequisites for making submarine telegraphs through the Atlantic feasible. To this day, submarine or, rather, fiber optic cables connecting Europe and North America follow that same path.[141] Yet, although the Telegraph Plateau served literally as the foundation for submarine telegraphy in the North Atlantic, it was the cable per se that "civilized" the ocean. According to *Punch*, the Atlantic telegraph's provision of instantaneous communication between the two continents broke the despotic reign of Neptune. The merchants remained no longer in ignorance of their cargo and price fluctuations once their ships had left the home port; they could now control transactions via cable. Only with the advent of wireless at the turn of the century would this control over the ocean be even further extended.[142]

A wave of industrialization and technological development spurred on the mastery of the material world, which Euro-America had gained decades earlier over all other societies. These increased Europe's and North America's superiority "exponentially in virtually all fields of science and technology."[143] The telegraph in particular, representing a technology that harnessed the lightning and put it to work for man's pleasure, embodied to many Euro-Americans the "ultimate symbol of man's power over nature."[144] Their ability to handle the electric telegraph put the telegraph operators in a class of their own. Time and again, the cable companies invited visitors to their cable stations, on board the cable ships, or to the manufacturing works, where the engineers, electricians, and operators acted as "showm[e]n" illuminating their guests on the "electrical mysteries."[145]

A sense of elitism played itself out, not only during these visits but also during the encounters of Euro-American operators with the "uncivilized" native population. At the time, contemporaries perceived electricity as the most capricious and least understood of nature's forces. The fact that Euro-Americans could still make it their servant seemed to them "a powerful indicator of the advance of civilization."[146] This notion was most vividly illustrated in the letters of George Kennan from his telegraph expedition to Russian America and Siberia on Western Union's 1860s scheme to establish a land telegraph route and beat the Great Atlantic Cable project. Kennan directed the telegraph surveying crew in Siberia. In an attempt to find the best route, an American team of surveyors explored over the course of two years almost six thousand miles of "unbroken wilderness, extending from Vancouver Island on the American coast to the Bering Strait, and from the Bering Strait to the Chinese frontier in Asia."[147] Almost en passant, a team of Smithsonian Institution scientists that accompanied Western Union's crew produced a detailed inventory of Alaska's natural resources that turned former "wasteland" into valuable property.[148]

In his letters, Kennan revealed common Western ideas of superiority. He described the indigenous population of Siberia as "gentle," "simple," and "childlike," in contrast to his believed "omniscience," which he based on his understanding of natural and scientific phenomena. In a letter to his mother, he relates his encounters with "those simple natives" who "thought, and still think, that there is not a single thing in the heavens or earth or under the earth that [he did] not know + [could not] explain."[149] Beyond his ability to explain natural and scientific phenomena, it was the telegraph machinery that established, according to Kennan's own understanding, his superior status among the native population and manifested the distinction between him and the others. To nourish his "stardom," Kennan performed tricks and used his knowledge of the telegraph against the indigenous population who had not yet seen such an invention. During his stay in Siberia, for instance, Kennan put up an electric telegraph and invited all the people in to see how it worked: "I can not [sic] describe all the tricks which I performed with that machine a battery + a few fathoms of wire." For the natives, being utterly unfamiliar with the technology, it appeared to be "alive," and subsequently, "they gave it the name of 'Ivan Machina' or 'Machine John' by which it goes to this day." As Kennan claimed in his letter, "the story of that machine, how it could talk + how [he] could understand what it said, was known to every native" and established "among all the wild natives [his] reputation . . . as 'schamán' or magician." Clearly amused, Kennan closed his letter: "Ha ha! You didn't know what eminence your son was about to attain when he left for Siberia. The civilized natives call me 'Yero weesokee Blagorodia' or 'his high Excellency' + the wild natives know me as 'the magician.' One title is hardly compatible with the other."[150]

The indigenous population's failure to explain and understand the science and working of the telegraph enforced Kennan's assessment of their naivety and childishness and furthered his notion of his superiority. The telegraph, which represented to Kennan and other Euro-Americans the "electric nerve of modern civilization," established a dichotomy and civilizational divide.[151] In comparison, the "Asian peoples had little to offer . . . in techniques of production and extraction or in insights into the workings of the natural world." Europe and North America seemed to be in a league of their own, distinct from all others as the polarities were numerous and obvious: metal versus wood, machines versus human or animal power, science versus superstition, and progress versus stagnation.[152] The transfer of Americans for resettlement to these areas appeared for those involved to be the remedy for bringing civilization and modernity. From his visit in St. Petersburg, Perry Collins wrote to Henry Seward, Secretary of State, that the Governor General of Siberia had suggested to him the possibility of recruiting 5,000 Americans for resettlement. Collins commented that this would give Russia "a population touched by American quickness, agricultural and mechanical ideals."[153] Although the idea was never brought to fruition, it illustrated the Americans' sense of superiority as well as the assumed transferability of its formula for success, the American frontier experience, abroad.[154]

Kennan's accounts were not the only ones relating the encounter between Western telegraphers (users of the technology) and the native population (nonusers of the technology) in such a racialized way. His interpretation of the latter's unfamiliarity with the technology as an argument for the former's superiority was common throughout the nineteenth century.[155] Already the discourses that emerged with the introduction of the telegraph in the United States in the 1840s reconstructed the relationship between mind and body and contributed to a racialized view of progressive civilization.[156] As a consequence, natives were generally considered to be unfit to perform any of the cable work other than the menial work of helping to bring the cable on shore after the cable laying or loading the coals.[157] The job of a telegraph operator working on the complex ocean cables was predominantly reserved for British (or Europeans) only. Even at Heart's Content, Newfoundland, there was a strict policy entailing a "distinction between 'natives' and men coming from England," although these "natives" were white and had once come from Europe as well.[158] At the time, scientific dominance of the world and the exploitation and oppression of "inferior" races by Euro-Americans were intricately linked. Telegraphy, as well as other technology, served "as metaphor and materialist basis for the domination of mind over body, capital over labor, and whites over Indians, blacks, Mexicans and Asians."[159]

Another means for the cable entrepreneurs to establish superiority was to take possession of their surroundings. The cable entrepreneurs decided where a cable station was to be built and so where "civilization" was to sprout. They named rivers, mountains, villages, and even children and left a distinct footprint of Euro-American power on native land, or in their understanding "empty," "waste," or "uncivilized" land, such as Cyprus. In 1878, as a consequence of the Russo-Turkish War (1877–1878) and the congress of Berlin, Cyprus came under British rule. In particular, the island's geopolitical setting made Cyprus valuable in Britain's policy of consolidating a world empire, because it added to other British bases in the Mediterranean at Gibraltar, Malta, and Suez. The Defensive Alliance between Great Britain and Turkey had been concluded on June 4, 1878.[160] Shortly thereafter, on June 13, 1878, Emma Pender reported of her son's departure to Cyprus on cable matters. Representing the Eastern Telegraph Company, Harry Pender was taking possession of the island: "In fact as Harry said of his commission," wrote Emma Pender, "he went out as a prophet to announce . . . where new cities would be raised over the land, where the earth would pour out riches, where the seas would cast up treasure."[161] Similarly, W. P. Granville, employee of the Eastern Telegraph Companies, expressed his conviction in his personal diary of his time on Cyprus that the British engineers brought civilization to the "Turks' mismanaged business."[162]

Stemming from a similar sense of superiority, there exist places all around the world that were named after telegraph engineers and entrepreneurs in the nineteenth century. Some examples are the village of Chauvin in Alberta, Canada, named after the telegraph engineer and Siemens Brothers manager George von Chauvin; Mount Field in British Columbia; and in Central Africa, Mount Gordon Bennett, which is today named Ruwenzori, as well as the Gordon Bennett River, named after the telegraph entrepreneur and newspaper publisher James Gordon Bennett.[163] In Cuba, some of the babies of female slaves on a tobacco plantation were named after the telegraph engineers Charles T. and Edwin Bright, who rested there while employed in laying cables in the West Indies.[164] It was the very act of naming geographical and human entities that implied a power over them.[165] In this sense, the telegraph agents represented a very powerful group of people at the time. Based on their belief in Euro-American superiority, they took possession of their surroundings and assimilated them into their understanding of the world until there was indeed, to use William Preece's phrase, not one "habitable spot on the face of the earth that [did] not bear traces of the presence of the engineer." The engineer's role appeared to be even more distinguished than that of the military, the traders, or the missionaries, who had in former centuries played the central role in first encounters. Now, as the Alaska Purchase illustrated, the engineer "not only immediately follow[ed], but . . . sometimes even precede[ed] the military conqueror." Thus, he distributed "peace and good-will without the accompaniments of fire, blood and famine."[166] This had its impact upon the self-understanding of the engineers at the time, who indeed acted as if they were the "great civilizers."

However, technology did not solely serve as a means of colonial suppression and an instrument of Euro-American superiority. It did so to a large extent, but modern technology also created new forms of subjectivity, not just new forms of suppression. In response to the close connection of technology and empire, many telegraph lines, particularly terrestrial lines, were destroyed by the native population and their operators attacked. In particular, the landline connections to India, such as the Siemens Overland Line, leading through the lands of what Charles Bright considered "barbarous and then unconquered tribes" were prone to, from a Euro-American perspective, native "vandalism" and "molestation."[167] Native opposition not only destroyed technologies but also put them to use, *their* use.[168] For instance, in 1866 in Great Britain, the Fenians, an Irish revolutionary group, celebrated the successful laying of the Atlantic cable because it accomplished "one great object of Fenian ambition, by uniting Ireland with America."[169] The same year, they staged their first aborted invasion of Canada from the United States starting a sequence of Fenian raids that were fought until 1871. The Knight of Kerry was so worried about Fenian takeovers of the Atlantic cable station that he made the inhabitants of Valentia proclaim their loyalty to the throne and their detestation of Fenianism. The Valentians in turn resolved "to take any steps [necessary] . . . for the security of the Atlantic cable and telegraph establishment."[170]

Contemporaries discussed the ideas of uplift, civilization, and development by means of telegraphy not only along the lines of an assumed setting of West versus the rest, but also within an Anglo-American setting. In fact, the category "West" and the idea of a common transatlantic model of civilization that presumed the cultural and political equality of Europe and North America did not come into use until after 1890.[171] According to *Lloyd's Weekly Newspaper*, the Atlantic cable would, by the benefits of a new proximity and speedier acquaintance and communication between the Old World and the New World, do great things to bring America "much closer to [Great Britain] than it ever did before."[172] With the use of the word *closer*, the author alluded to meanings beyond mere spatiality. In the mid-nineteenth century, the United States of America still represented to many Europeans, and particularly the British, the uncivilized cousin. This view was nourished through British travelogues on America, such as Frances Trollope's bestseller *Domestic Manners of the Americans* (1832). Hardly any other travel book had been more wildly debated and hardly any expressed the notion of English superiority better.[173] The episodes from social life, observations on eating habits, dress, and behavior in public places, and criticism on subjects ranging from slavery to popular idioms "provoked amused curiosity on the one side of the Atlantic and hot resentment on the other."[174] According to the *Times*, the Atlantic telegraph was meant to "Europeanize America more than anything yet has done." It would effect in "taking away the remoteness and the strangeness—if we may say so, without giving offence, the rawness of the New World, and it will in short bring the New World into the Old World."[175]

The Americans had quite contrary views on their own civilized superiority over old Europe. From the early nineteenth century onward, the country was changing rapidly: it industrialized, became a transcontinental power, and began to expand overseas on a sustained basis.[176] Poets, philosophers, and statesman as diverse as Henry David Thoreau, Walt Whitman, Thomas Hart Benton, and William Gilpin agreed with Ralph Waldo Emerson, who portrayed America as "the country of the future." America was steadily progressing "into a new and more excellent social state than history has recorded."[177] In particular, the frontier expansion was fundamental to the saga of the rise of the nation and its concept of

American exceptionalism. Well into the twentieth century, schoolbooks described this "hegemonic narrative of pioneers taming the 'Wild West'" as an "epic struggle by sturdy yeoman farmers to civilize a continent thinly peopled by savage Indians." America's manifold expansionism in the nineteenth century had become a defining source of Americans' sense of themselves as exceptional people and of the ways they conceived their relationship to the rest of the world. It lent legitimacy "to the ever more widely held conviction that the society fashioned from the progressive mastery of the North American wilderness ought to serve as model of modernity of all humankind."[178]

Numerous landscape paintings of the era, such as John Gast's *American Progress, or Manifest Destiny* (ca. 1872) or Frances Palmer's *Across the Continent, "Westward the Course of the Empire Takes Its Way"* (1868), celebrated the pivotal role of industrial technologies—from farm machinery and canal locks to the telegraph and the railway—in the spread of settlement westward and on to the Pacific coast.[179] With the spread of steam power and industrial production across the young republic in the mid-century, a revolution in transportation and communication in combination with major improvements in machines for processing raw materials and cultivating soil enabled even higher levels of resource extraction. Advances in ironworking for locks facilitated the construction of canals that linked newly settled areas with steamboat carriers on the Mississippi and other centers of commerce and industry on the Great Lakes and the Atlantic coast. Steam engines drove mechanized "donkeys" that skid cut timber to rivers and railway lines, which carried it to steam-powered sawmills, or drained western mines and carted the ores to the trains.[180] Machines were essential to transform the western prairies that had beforehand been dismissed as "sublime waste" into one of the most productive agricultural regions of the world; the previously mentioned paintings graphically confirm technology's vital contribution "to the fulfillment of America's divinely appointed civilizing mission, which would ultimately benefit all of humanity."[181] With his invention of a telegraph, Samuel F. B. Morse had added another chapter to this narrative. In the popular imagination, electricity was a mysterious power that originated in the heavens and had for millennia been linked to the divine. That an American had transformed an elemental force of nature into a species of private property seemed to Americans to fit their nation's progressiveness as well as its citizens' moral purity.[182]

This context of contesting notions of superiority in the transatlantic world is vital for the interpretation of Constantino Brumidi's 1862 painting *The Telegraph*. The Italian frescoist had been hired to paint some of the rooms in the U.S. Capitol extension in Washington, DC. He took a classical European theme and adopted it to New World achievements. Instead of being carried to Crete, Europa has been conveyed across the Atlantic where she is greeted by America, who is wearing the Phrygian or freedom cap and holds the caduceus (ancient emblem of the messenger of the gods), while resting her arm on an anchor representing hope. America's strength is represented by the cannon lying behind the anchor, her mechanical invention by the gear wheel beside her, and her generosity by the cornucopia. Among the fruits are giant grapes, which signified the Promised Land in the Old Testament, and the pineapple of hospitality.[183]

The journey's purpose is for Europa to receive the Atlantic telegraph wire from America; the telegraph line and a pole can be seen on the right-hand side. Contrary to the actual cable-laying route, the cable will be taken back to Europe. Although the meeting of the two continents is amicable, their attitudes clearly signify "that Europe is the suppliant and America the generous benefactor in this exchange."[184] America's superiority over

3.2 Constantino Brumidi, *The Telegraph, 1862*, U.S. Senate Collection.

Europa is expressed in its form of government (the Phrygian cap), its military strength, and its technological inventions. The location of the fresco at the heart of U.S. political power, where every foreign statesman and diplomat would be welcomed, was of utmost significance. The fresco symbolized every visitor's take-home message. This contest between the Old World and the New World remained simmering throughout the nineteenth and early twentieth centuries and expanded beyond the underlying tension between two national narratives of invention and progress. With anxiety, the British watched the rise of the United States to become an industrial power and later a world power.

In 1864, during an Atlantic cable banquet, John Pender pointed to the undertaking's cultural promises. The Atlantic cable's importance was not to be measured "by the mere standard of capital or prospect of commercial gain." Rather, "scope and objects of the new organization were calculated to bring the whole of the civilized world within their influence."[185] It provided not only for the movement and transmission of information, but also for the spread of values and ideas. Probably everybody present agreed with Pender that the Atlantic cable entailed an inherent promise of a culturally unified globe; they differed, however, on the explanatory model for the electric union and "peace and goodwill."

Throughout the nineteenth and twentieth centuries, contemporaries harbored a variety of global imaginaries of universal peace and civilization. These were nourished from different local, national, and ideological contexts and changed over time. All shared an unwavering belief in progress and development inextricably linked to the technological developments of the age. In the 1850s and 1860s, the mermaids' image and their wish for "peace and goodwill" turned into a household message, as did their promise that instantaneous communication would rectify all misunderstandings. The "annihilation" of time and distance seemed to be a crucial structure for the development of universal peace, whereby the Great Atlantic Cable was only one of many development projects at the time to achieve such a state. Its peculiarity was its incredible speed, which brought a new sequence in the chronology of events, as well as its implied global reach.

In the 1850s and 1860s, actors differed in approaching the peace concept from two different national and economic contexts: Manchester Liberalism and the idea of a *Societas Christiana*. The British protagonists were influenced by Manchester Liberalism and the idea that the spread of markets and the resulting international dependencies would result in the spread of peace. Key representatives of this economic philosophy, such as Richard Cobden and John Bright, supported the cable enterprise because it exemplified their theory on the correlation of international trade and peace. Although religion had little influence on John Pender and James Anderson, it played a tremendous role in the agency of the American protagonists. For Samuel F. B. Morse, Cyrus W. Field, and Peter Cooper, the explanatory principle for the cable's moral influence was not a secularized notion of peace but God's providence and the idea of a global Christian union. Time and again, they connected their cable work with religious thinking and even missionary expeditions. In both concepts, "global" electric union did not mean "universal." Electric unity was employed within a Eurocentric rhetoric of imperial and industrial power relations and was only valid as such. All others, that is non-Euro-Americans, non-Christians, and nonelites, were excluded from "peace and goodwill" as well as all the other advantages that global communication brought.

From the 1870s on, a third explanatory model became important and was institutionalized: internationalism in the form of a codified world, bound by universally applicable laws and an international system. The two law reformers, David Dudley Field and Louis Renault, represented the key players in attempts to codify the global media system. Their focus lay on the protection of submarine cables in times of peace and, more importantly, in times of war so as to not disconnect the world. Both believed that the codification and regulation of the international movement of goods and people as well as knowledge and information would result in the creation of an international system that was inherently peaceful. Their greatest breakthrough was the International Telegraph Convention for the Protection of Submarine Cables of 1884, the rules of which were modified again in 1907 by the Hague Conventions of land warfare. The outbreak of World War I and the speed with which "German" Atlantic cables were cut that very summer illustrated that, in the end, the ideas connected to the implementation of a Manchester liberalism, the *Societas Christiana*, and *l'esprit d'internationalité* could not prevent the entire world from going to war with each other.

Finally, the idea of a unified market of morality via means of global communication entailed the notion of universal peace, which, after all, only concerned the major industrial powers of the time. However, its reach also expanded beyond the Euro-American realm of power relations because it also entailed the concept of "civilization." At the time, technological progress was strongly intertwined with notions of civilization, and development and technologies, such as the railways, steamships, and electric telegraphs, were contemporaries' proof of their own superiority over nature as well as the "rest," or the "uncivilized." But the West's "mastery of the material world" served not only as a "justification of imperial dominance" but also as means and instrument of its diffusion. Submarine telegraphy was not only argumentative proof of, but also instrumental to, the civilizing project per se.[186] As a result, the cable protagonists found themselves as equally in the positions of engineers, electricians, or cable operators as they were explorers, travelers, and "civilizers." We may not forget, however, as the examples of Mrs. Trollope and Brumidi's fresco illustrated, that the demarcation line of "civilization" not only ran between the "West" and the "rest" but also sometimes right through it.

4 | WELTCOMMUNICATION

When the Atlantic Cable is completed, it is a fact, that a message will be received in America five hours before it leaves England.

Punch, "Something Like a Telegraph," August 4, 1866.

WITH THESE words, an article in *Punch* expressed a general perplexity about a new compression of time that the Atlantic telegraph seemed to inaugurate. News would arrive not only faster, as the submarine telegraph "eliminated" the twelve days a mail steamer took from Europe to North America, but also more current or "newsworthy."[1] The electric telegraphs dematerialized global information flows.[2] For the first time, the message was separated from its material messenger and could reach its destination at unprecedented speed.[3] Undoubtedly, the globe-spanning telegraph network was essential for the development of entirely new communication patterns.

With his sketch of a theory of *Weltcommunication* (world communication), German philosopher Ernst Kapp envisaged the ocean telegraphs as electrical nerves of a global body and as the innervation of the *Weltgeist* (world spirit). This notion soon morphed into the current historiographical perception of a larger global entity organized in a global media system.[4] Some scholars see the development of a global submarine cable network as the first step in the history of telecommunications from a *Victorian Internet*; that is, they see the telegraph as the equivalent of the Internet for the late nineteenth century, and via a "global public sphere," as the first step leading to the world as a *global village* in the information age.[5] Yet, the assumption of a unidirectional progression of global communication densification is, at least for the nineteenth century, erroneous. The promise of an all-inclusive *Weltcommunication* was as empty as the analogy of a Victorian Internet is misleading. Submarine telegraphs never were a means of social or mass communication but remained a specialty service for an exclusive clientele until they were displaced by wireless and radio telegraphy as the dominant means of global communication in the 1920s.

The second half of the nineteenth century witnessed a continuous battle of contesting theories of communication, each attempting to give context to the concept of *Weltcommunication*. At the heart of these battles, we find those running the global submarine network, men such as James Anderson, William Siemens, and John William Mackay, as well as the global media reformers like Henniker Heaton, Sandford Fleming, and George Squier. From the outside, the disputes played themselves out in commercial competition, price wars, or the demand for a nationalization of international communication. Especially during the time of the media reformers, between the 1890s and 1910s, the driving question was one of more or less state involvement. On the inside, the struggle about access to cable and global communication revealed competing concepts of modernity itself. For most contemporaries, technology and media as means of global awareness marked moments of liberation and emancipation from nature and the triumph of culture.[6] Ironically, these very processes sharpened the rift between different modes of civilization and (un)civilization, urbanity, and global connectivity. In retrospect, the ever increasing geographical expansion of communication networks went hand in hand with an increasingly disunifying development of the world. This was furthered not only by tariff policies but also by how global news was made. Cable and media moguls like James Gordon Bennett Jr. depended on a "shrinkage of the globe." At the same time, with the instantaneity of their news coverage of events such as George DeLong's expedition to the North Pole or the "discovery" of David Livingstone, they created the notion of the "dark continent" or faraway places that lay outside of the system.[7]

PROVIDERS, USE, AND THE CONFUSION OF TELEGRAPHIC COMMUNICATION

With the opening of the Great Atlantic Cable in 1866, Europeans and Americans seemed to believe that a new era of communication was opening up before them. A large number of people wanted to be among the first to use the new tool for *Weltcommunication*. Already before the successful completion of the cable, telegraph clerks brought an enormous number of applications for priority of messages before the managing director of the Anglo-American Telegraph Company.[8] The cable agents furthered this telegraphic frenzy. During the telegraphic soirées they held in celebration of the Atlantic telegraph, they set up apparatuses from which guests could telegraph at any length and to any place they liked, free of charge. While this was a nice promotional trick to accustom people to the new instantaneity of communication, it also projected a false image of the technology in use, which was simultaneously countered by the high costs of the cable entrepreneurs' tariff policy. This initial submarine enthusiasm resulted in a number of extraordinarily long and expensive telegrams that showed that many customers misunderstood the new medium as a faster version of the letter. Newspapers made headlines with telegrams extraordinaire, which were also used to ostentatiously display power and wealth, as for example those passed between Emperor Maximilian of Mexico and his wife, Charlotte of Belgium. As the *Birmingham Daily Post* reported, "[a] dispatch of 478 words in cipher" had passed over the Atlantic Telegraph between the two of them. The cost of transmission was over £726.74.[9] This record was soon beaten by a dispatch from the U.S. government to the American ambassador in Paris in December of 1866, which consisted of more than four thousand words, cost over £2,000, and occupied ten hours in transmission.[10] Because government dispatches had priority in transmission, the message blocked all other telegraphic communication for hours.

Although in the first frenzy of technical enthusiasm, a relatively large number of such telegrams extraordinaire were sent, in everyday use, telegrams in the form of fast-written letters remained the exception. Rather, the transatlantic communicational space played itself out as one where time, money, and brevity ruled. Soon after the establishment of transatlantic traffic in 1866, it became clear that its use was absolutely exclusive, bestowing the benefits of instantaneous communication only upon those "who can pay."[11] Decisions in this regard were primarily made by the cable companies whose actors enforced their understanding of *Weltcommunication* upon its users. Tariffs between Great Britain and New York started out at £20 for a minimum of twenty words and, in November 1866, were reduced to £10 for a minimum of ten words. Prices fell further in December 1867 to £5 for ten words plus the costs for each additional word according to destination and, from 1869 onward, to £2 for a minimum of ten words. For the cable entrepreneurs, this extreme price reduction in the first three years of the cables' service was a balancing act between recouping the initial investment by paying high enough dividends to the shareholders and attracting enough users willing to pay a certain price. In 1867, the Anglo-American Telegraph Company tried a word rate of £1 for the traffic on their 1866 and 1865 Atlantic cables, but it was not until 1872 that traffic manager Henry Weaver instituted a regular word rate system of four shillings per word.[12] These rates only covered the connection between Great Britain and New York or Boston; messages beyond these two points were additionally charged for the local landline connection as well as the various national taxes. A message from London to Austin, Texas, for example, cost £6.67 in 1867 at an ordinary ten-word rate of £5.00.[13] People were even worse off if they attempted to send a message to a place "beyond the range of the Telegraphic System," such as Fiji, German New Guinea, or the Marshall Islands.[14] These messages were then sent on via ordinary mail.[15] Due to such a tariff policy, the ocean telegraphs did not intend to be a medium of mass or social communication and did not supplant ordinary mail. Instead, they furthered direct, point-to-point communication between the respective centers of trade.[16]

Once a balance between the interests of shareholders and users, or between dividends and tariffs, had been established, only two things could influence the tariffs: a cable breakage, and thus a shortage of transmission routes, or a new cable, and hence additional transmission options. Each time a new competitor entered the field, the telegraphing public hoped that charges would be reduced. Such hopes were not irrational, as all new companies claimed to "inaugurate a new era in the history of Atlantic telegraphy by . . . abolishing for ever [sic] the prohibitive tariff" of the existing companies.[17] In the 1870s and 1880s, fierce price wars, which brought competitors to the edge of bankruptcy by lowering the tariff to so-called "nonremunerative rates," were the result of this competition. As a rule, the outcome of this "system of competition ending in an amalgamation" was that each new competitor joined, voluntarily or by market force, the working agreement of the Atlantic pool, headed by the Anglo-American Telegraph Company.[18] Thereafter, tariffs were usually brought back to or near the old level before the price war. There were four major price wars in the 1870s and 1880s on the Atlantic connections. The first of them was the Siemens-Pender controversy from

1875 to 1877. A second one soon followed, when the French enterprise La Compagnie Française du Télégraphe de Paris à New York (or PQ, as the company came to be called) entered the market. The cable war with PQ illustrated how quickly and effectively such an amalgamation could be achieved. PQ opened its cables for business on June 2, 1880, at a rate of two shillings per word, undercutting the Atlantic pool by one shilling.[19] The pooled companies immediately followed suit. While Direct United States Cable Company (DUSC) drew level at two shillings and on June 19 lowered their price to one shilling and six pence, the Anglo-American Telegraph Company dropped their rates to an unprecedented six pence.[20] PQ was not able to follow at a rate that, according to the *Pall Mall Gazette*, would "pay working expenses, but [would] leave little or no dividend on the respective capitals of the various cable companies."[21] Four months later, in September 1880, an agreement was formed between the Atlantic pool and PQ, amalgamating the latter into the working agreement. It was ratified and finalized in January 1881.[22] From October 1, 1880, tariffs were back at two shillings per word.[23]

Almost immediately upon the amalgamation of PQ into the pool, another tariff war was in sight. This time, an American competitor, Jay Gould, railroad magnate and main shareholder of the largest U.S. landline service provider, Western Union Telegraph Company, entered the market. His entry marked an important change in the structure of the global communication system. It not only connected the two single most important communication markets of the time, the ocean cables and the American terrestrial service, but also combined the organization of the message via cable service with the organization of its content via news service. In December 1880, John Pender unsuccessfully attempted to discourage Jay Gould from laying his own Atlantic cables. He warned him that "to spend more money in laying new cables at present would be equally disastrous to the new as to the present system."[24] Pender seemed convinced that the market could not sustain another rival. Gould's focus, however, was not necessarily set on the ocean market, but the American national system. Recognizing Pender's advice as a hoax, he ordered, via telegram, Siemens Brothers to manufacture and lay two cables across the Atlantic. After a brief price war, both combatants signed a working agreement, giving Western Union 12.5 percent of the revenues as long as it had one cable and 22.5 percent once it had two. In return, Western Union would pass on all outgoing transatlantic telegrams to the Anglo-American

Telegraph Company.[25] The final major tariff war occurred between 1886 and 1888 and came with another American contestant: John W. Mackay and James Gordon Bennett and their Commercial Cable Company. They managed to break the Atlantic pool's monopoly and stayed independent. On September 1, 1888, the Commercial Cable Company and the Atlantic pool entered a working agreement, fixing the cable tariff at one shilling per word for the remainder of the nineteenth century. On the transatlantic line, this essentially remained the standard until 1923.[26]

Although most telegraph companies claimed that their motive in price reduction was "to extend their business in what they call[ed] 'social messages,'" they did not aim at the working or middle classes. The Atlantic cables were, to use the words of a critique, "a golden bridge, to be used by the possessors of gold only."[27] Despite the enormous reduction from £20 for twenty words in 1866 to one shilling per word in 1888, one shilling per word was in no way a social (i.e., generally affordable) tariff. In 1873, an agricultural laborer in Great Britain earned twelve to fourteen shillings a week, and a skilled artisan earned between twenty-eight shillings (1£ 8s) and £2 a week.[28] In 1909, Henniker Heaton, British parliament member, journalist, and one of the most ardent supporters of a penny post system for the British imperial telegraphs, pointed out that the cable rate for one word, which was eight pence at the time, still ranged from one day's to six days' wages for a farm laborer.[29] Those running the

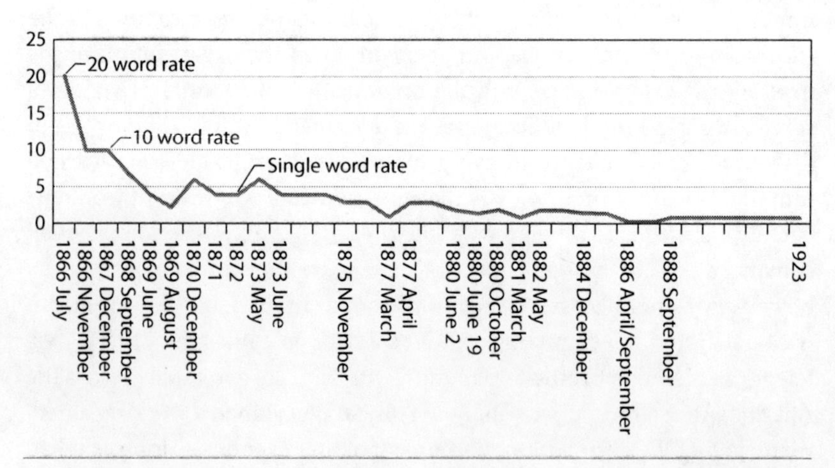

4.1 Simone M. Müller, "Atlantic Tariffs in Shillings/Word (Not Considering Press Rates)." Data source: Atlantic Pool Cable Companies Tariff Books, 1866–1923.

global media system established a two-class system of communication in which "those with long purses, and engaged in large transactions" were in possession of intelligence from the opposite sides of the ocean "days in advance of their neighbours."[30]

This system did not remain uncontested. Users found new ways to circumvent the hefty tariffs. Soon, transatlantic telegraphy was characterized by the messages' brevity. Users followed the rule that "[t]he wordier a message and the greater the distance, the higher the charge."[31] The pressure to be cost-efficient led to shorter telegrams and eventually to the development of the "telegram style," which left out anything that was redundant or not essential for the message. Additionally, codes and cyphers were used and codebooks developed for the different industries and purposes. Between 1866 and 1869, J. Wagstaff Blundell, accountant and author of a *Manual of Submarine Telegraph Companies*, meticulously noted the average number of daily messages and the Anglo-American Telegraph Company's daily revenues: while the numbers of messages rose steadily, the company's per-word output remained almost the same. Starting out with 29 messages a day and an average daily amount of tariff income of £757 in 1866, this climbed to 64 messages and £868 in 1867 and to 131 messages and £952 in 1868. Finally, in 1869, on average, 219 messages a day were sent, bringing in £975 a day.[32]

According to Blundell, not only did the number of messages increase across the Atlantic connection, but they also became shorter. In 1872, the word rate had been introduced in the Atlantic traffic, and in commercial communication between established business partners, often a telegram of two or three words was sufficient: buy, sell, or order. Since the late 1870s, up to a third of the telegrams leaving Germany, for instance, did not contain more than five words. Only one out of ten telegrams was longer than twenty words.[33] On the Atlantic cables, the trend was similar. In 1877, transatlantic messages had an average of 11.4 words according to the Anglo-American Telegraph Company's statistics.[34] This was in no way comparable to the initial extremes of dispatches of 400 or even 800 words during the first global communication frenzy. People had accommodated their way of communication to the medium and its tariffs and to the cable companies representatives' idea of *Weltcommunication*.

To fit as much content into as little telegraphic space as possible, the "packers," senders who abbreviated their messages, adopted highly creative methods.[35] Aside from the usage of shortened spelling such as "immidiatly" instead of "immediately" at the time when letters and not words were counted, the use of foreign languages was common. As the *Electrical World* reported in 1884, once the simple word count had been introduced in the late 1860s, often two or three words were run together in a foreign language for the benefit of brevity. Yet these "evasions" helped little "to carry the message unquestioned out of the originating country."[36] Indeed, a duel was going on between the packers and the companies. Once the evasion was noticed, extra toll was collected. Furthermore, the companies turned to the International Telegraph Union to obtain "a definition of what a word, telegraphically speaking, is."[37] Ever new rules as to the word count were introduced by the various companies or by the International Telegraph Union, thereby making, for example, *c'est-à-dire* four words instead of one.[38] It illustrates the power of the ocean cable companies that different rules for word count were applied for European traffic (primarily the terrestrial system) than for extra-European traffic (primarily the ocean cable system). Within Europe, a word could contain as many as fifteen Morse characters; for extra-European traffic, this number was only ten. This made the word *responsibility* one word in European traffic, but two beyond it.[39]

The effect of such packing meant that it was not uncommon to see messages from correspondents asking for more definite instructions or information, "as the former abbreviated message was unintelligible."[40] Such failed communication is vividly documented in the correspondence of Lady Emma Pender, wife of Sir John Pender. Emma Pender's letter writing to faraway places picked up with her daughter Marion's marriage to William des Voeux in 1875, an official of the British Colonial Office, and the couple's subsequent stays abroad. Lady Pender developed the habit of corresponding via telegram, often using the transatlantic connection. Each of the telegrams, however, was accompanied by an extensive and often explanatory letter. This letter followed the telegram via the much slower steamships arriving many weeks later, a habit that soon became common practice when using telegraphic communication.[41] One of the most common misunderstandings in the cablegrams was the phrase "All well," which turned into a vacuous expression in telegraphic correspondence. Upon their daughter's departure from a visit in England back to her husband in the colonial service at St. Lucia in the Caribbean, John Pender himself struggled in a telegram with the very expression "all well" as the following letter from Lady Pender exemplifies:

> Your Father telegraphed your departure to William. He wished to tell him somehow that you were well but far from strong. This however considered might alarm so at last he left the message ending with "All well." This does not strengthen my faith in telegrams.[42]

Their correspondence failed not only in the attempt to communicate Marion's state of health, but also with regard to William des Voeux's career in the Foreign Service. Probably induced by his wife's wish to have her daughter closer to home, John Pender used his close ties to the undersecretary of the Colonial Office, Robert George Wyndham Herbert. He negotiated not only cable policies but also the career of his son-in-law. Several times, Pender's telegrams to his daughter and son-in-law, advising them to return home, had them wait with packed bags before the explanatory letter shattered all hopes. Surely, so William des Voeux wrote in a letter to Emma Pender, such a telegram could only mean one of two things: "either that an appointment had been actually promised, or that he [John Pender] considered it so certain as to be prepared to pay our expenses."[43] As successful as John Pender was in the cable business, so unsuccessful were his attempts with des Voeux. Soon after that last fateful telegram, a new appointment indeed came for his son-in-law: he was promoted to become governor of Fiji. Although this was certainly a rise in rank, it did not help Emma Pender in bringing her daughter closer to home. At the time, Fiji was a telegraphic "wasteland," and this left Emma Pender "very angry with Fiji for not possessing a 'cable.'"[44]

Because messages were always going through the hands of a clerk, for reasons of secrecy but also brevity, users soon employed ciphers and codes. These were developed to meet the needs of particular industries or user groups and could be used on all lines, unless the respective government vetoed its use. In 1867, for example, the Spanish government mandated that messages to Cuba had to be written in "plain ordinary language," with code and cipher not being allowed.[45] Sometimes the cable companies also developed their own code system, which they then provided for the telegraphing public.[46] Early on, wealthy transatlantic travelers and tourists were identified as possible cable users; predominantly Anglo-Saxon white American males joined their British and Continental European counterparts on their "grand tour" through Europe. Although in the early decades of the nineteenth century, the number of overseas travelers among Americans was likely less than 2,000 per year, numbers spiked upward after 1849. Aside from a short drop during the Civil War, numbers increasingly went up, surpassing the 100,000 mark in 1885. By the start of World War I, nearly a quarter of a million of Americans were traveling abroad.[47] Already in 1880, J. E. Palmer edited his first *European Travelers and Telegraph Code* book; a second edition followed in 1884. In 1887, Golder Dwight published his *Official Cable Code and General Information for European Tourists Including French and German Phrases with English Pronunciation*.[48] With the use of a cable code, travelers could send "almost any information they wish[ed] (at a comparatively small expense) to friends at home," such as advising them of their safe arrival, their state of health, what sort of voyage they had, or where they intended to go first.[49]

The cable companies were quite biased against any code and cipher usage. As James Anderson revealed in a speech in 1873, they considered codes to be unfair and defrauding them of their lawful dues. Through the use of codes, they were "deprived of [their] anticipated profit by the economy of the commercial world, which has learned a new condensed language by which they can express all their wants."[50] Laying a cable was one thing, but making it pay was another. The frequent changes in tariff regulations and regulations concerning the form of a telegram illustrate vividly the constant battle between those coming up with new codes, the packers, and the cable companies coming up with new rules on how to charge even a coded message its proper worth. Between 1866 and 1869, coded messages were, as a rule, charged twice the actual tariff.[51] Thereafter, codes had to form a known or dictionary word; letters, grouped or otherwise, that did not obey this rule were charged by letter. Numerals were also charged each as a word, making it harder for cipher usages. Throughout the century, the cable companies used the International Telegraph Union (ITU) conferences to counter all attempts at facilitating telegraphic communication by means of artificial codes. The telegraph users, however, soon adapted new ways for packing as much information as possible in as few words as possible.

In 1884, facing new competition on the Atlantic market through the Commercial Cable Company, James Anderson, spokesman of the Eastern and Associated Companies, raged against the code and cipher users. He sent up a warning to the commercial public that, sooner or later, "the present abuse of the code system" would be stopped "to the great inconvenience of those firms who are building up a huge mass of made-up words, syllabic combinations having the appearance of words only, and unknown to any language." This growing practice was, according to Anderson, "*unfair*, if not illegal, and contrary to the convention by which all telegraph administrations [were] bound." It would have an abrupt ending, especially, if there was ever "government control of any kind," to which Anderson was completely opposed.[52] Anderson's broadside against nationalizing the ocean telegraphs is of particular interest when seen in light of the actors' relation to the nation-state. Up to the First World War, national monopoly theory, meaning government ownership of ocean cables, was the panacea for media reformers and those urging lower rates. *If only* the governments would buy out the cable companies, rates would unquestionably be lower, reformers argued throughout the Euro-American world. Time and again, the companies also threatened to sell out, knowing quite well that no government would make that investment.[53] Particularly fierce debates surrounded the issue of code cablegrams and the adoption of an official ITU vocabulary for code messages between 1890 and 1908 at four of the ITU conventions at Paris, Budapest, London, and Lisbon. Each time, the results of these debates exemplified the power of the submarine cable companies against the ITU, government representatives, and telegraph users. Although such a standardized European code language had already been adopted at the Paris convention, it was never enforced.[54]

Upon the completion of the first Atlantic cable in 1866, the *Belfast News-Letter* enthusiastically exclaimed how it now became possible "within a few brief minutes, not merely to telegraph from London to New York, but, by a process of easy re-transmission, the golddigger [sic] at California may, if he wishes, communicate within an hour or two with a Parsee merchant in Bombay."[55] This statement has to be re-read critically. The stress has to be on *merchant* as well as *gold* (less on digger) because the cable actors created (submarine) telegraphy as a communicational tool of exclusivity. The telegraphic "girdle around the globe" did not connect people, but markets.[56] Still, the tariff system was not stagnant. Tariff wars between companies brought the costs down, and the users found ways to circumvent the exorbitant tariffs by means of packing or the use of code and cipher. In the end, however, submarine telegraphs highlighted the distinction between those within and those outside of the globally communicating community.

NEWS AGENTS AND THE CABLE MARKET

From its beginning, the transatlantic telegraph market was of great interest to news agents. Already with the introduction of the first landlines in the 1840s, the advantages of the medium's speed and instantaneity for making news more current, albeit not necessarily more "newsworthy," had become transparent. Before the Atlantic cable, news gatherers employed optical telegraphs, hired spotters to locate ships en route, or employed rowboats to contact incoming boats before they docked to obtain news from across the ocean as fast as possible.[57] Now, the telegraphs helped them spread information, their commodity, even more rapidly. Among the first newsmakers to make extensive use of the Atlantic cable was the *New York Herald*, run by James Gordon Bennett. Within a day of the opening of the 1866 cable to the public, the newspaper had King William of Prussia's speech on his victory at Königgrätz cabled in full. The transmission cost 36,000 francs (roughly $7,059).[58]

Often, developments in the organization of news and the organization of news transmission via telegraph were interconnected. In the 1880s, they became interdependent as key actors, such as Jay Gould or James Gordon Bennett Jr., took on positions in both systems. Given high transmission rates, news agents followed the wave of merchants with transatlantic ambition, such as John Pender and Augustin Pouyer-Quertier, and pushed into the Atlantic telegraph market. Because the question of "What is news?" directly related to the policy and costs of sending a telegram, they took on the role of cable entrepreneurs who attempted to secure their own means of transmission and so became important actors of globalization.[59]

In the late 1840s, Charles-Louis Havas, Bernard Wolff, and Julius Reuter recognized the importance of the "latest communication technology," in the words of Wolff, for the transmission of news.[60] Each of them set up a telegraph agency in Paris, Berlin, and London dedicated to the transmission of primarily commercial and financial news. Soon the triumvirate took over global news reporting, and only twenty years after the foundation of their first press offices, Wolff, Reuter, and Havas "were talking quite confidently in terms of empires."[61] One of these news agents had his eyes not only on the messages but also on the medium: the telegraph lines. In addition to Reuter's Telegram Company, Julius Reuter established Reuter's Telegraph Company in 1851 (switching from the message to the medium) and built telegraph lines across Ireland and between Great Britain and the European continent. His was not a public telegraph company, but similar to the line of Colt and Robinson in the United States, solely a news-gathering one. It predominantly served banks, brokerage houses, and leading business firms.[62] With the Telegraph Act of 1870, nationalizing British landlines, cables and cable companies passed into the hands of the British government. Reuter's venture into the telegraph service greatly influenced his news business. The Irish overland connection alone secured him an eight-hour advantage over his rivals on transatlantic news.[63] As costs for obtaining transatlantic news rose from an average of £67 per month in 1865 to £424 per month in 1867, Reuter decided in 1868 to also enter the Atlantic telegraph market. Together with Baron Emil d'Erlanger, he launched the French Cable Company. Yet, soon after its opening, the company had to bow to market pressure. It was forced into a working agreement with the Anglo-American Telegraph Company.[64] After this failure, Reuter withdrew as a cable entrepreneur. His sole connection to the cable business remained his directorship in Pender's Globe Trust in 1872. Still, Reuter was the first to establish the structural link between medium and message.

Reuter was not the only one who saw the connection between the cable business and the news business. Cyrus W. Field, for instance, owned interests in the *New York Evening Mail* and the *New York Evening Express*, two leading New York City newspapers.[65] However, it was in the 1880s with the appearance of Jay Gould, who controlled Western Union and several Associated Press newspapers, such as the *New York Tribune*, the *New York Sun*, and the *World*, and James Gordon Bennett Jr., the owner and main editor of the *New York Herald* and the *Paris Herald*, that news came to play a central role in the cable business. In the 1880s, Gould and Bennett entered into a cable and news war with each other and redefined the parameters of the global cable system. With them, American agents took on a much stronger role in contrast to their British counterparts. The strong antagonism between these two American cable entrepreneurs marked a shift within the cable business from a London-based center of power to a more broadly defined Anglo-American center of power. In addition, it signified a struggle within the United States concerning news distribution as well as the question of social status of various immigrant groups coming from Europe. In a way, the Gould-Bennett controversy of the 1880s was similar to the telegraph war between John Pender and

William Siemens. Again, personal animosities were entangled with business interests. And, like the earlier war, its outcome marked a milestone in the development of the ocean cable system.

A so-called "robber baron," Jay Gould originally came to the telegraph business from the railroads, taking a route similar to that of many of the early telegraph engineers, such as Daniel Gooch or Samuel Canning.[66] The "Napoleon of finance," who had made his fortune as part of the powerful Erie triumvirate crisscrossing the United States with rails, added the telegraph market as a symbiotic acquisition to his vast enterprises.[67] In 1874, Gould commenced taking over the American telegraph market targeting the market leader, Western Union. In 1875, Gould launched his first raid on Western Union by challenging its patent rights on quadruplex transmission. It ended in 1877 with Western Union taking over Gould's rival network provider.[68] Gould commenced his second raid on Western Union in June 1879 when he bought up the insignificant American Union Telegraph Company and thence steadily started to amalgamate all of Western Union's rivals. In January 1881, Gould swallowed the top of the class, Western Union, and so controlled 90 percent of the telegraph market in the United States. This utterly changed the public debate over federal telegraph legislation and state ownership. Within a week of the takeover, Western Union was at the center of attention of antimonopolists.[69]

Gould's interest in the ocean cable market was, on the one hand, part of a strategy within the Western Union takeover. The company already owned interests in ocean cables and had in the 1860s pursued the Collins Overland Route across Alaska and Siberia as an alternative to the Great Atlantic Cable. On the other hand, it was part of a great financial scheme.[70] In late December 1880, Gould decided to have two Atlantic cables manufactured and laid by Siemens Brothers. Although he called for 70 percent subscription for his American Telegraph and Cable Company from external sources, the capital of £2,000,000 ($10,000,000) was raised within 40 hours.[71] The price war that took place in the spring of 1882 was brief and vigorous. Three months after opening his cables to the public, Western Union, which in fact leased the Atlantic cables from Gould's cable company, signed a joint-purse agreement and entered the Atlantic pool. Charles Bright suspected that this was, after all, "the happy destination for which . . . it was originally launched into existence."[72] This agreement set new standards for the business of submarine telegraphy because it formally combined land and ocean telegraphs. The market advantage on the Atlantic was no longer determined by vertical business networks securing independent means of cable manufacture, laying, and operating, but was increasingly set via customer access. This raised the importance of the landlines. During the initial period of point-to-point connections between cities such as London and New York, the cable companies' offices there had been sufficient to serve and reach their cabling customers. The inclusion of Western Union's far-reaching network showed that by 1880 the use of the Atlantic telegraph extended far beyond these points.

The involvement of Gould was the first instance amplifying the growing interdependence of news and cable business in the 1880s. This linkage became even more defined with the appearance of James Gordon Bennett on the cable market. Whereas Jay Gould represented the businessman, financier, and cable entrepreneur, James Gordon Bennett represented the news agent through and through. In the late nineteenth century, Gordon Bennett, as he came to be called in distinction to his father, James Gordon Bennett Sr., was one of the most flamboyant figures of Anglo-American public and news life. Backed by his enormous wealth, he embraced the possibilities of the new technologies for a "time-space compression," especially the telegraph, the telephone, and later on wireless. He ordered his correspondents to and fro across the entire globe, always on the lookout for a story in Greece, the Caribbean, or the North Pole.[73] Gordon Bennett did not just make global headlines, he was global news.

Gordon Bennett, son of Scottish and Irish immigrants, entered the news market in 1866, the year of the Great Atlantic Cable, when his father, who was one of the most influential journalists of mid-nineteenth-century America, partially turned the business of the *New York Herald* over to him.[74] By 1866, Bennett's father had revolutionized American journalism. With his cheap penny press, he brought "newspapers to the masses," which he entertained with a sensationalism to reporting that "startled readers, but drew them back again and again."[75] In 1877, Gordon Bennett's behavior at a New Year's soiree of his fiancée's father in New York made him a social outcast from American high society and a permanent émigré to Europe.[76] From then on, he spent most of his time in Paris or on board his yacht roaming the world's oceans. He never permanently returned to the United States and did not get married until the age of 73.[77] He died four years later at his Beaulieu villa in France.[78]

The cable business came into Bennett's focus in the early 1880s upon his acquaintance with the "Bonanza King," John William Mackay. Mackay was an Irish immigrant to the United States who had, in the 1870s, made an enormous fortune with silver mining in the American West.[79] After his mining career, Mackay had remained in the American West and only intermittently joined New York's high society; his wife and son, in contrast, returned to Europe and set up a family residence there. Allegedly, it was his wife's exorbitantly high transatlantic cable bills that aroused Mackay's interest in setting up an ocean cable enterprise.[80] At the same time, Bennett's *New York Herald* was waging war against Jay Gould's Western Union telegraph monopoly, which dictated prices of national and, because it owned two Atlantic cables, international news transmission. Mackay estimated that the prospect of federal antimonopoly legislation would aid the scheme in its fight against Western Union and that Bennett, who ran the *New York Herald* from Paris, would be the transatlantic connection's best customer.[81] On December 10, 1883, John W. Mackay and Gordon Bennett launched the Commercial Cable Company. The following year, they visited Siemens Brothers in London and ordered two transatlantic cables. These were opened for public traffic on New Year's Eve 1884 with a tariff of one shilling and eight pence a word between Great Britain and New York.[82] What followed was the almost mandatory cable tariff war between the strongly entrenched Atlantic pool monopoly and the Commercial's new lines, which brought both Mackay and Bennett to the brink of bankruptcy. As Ludwig Löffler, managing director of Siemens Brothers, reported to Carl Siemens in St. Petersburg, the Atlantic pool companies declared war on the Commercial early in 1886 when they reduced their tariffs to six pence per word. They hoped that thus ruining the Commercial's income (and everybody else's as Löffler sharply noted) they could force Mackay to return to a tariff of two shillings and six pence or even three shillings a word in order to recoup his investment. Yet, as Löffler concluded, "there was no way Mackay would give up."[83]

In response to this attack, Bennett and Mackay followed a threefold strategy: They took over a successful landline network competing with Western Union. They attempted to lure away PQ, the newest acquisition of the Atlantic pool. And lastly, they enjoyed the benefits from financing their cable predominantly from their own fortunes. As Gould complained, if Mackay ever ran low on money, "he could simply go back to

Nevada and dig up some more."[84] This financial independence brought one important advantage for Bennett and Mackay: they cared little about dividends and shareholders' interests. When in 1886, the pooled companies reduced their tariff to six pence, the Commercial could no longer follow. Instead, they officially appealed to the telegraphing public "to make a temporary sacrifice" by supporting them at a one shilling tariff "on the assurance that if [they were] vigorously backed up now the tariff [should only] return to the original figure," one shilling and eight pence, and not go back up higher.[85] The deciding movement was the establishment of the Postal Telegraph Cable Company to counter Western Union's monopoly on landlines in 1886, as well as the support of Mackay's mining colleagues to maintain liquidity. In effect, the cable war was waged over the American customers and won on the American landline market. By 1887, the Commercial Company had secured 50 percent of the transatlantic cable traffic and established itself to stay independent.[86] In August 1888, an agreement was concluded with the Atlantic pool, whereby the Commercial secured its independent status from the pool. Additionally, a word rate of one shilling, commencing from September 1, was arranged.[87]

The battle between the Commercial Cable Company and the Atlantic pool had "commanded the attention and admiration of the entire business world."[88] People were aware that for the ocean cable business, the Commercial's success marked the end of an era. After more than two decades and several attempts to break the Anglo-Americans' grip on transatlantic communication, a duopoly was established.[89] With it, the importance of the Class of 1866 dwindled and a transition occurred from a solely Anglo-European market control to one with a broader American basis: Jay Gould, James Gordon Bennett Jr., and John W. Mackay were the leading protagonists. In effect, the cable war was decided on the American landline market and over the question of access to the American customers in the "hinterlands." Moreover, whereas in the 1850s and 1860s, Cyrus W. Field had had to apply to the world's bank, London, for financial support, two decades later, the American capital market had developed so strongly that it was able to finance an Atlantic cable entirely by itself. This process coincided with the United States' rise to a global player.

The dispute between Gould and Bennett was not solely a question of cable business. It also concerned control over the media and, in a way, control over the Associated Press (AP), one of the leading news agencies in the United States. In fact, Gould's "big haul of plunder," as the *New York Times* christened Gould's 1882 coup with the Atlantic pool, caused great agitation.[90] Thereafter, he controlled and influenced the costs of cable transmission within, as well as to and from, the United States. In response, an entire series of new Atlantic cable proposals were thought through in the United States. People worried not only about their cable bills but also about the news market itself, as Gould seemed to be about to land his next coup. In the early 1880s, the *New York Times* and the *New York Herald*, both members of the AP, had for some time been fearfully watching Gould's increasing control over the communication and media sector. Gould owned large parts of the *New York Tribune* and the *New York Sun*, which were looked upon as his "personal organs," and in 1881, he added to his assembly the *World*, another New York paper, which replaced the *Courier and Enquirer* in the AP.[91] It was further rumored that Gould was behind the purchase by Cyrus W. Field, his cable business partner, of the *Evening Express* in December 1881, which also was part of the AP.[92]

Supposedly, Gould's move was part of a scheme targeted at controlling four out of seven of the AP newspapers to force them into an exclusive working agreement with Western Union and the Western Associated Press, a rival news agency in the American West with a reach only second to the AP. This would have secured Gould an absolute monopoly over news transmission and distribution in the United States.[93] Moreover, if Gould would have succeeded in controlling the AP, he could have told the public "exactly what he wanted them to hear."[94] For the *Times* and the *Herald*, Gould's increasing influence on the AP concerned them directly, as he might have gained control over their news reporting as well. In February 1881, the *New York Times* mobilized against Gould's attempt to monopolize the flow of news and his taking control of what once "ha[d] been the free press of America."[95] The *Herald* warned that in "improper hands" the AP might be not only "unfaithful to public interests," but also "a very powerful engine for mischief."[96] Although Gould did not take as much control over his newspapers as the *Times* and the *Herald* feared, his influence should also not be underestimated.[97] With the 1882 Atlantic pool agreement, he established a system of censorship on all telegrams processed via the transatlantic lines. Upon inquiry, John Pender admitted to John Garrett, president of the

Baltimore and Ohio Railroad, a Western Union rival, "that whenever the message contained anything affecting Western Union Company in any way, it would be the privilege of that company to inspect them."[98]

For Gordon Bennett, who ran the *Herald* from Europe, such control had an immediate effect on his news business: anything Bennett cabled from Europe went through Western Union's hands and might fall under its control. Additionally, Gould's expansion into the American news world coincided with the slow descent of Bennett's newspaper. The previous decade had been the most profitable for the *New York Herald*, and between 1873 and 1883, it was "potentially the most powerful newspaper in the world."[99] This changed in the early 1880s, mostly due to Joseph Pulitzer, who took over Gould's *World* in 1883 and made it into one of America's most important and influential newspapers.[100] The *Herald's* dominance was contested, and Bennett's interest in the cable business was connected to his need for speedy and cheap transmission but most of all to "the desire to escape from the necessity of using lines even indirectly controlled by Mr. Gould."[101]

As a consequence, it was not only an Atlantic cable war but also an American news war. It was not long before Bennett became "Jay's most persistent tormentor."[102] All of Bennett's newspapers, the *New York Herald*, the *Paris Herald*, and the *Evening Telegram*, did not miss one chance to criticize Gould's way of pursuing business as dishonest and fraudulent. Bennett particularly condemned Gould's entry into the Atlantic pool in 1882 as "the most gigantic system of organized robbery in existence in any civilized country."[103] Simultaneously, the *Herald(s)* ran headlines in Europe as well as in the United States to justify John W. Mackay's conduct as the Commercial's manager. Additionally, the Commercial's regular advertisement was most prominently displayed in Bennett's newspapers, usually on the front cover right underneath the weather forecast.[104] In response, Gould used the *Sun*, with which Gordon Bennett had, to make things even more delicate, sort of "inherited" a news feud from his father, as well as the *New York Tribune*, for stark statements concerning the personalities of his antagonists.[105] In 1888, matters climaxed when the *Herald* waged an all-out offensive against Gould, lunging at the Gould-Sage Trust and its alleged misappropriation of Union Pacific money. Although the grand jury decided for Gould, the *Herald*, as well as other newspapers, "convicted him" of fraud.[106] As a response, Gould composed a bitter personal attack against

Bennett, which he sent to all American newspapers, except the *Herald*, in April 1888.[107] Gould portrayed "Bennett, the libertine" as the social derelict of New York's respectable and honorable men, whose entire life had been "a succession of debouches and scandals" and essentially "one of shame." He argued that no respectable man would invite Bennett into his residence "where virtue and family honor [were] held sacred" as his "very touch in the social circle [was] contaminating."[108] Although Gould's letter was unfavorably received by most of the American newspapers, it put an end to the Bennett affair.

Bennett's armistice was not solely caused by this letter, but was also influenced by Mackay's business negotiations with Gould. Simultaneous with the American news war, the Atlantic cable war also came to a close; during the summer of 1888, Gould and Mackay negotiated an end to their cable war and, in September, fixed the cable rates at one shilling per word.[109] At the same time, the general excitement about the robber baron Jay Gould, Western Union, the Associated Press, and national monopoly theory in the United States died down. It seemed as if with time, the public's fear of Gould's control over the press was laid to rest.[110] For the Atlantic Ocean cable market, the Gould-Mackay agreement marked the beginning of a firm duopoly, which lasted until the 1920s. All future rivals, primarily the French and German state cable ventures, aligned themselves with one of the two. In 1911, upon Western Union's refusal to continue to collect telegrams for Atlantic pool companies, those companies were forced to enter into an agreement with Western Union. It was decided that from 1912 on, all the company's cables would be leased to Western Union. The Commercial Cable Company was by that time already operating the "German" Atlantic cables. As a consequence, the American-based companies, Western Union and the Commercial Cable Company, controlled all thirteen transatlantic cables by World War I.[111]

The mounting ferocity of the Gould-Bennett controversy suggests "motives that transcended the cable war or even the desire to sell papers."[112] Indirectly, the dispute also responded to inner-American processes of national and social integration with regard to the massive waves of immigration, particularly from Ireland; about 1.5 million people immigrated to the United States from Ireland during the famine years of 1846–1855. Throughout the nineteenth century, the Irish underwent a process of "becoming white." Upon their arrival in America, they were treated as social or, as historian Ignatiev claims, even racial inferiors.[113] The "Irish" Mackay and Bennett had to face similar resistance from New York's high society, the entrance to which their money had bought them.[114] Yet, although money could make them upper class, it could not make them accepted. Gordon Bennett was literally driven from the country. Although William Mackay remained in the United States, his wife and family had taken up permanent residence in European exile. Jay Gould, in contrast, was of an old and distinguished Anglo-American family that had come to America in 1647.[115] The Gould-Bennett battle contained the "classical ingredients of Catholic/Protestant, American (Irish)/British antagonism."[116] Beyond this antagonism, this conflict also shows how fluctuating social determinants of class and belonging were within a Euro-American group. These social divisions broke down due to migration, and they could be challenged through means of technological progress and capital. The case of Bennett and Mackay contains a strange parallel to the Class of 1866 and its members from Britain's North. In contrast to the two Irish-Americans, however, these Scotsmen and northern English capitalists had found acceptance within Britain's peerage system.

After the disputes had been settled, Gordon Bennett emerged as the most visible representative of a "global" news modernity based on technological speed. The survival of the Commercial Cable Company as well as the establishment of the Postal Telegraph Cable Company secured Bennett a great level of independence for his newspapers unparalleled by any other American newspaper owner. The instantaneity and speed of the ocean cables and their linkage with landline systems allowed him to run his New York newspapers from Paris, Nice, or wherever he happened to be, sometimes even on his yacht touring in the Mediterranean as he sent orders by cable. While Bennett was touring the world, his executives were required "to sit at the end of a cable." Continuously, "a never ending stream of editors and reporters was kept shuttling back and forth across the ocean."[117] In Bennett's way of running the papers, geographical distance was of no importance. It could be easily bridged by modern means of transport and communication. Indeed, technology was key to Bennett's sort of obtrusive modernity. In particular, the submarine telegraphs expanded his work space and "annihilated" the space in between. The global ocean cable system helped him run his paper successfully despite "the apparent disadvantages of remoteness and absence [that were] always omnipresent in *The Herald* office."[118] Bennett was equally here as there.

In his news reporting, Bennett catered to a particular class of cosmopolitans to which he himself belonged. With the *Paris Herald*, which Bennett started in 1887, he primarily targeted the newly rich Americans in Europe, who are so vividly depicted in the works of the American writer Henry James.[119] They made the long ocean crossing "in the highest style, lived and traveled on the continent in the greatest comfort, ordered their custom-tailored wardrobes . . . and then made the rounds of major cities, spas, racetracks and ports where they were sure to meet other people much like themselves." Newly arrived travelers in Paris were invited to register at the *Herald* so that their presence could be reported in the paper.[120] As Bennett was himself, the readers of his Paris edition were part of a new cosmopolitan class nourished through the means of modern communication and transport.

In the same way that Gordon Bennett traversed the entire globe via cable to organize his paper, he also saw the entire globe as a source for news. As one former employee recalled, Bennett always "kept watch for rising storms all over the world, and whenever he thought a war, a revolution, or some other great event was impending, . . . he would send a correspondent there."[121] Bennett had a sense for sensationalism, which he connected to global "discovery" stories. Already in the 1870s, Bennett used the ocean telegraphs "to break one of the most sensational stories of his time": the search for the English missionary, David Livingstone.[122] Readers of the *Herald* were taken aback when they read on July 2, 1872, that the famous missionary, who supposedly had been lost in the "dark continent" of Africa for three years, was alive and well and moreover had discovered the source of the Nile River.[123] Livingstone was not his only coup. In 1879, Bennett supported U.S. Navy Lieutenant George DeLong's ill-fated voyage to reach the North Pole, providing him with his own boat, the *Jeannette*, which he had explicitly bought for that purpose. In 1881, DeLong discovered several islands that he claimed for the United States, among them Bennett Island. When DeLong and the ship went missing, Bennett sent out a search team for him. It returned without a trace of the *Jeannette*. Later, people learned that the *Jeannette* had been closed in by the ice and sank. On their attempt to reach home, nineteen of the crew members, including DeLong, died.[124] Bennett never lost interest in the Arctic. In 1909, he paid Frederick A. Cock $25,000 for his exclusive story of how he had reached the North Pole for the *Herald*.[125]

In the 1900s, Bennett shifted his focus from stories of global discovery to stories of global speed. He tendered the Gordon Bennett Cups for international yacht, car, or air balloon races across Europe or the North Atlantic.[126] Moreover, just as Bennett embraced the ocean cable technology, he also took on the telephone and wireless upon their appearance. For years, his newspapers marked certain articles as "by the *Herald's* special telephone."[127] Upon learning of Guglielmo Marconi and his experiments with wireless, Bennett brought Marconi to America to use his device in reporting the America's Cup yacht race in 1899.[128] With this change in news reporting from stories of discovery to stories of speed, Bennett recognized important trends with regard to technological developments as well as visions of the world. By 1900, the age of discovery was essentially over. The central question was no longer whether a place could be reached across large distances, but how fast.

As the *Herald's* news making turned to stories of speed, the Commercial Cable Company did everything possible to annihilate any kind of distance. It even ignored what had long been considered to be the business's safety rules with regard to the cables' landing places. As the *Birmingham Daily Post* announced, the Commercial's 1884 Atlantic cable was to land "almost at the gateway" of New York only to be continued "underground to Wall Street, within two doors of the Stock Exchange."[129] For the first time, ocean cables were "landed in the heart of a great commercial center." Its telegraph offices were, perhaps mindful of Bennett's constant shift in time zones, "open day and night."[130] Additionally, Muirhead's duplex system, allowing for messages to be sent simultaneously from both sides of the cable, was applied to the two Atlantic cables. This made them the fastest means then available for transatlantic communication.[131]

In the 1890s, Bennett became obsessed with increasing the speed of his news reporting. He staged events, such as an interview conducted via cable between the Old World (Europe) and the New (America) and Anglo-American cable chess matches across the Atlantic. The players were members of the House of Commons and the House of Representatives or the members of the London and Brooklyn chess clubs. The Commercial Cable Company provided the connection, and the *Herald* was the first to report upon the events.[132] In 1886, a telegram message between New York and London took twelve minutes from the moment it was posted to the moment it was handed over to its receiver. This marked a record at the time, but it was soon beat by the Commercial itself.[133] In 1895, reporting on the America's Cup, the Commercial Cable Company had its cable steamer *Bennett-Mackay* stationed at the starting point for the yacht race and connected with its cables to France and England. This gave the public "a telegraph office on the racecourse in the Atlantic Ocean from which messages may be forwarded to all parts."[134] In the end, the Commercial reported the result of the America's Cup within two minutes on the Atlantic connection, beating its own previous record by one minute.[135] On July 6, 1903, many newspapers ran headlines of the Commercial's successful attempt to encircle the earth in nine minutes. The completion of an around-the-world cable system enabled the *Herald* to receive the latest news directly from points as far from its New York and Paris bases as the Philippines, California, and Asia.[136]

Bennett was obsessed with speed, the annihilation of time and distance, and the technology that allowed him to experience such a global placelessness. Yet such acceleration of communication had one paradoxical result. It widened the gap between those places connected to the global communication system and those outside of it. This intensified a "relative backwardness of those parts of the world where horse, ox, mule, human bearer or boat still set the speed for transport." This was most strikingly illustrated in the case of the missing David Livingstone. In an age when New York could telegraph with London, Buenos Aires, or Tokyo in a matter of minutes, it was striking that the *New York Herald* could not receive Livingstone's letter to that newspaper in less than eight to nine months and even more striking when the London *Times* could reprint that letter the very next day. As Hobsbawm argues, "[t]he 'wildness' of the 'Wild West', the 'darkness of the 'dark continent', were due partly to such contrasts."[137] *Weltcommunication,* in the form of news transmission brought some parts of the globe, such as the Euro-American realm, into a closer imagined community, making Euro-Americans into world citizens. Speed seemed to be the deciding factor. Different modes of speed translated into different levels of "civilization" or "development," whereas the holes within global telegraphic communication disclosed power relations within an imperial-colonial setting, as well as within an urban-rural setting.

So far in popular history, Bennett has been treated as a phenomenon rather than a noteworthy newsmaker. Nevertheless, Bennett needs to be critically reconsidered. Despite all his eccentricities, he was an important figure in the process of "globalizing news." He did so in two ways, namely in globalizing the system and in globalizing its content. First, the business merger of a globally active news distribution system, the ocean cables, with a news-making system allowed Bennett to successfully operate several newspapers on two continents. Second, with his news reports on expeditions through Africa and to the North Pole, he brought these faraway regions home to a Euro-American public, who thus experienced the shrinkage of the globe through media coverage. In fact, his entire conduct of business was globally oriented. Bennett had one advantage: his wealth allowed him a degree of independence in news making that thereafter has never been accomplished again. As the *New York Times* pointed out in his obituary, Bennett ran all his papers "solely for his own personal satisfaction and according to his own sweet will."[138] Still, Bennett's newspapers were not out of the ordinary and able to compete with other papers. They had high sales numbers and belonged among the leading journalistic organs of the time. Moreover, in his shift from news of discovery to news of speed, Bennett showed the instinct of a shrewd businessman paying tribute to a changing world. The independence Bennett bought himself through his wealth was not necessarily one from public opinion but from the news agencies, such as Reuters, Havas, and Wolffs Telegraphisches Bureau, simply known as WTB, that were providing their "empires" with global news. Bennett's wealth allowed him to be his own global news supplier and eventually an alternative model to a global news system run by the news agencies. In fact, Bennett's business model was no different from what people had accused Gould of in the early 1880s: the control over a vertical business network that encompassed the process of news making, news gathering, and news transmission exercised on a global scale.

SOCIAL MESSAGING AND A UNIVERSAL PENNY PRESS

Apart from the commercial and news community, the general public was also interested in what the cable entrepreneurs came to call "social communication," and they complained about the system of ocean telegraphy the Class of 1866 and others had set up. The first protests against the company's "extortionate tariffs" and calls for a global media reform set in shortly after the laying of the Great Atlantic Cable. These attempted to change the nature of the system such that mass communication via ocean telegraph could be an option for more people than just the wealthiest Euro-Americans. Nevertheless, only from the 1890s on and with the increasing interest of national governments in control over ocean cables were the "global media reform[ers]," such as Henniker Heaton, Sandford Fleming, George Squier, Ernst W. H. von Stephan, and Edward Sassoon, heard and their ideas of a nationalization of ocean cables and government monopoly taken seriously.[139] The Australian Henniker Heaton was the most forceful spokesman for a universal penny press, that is, a cheap telegraph communication all around the world. He was also the most important opponent of the cable system's general manager, James Anderson. To both men, the term *Weltcommunication* had entirely opposite connotations. While Anderson, representing the entrepreneurs, aimed at a select circle of patrons, Heaton argued for the telegraphs' application to the masses.

Initially, "social messages" neither played an important role in the business strategy of the cable entrepreneurs nor made up a large percentage of the actual cablegrams sent. Only in 1869, three years after opening Atlantic cable traffic to the public, did the Anglo-American Telegraph Company adopt a resolution proposing a reduction of tariffs in order to extend their business in social messages.[140] According to the response in the media, such a move was long overdue as costs for ocean cabling did not invite people to choose a cablegram over a letter. "Many a friendly message" so *The Era* complained, "would be sent if it could be managed for a souvereign, but when we come to Two Pounds for name and address only people recollect that a letter is delivered in nine or ten days."[141] Between 1869 and 1872, the tariff remained between £1 and £3 for a message of at least ten words. With the introduction of the word rate in 1872 and the possibility of sending cablegrams of less than ten words, ocean telegraphy still did not convert into a technology that was widely used.

In February 1873, the British *Daily News* led off a discussion on the demand for long-distance communication and the need for a social tariff. James Anderson, acting as spokesman of the Atlantic telegraph companies, debated an anonymous, yet very cable-knowledgeable correspondent, H. L., who had initiated the dispute with his letter to the editor. In the process, Anderson revealed the cable entrepreneurs' understanding of *Weltcommunication* and the motives behind their price policy. H. L. argued that since there were only two means for bringing down prices, government purchase of the Atlantic cables and competition, of which the former, government purchase, was not on option, competition in cable traffic was the only option for reducing prices to a social tariff.[142] In an open response to H. L., James Anderson opposed all suggestions for a social tariff, which at the time was discussed at one shilling a word. In an earlier manifesto, Anderson had argued that tariffs could only be lowered under three conditions: the government purchase of lines, the granting of monopolies, or amalgamation of competing companies. Furthermore, he promised that if the cable companies were unencumbered by state involvement and competition, prices would slowly come down on their own.[143] In his response to H. L., he claimed that a social tariff would be utterly unprofitable on the transatlantic connection, as there was no market for social messages in Europe or the United States, neither of which, so he claimed, possessed "a sufficient number of persons ready to spend 1s. per word upon messages which [were] not commercial, or of serious importance." He based this statement upon the conviction that "people separated by a great distance [did] not either write or telegraph frequently to each other, and, as a rule, the greater the distance, and the longer the period of separation, the less frequent would the interchange of communication become." From this, Anderson concluded that "[o]ne shilling per word would not be a social tariff low enough to encourage travelers to bother their friends with anything but the most important affairs, and if they had important affairs to communicate they would not be deterred by 4s per word."[144] At the time, Anderson was not the only one to hold such a conviction. Similar statements were made by representatives of American landline telegraph providers, such as William Orton and Norvin Green, who were convinced "that the telegraph would always remain a specialty service" and that "the telegram would never supplant the letter."[145] While Orton and Green argued that the telegram was ill suited to link sender and receiver by a similarly intimate bond as the letter, Anderson's argumentation entirely defied the image of a world united other than by trade. His sole emphasis was on profitability and cash flow.

In the days following Anderson's response to H. L., the *Daily News* received several commentaries on the ongoing debate. One writer called Anderson a "veritable Balaam"—a wicked man who was only interested in his company's revenues and ready to exploit the public.[146] Others challenged Anderson's theory of communication. Both had their point. Considering the underlying critique of telegraphy as an "exploitive" capitalist system, it is undisputable that submarine telegraphy was a business of immense initial set-up expenses, great risks concerning cable breakage, and high costs of cable maintenance. In the same instance, however, once the business was running, it was also one of large profits.[147] Anderson's response shows that for the cable entrepreneurs, the shareholders and their revenues were closer to their hearts than the users. After all, many of the entrepreneurs themselves held large interests in the various cable companies. Even H. L. admitted in his response to Anderson that "shareholders in existing companies [were] not philanthropists, but investors."[148] They would not lower prices if it lowered their revenues. Captain Henry A. Moriarty, former navigator on board the *Great Eastern*, expressed a similar sentiment in a letter to *The Standard* in 1870. As an investor, he simply possessed a right to siphon off large profits. After all, the cable pioneers had carried great risks, and now they should have the right to their proper pay.[149] Certainly, not everybody would have phrased it as undiplomatically as Moriarty, but his understanding of cable capitalism was not uncommon.

In particular, Anderson's thesis that the demand for social communication lessened in proportion to distance was hotly contested. Distance, one reader of the debate agreed, did play a role in social communication. However, it was distance created by time, the days and weeks it took a *letter* to arrive at its destination thus rendering all information outdated, and not distance created by geographical space. It was only "pure assumption on the part of Sir James Anderson that people separated by a great distance do not care either to write or telegraph to each other frequently"; rather, it was the great expense connected with transatlantic telegrams that hindered traffic.[150] Although this respondent mistook a cablegram as a faster version of a letter, Anderson's belief in lacking global social communication is unconvincing. Rather, at a time when the world was in migration, the cable entrepreneurs were deliberately closing themselves off from an enormous market—the migrants—thereby manifesting the telegraph as a medium for the elites.

Between 1815 and 1914, at least 82 million people voluntarily, not counting slave trade movements, migrated from one place to another. One of the most important migration movements at the time was that of Europeans to North America, which peaked in the 1850s, 1880s, and 1900s. Annually, between 260,000 people in 1850 and 1,000,000 people at migration's height in 1911 immigrated to the United States.[151] Calls for a national transatlantic German cable in the 1890s were in part induced by the large number of German immigrants.[152] Also rising traffic revenues that followed tariff reductions mirrored the demand for cheaper telegrams. Every reduction brought an increase in cable traffic. When in the early 1880s, during the tariff war with PQ, charges were reduced from three to two shillings, receipts increased considerably.[153] Still, the cable entrepreneurs based the tariff policy on the conviction, in the words of Anderson, "that a high tariff and a few messages pa[id] better than a low tariff and many messages."[154] Such a policy seemed to be confirmed by further price reductions. During the price war with the Commercial Cable Company between 1886 and 1888, the tariff drop from two shillings to six pence a word more than doubled the volume of traffic, yet could not, in the eyes of the cable companies, develop "a remunerative revenue."[155] The stagnation of the telegraph charges at one shilling a word rested on the entrepreneurs' claim that within the balance of supply and demand on the one side and revenues on the other, they could go no lower. The companies' user target group was Euro-American, white, and well-off; a social tariff remained an experiment not worth pursuing.

At the turn of the century, a number of different figures revived the debate about cheaper global communication, global media reform, and national monopoly theory. Among them were the Canadian railroad and telegraph engineer Sandford Fleming, the Australian newspaper proprietor Henniker Heaton, George Squier of the American Signal Corps, and the British politician Edward Sassoon. Their crusades fell into a period of growing nationalist interests in ocean cables that translated into various governments' desire for national submarine cables, such as the German Atlantic cables of 1900 and 1902 and the British and American Pacific cables of 1902 and 1904. In such a context, the media reformers' main argument of more state involvement hit a nerve. Some of them argued for a nationalization of all existing ocean cables, whereas others argued for a national entrepreneurship in the laying of new cables. While the ambitions of Fleming and Squier and their calls for government involvement have to be seen in the context of a British and American Pacific cable discourse under the heading of national security, the Australian Henniker Heaton's primary goal, indeed, was a social tariff.

In his writings, "the apostle of cheap communication," as Heaton came to be called, reflects on central aspects of the time, such as the interrelation of state and markets, rising (cable) nationalism and a changing economic system, and the reach of globalization-in-use.[156] Finally, Heaton's advocacy is another example of nationalism as a strategy and not a goal. Indeed, in his position as a spokesman "on behalf of the poor and middle classes," Heaton was not a singular appearance or a phenomenon;[157] he was also a spokesman of his time, employing the right strategy, i.e., national monopoly theory, at the right time. Furthermore, with his critique, Heaton expressed not only a particularly Australian point of view on state entrepreneurship, but also a more general viewpoint within the Anglo-American public against trusts, monopolies, and the exploitation of the public. By the beginning of the twentieth century, Australia had successfully nationalized its railways as well as its landline telegraphs, spurring on similar debates in Canada, New Zealand, and Great Britain.[158] In the United States, antitrust movements had already come to the fore in the aftermath of Gould's Western Union takeover in 1881 and found famous expression in the muckraking movement of journalists investigating (alleged) corruption and exploitation and much antitrust legislation during the Progressive Era (1890–1930), such as the Sherman Anti-trust Act of 1890.[159] Before 1900, a congressional committee examined the issue of government versus private monopoly in telegraphy nineteen times.[160] Heaton's antitrust and antimonopoly schemes were not an uncommon occurrence in the Anglo-American world. Heaton's writings, however, mirror not only economic and political debates, but also social concerns and the demand for social messaging in the global realm. In fact, his private papers and publications are some of the very few sources through which those acclaimed 90 percent unconnected to global communication speak.[161]

Heaton, an Australian newspaper publisher and member of the British Parliament, first entered the stage of ocean cable policy in 1885 when he represented Tasmania at the Berlin International Telegraph Conference. Heaton thereby opened his battle against the world's global monopolies with an initial success: he managed to secure lower cable rates between Great Britain and Australia.[162] From then on, he was set on the topic of a global penny post, equally devoting his energies to ordinary mail service and the ocean cable service.[163] His two favorite enemies were James Anderson, who countered all of Heaton's proposals concerning cable management as "impossible" or "ineffective," and the respective postmaster generals of Great Britain, whom he badgered in Parliament with his constant enquiries.[164] In 1898, he enjoyed his first breakthrough with the introduction of an imperial penny post, followed in 1908 by an Anglo-American penny post for mail sent between the United States and Great Britain. In the decade prior to the First World War, Heaton concentrated on the monopolies of the ocean cables, arousing great controversy with his daring essays on "How to Smash the Cable Ring."[165] Until his death in 1914, Heaton used every means possible, including pamphlets, essays, and letters to the press, as well as his position and influence in the British Parliament, to fight for a universal cable penny post. In contrast to his achievements concerning imperial penny postage, Heaton failed in his crusade against the cable monopolists. A conference held by the British Royal Colonial Institute in 1908 and an "influentially attended" meeting of London merchants and politicians presided by the Lord Mayor in December 1908 to consider the feasibility of an ocean penny press throughout the British Empire remained his only triumphs.[166] In 1888, tariffs on the transatlantic route were one shilling per word, and they stayed so until 1923.

Heaton opened his battle against the "tyranny of capital" of the cable companies with a letter to the British Postmaster General in which he fervently urged the reduction of cable tariffs.[167] Therein, he painted the picture of a two-class society across the globe falling sick from a technology that had an increasingly disunifying character. Although the telegraph had "annihilated time and space," it increased the gap between those who could communicate with it and those who could not, thereby exposing social fragmentation.[168] He argued that the bulk of trading negotiations were still conducted in writing "just as they were between Assyria and Egypt thousands of years ago." This circumstance not only produced "a lamentable waste of time," but also disadvantaged those not communicating via cable.[169] The gap was such that those unconnected "might be living in another planet" for all the use they could make of "the great invention."[170] This reality of an increasingly fragmented world, foreshadowing current debates about a digital divide, remained one of Heaton's main arguments throughout all of his writings.[171] As Heaton disclosed in his speeches and articles, the disunifying character of the modern age expressed itself not only in the medium of space, as it mattered greatly whether you were born in Africa or in Europe, but also in the medium of time. The speed and inherent instantaneity of the ocean cables produced an entirely different time sphere in which people with money could move and govern the others.

Heaton admired the work of the French philosopher Pierre-Joseph Proudhon, and his critique was influenced by Proudhon's socialist writing. Considering Heaton's conservative political background, this comes as a surprise. In 1884, Heaton settled in London and returned to Australia only for brief business visits. For the next 25 years, from 1885 on, he occupied the conservative seat for Canterbury.[172] Neither his contemporaries nor his biographers called Heaton a socialist, but he socialized with radical sympathizers such as the American Mark Twain. Moreover, in Heaton's writings, one finds striking similarities in tone and argument, clothed in a more general notion of entrepreneurs' social responsibility. To Heaton, electricity represented a "common heritage of humanity," which had to be made available to the entire global public. Private property, as expressed in the cable monopolies, and restricted access to global communication were, in accordance with Proudhon's script *Qu'est ce que la propriété?*, theft.[173] Prohibitive transatlantic cable tariffs helped to "accumulate business in the hands of a few . . . and to accentuate the inequalities of distribution." The cable ring, personalized by John Pender and James Anderson, "cut off all electric communication between the masses" and made them into illiterates.[174] Ocean telegraphy was nothing but a "plaything . . . of millionaires."[175] Heaton's was not the sole voice claiming telegraphy as people's property. As early as 1866, an article in *Reynold's Newspaper* advocated Marxist argumentation: the Atlantic telegraph represented, in the same way as railroads and steamships, an achievement of the working classes, and this fact grounded their right to political participation.[176] In 1877, Friedrich Engels proclaimed in his *Anti-Dühring* essay the necessity of state ownership of mail, telegraph, and railroad.[177]

It remains unclear whether Henniker Heaton had read any of the writings of Marx and Engels. Nevertheless, he reached the same conclusion: the only solution to counter "the exclusive possession of a right by an individual" and the "deprivation of essential or valuable privileges to the entire community" was state ownership.[178] After decades of the failure of competition on the ocean telegraph market to lower prices enough to create a social tariff, the nationalization of all already existing ocean cables seemed to be the ultimate solution to cheap global communication.

Nevertheless, in Heaton's militant rhetoric, which had a distant sound of expropriation and emancipation of the masses, he left a loophole for the cable entrepreneurs, appealing to their philanthropic side and their social responsibility. Heaton claimed access to ocean telegraphy for everyone as a "precious birthright" that had been taken away from the millions by "unenlightened management," but in the same way, it could be returned to them through enlightened state management.[179] In 1895, he publicly urged John Pender to fulfill the technology's original messianic promise of an "immense increase in the happiness of the masses" by linking "the hearts, and not merely the pockets," of people across the Atlantic. To facilitate communication "for the millions who [could] never pay high tariffs, but who none the less long[ed] for the means of communicating in a moment with those who are dear to them" would certainly "crown [his] work" in global communication.[180]

Nevertheless, as H. L. had already detected in 1873, cable entrepreneurs and their stockholders were not philanthropists, but calculating investors and capitalists. Throughout the period, they doubted the technical feasibility of a universal penny post and based their tariff system on a communicational model that postulated that "[t]he social element which justifies the penny postage and one shilling or six penny telegrams within the limits of a State does not exist outside these limits and cannot be created."[181] Despite the increasing number of transatlantic letters, they argued that "the two [English-speaking] peoples, numbering more than 100,000,000 of the same blood and speech, ha[d] nothing to say to each other and no desire for more frequent, rapid, and intimate communication," at least nothing that could be translated into cable messages.[182] Such differentiation between the national and the international communication market is also important to note in the light of the emergence of publics and communities. Contemporaries, represented by the cable entrepreneurs, assumed around 1900 that, speaking in social terms, a national community, but not a global community, had developed. Their view was certainly supported by the popularization of the telephone and the national telegraph systems. In the United States, for instance, both became a means of social and mass communication within the national realm around 1900 and 1910, respectively.[183]

This was not the case with the ocean telegraphs. In response to the entrepreneurs' unwillingness to take on social responsibility, Heaton's claims for state ownership became more forceful. The year 1908 saw the peak of Heaton's cable reform ambitions. He managed to mobilize a large number of influential spokesmen from politics, finance, and trade on behalf of his cause during a meeting held by London's Lord Mayor; he also organized a conference on global communication held by the Royal Colonial Institute, and the press even reported him spearheading a Cable Rate Reform Party.[184] In his work, Heaton found support from other cable reformers such as the Canadian Sandford Fleming and the American George Squier, who had also engaged in the quest for a national Pacific cable, employed nationalist rhetoric, and argued for state ownership. In particular, between Fleming and Heaton, there existed mutual support in cable matters. On closer look, however, Heaton's cable nationalism was different from Fleming's or Squier's. Fleming and Squier had little in common with Heaton's internationalist approach to national intervention. While other cable reformers emphasized aspects of national security and imperial control as the main reasons for government intervention, Heaton demanded state intervention only as a preliminary step leading up to the abolition of national frontiers and the international regulation of global communication. While the broader discourse of government control over ocean cables was targeted at reaching nationalist aims, its most outspoken representative turned out to be a proponent of *l'esprit d'internationalité* as proclaimed earlier in the century by Louis Renault and David Dudley Field. In his speech before the Royal Colonial Institute in 1908, Heaton fervently argued in favor of a system of global communication without political boundaries, hence without state taxes and national telegraph regulations. These were "[t]he chief obstacles" before them in their fight for affordable global communication for all. Consequently, the conference delegates' objective should be "to abolish political frontiers . . . in [their] communication with every part of the earth," as far as telegrams were concerned.[185] Heaton's biography is similar to that of many cable engineers and operators in the sense that he constantly moved back and forth between various parts of the Anglo-American world. He was convinced that "[b]etween man and man these political frontiers should not exist" and that, as a matter of fact, "to the travelled individual who has friends all over the world they do not exist, except on paper."[186] Although he does not make this point explicit for the ocean telegraph system, this conviction meant the abolition of national borders in the form of different state taxes and regulations concerning the use of cipher and codes. Heaton probably endorsed a system of communication that charged by distance and not by route and that was internationally regulated. The national element only represented a means to achieve his end of a social communication system on a global scale. With this strategic nationalism, Heaton was not far from the cable entrepreneurs, who similarly employed it to uphold their global economic system.

Although their strategy was similar, Anderson's and Heaton's ideas and perceptions of *Weltcommunication* and its importance and implications could not have been more different. Although those directly involved in the cable business considered the ocean telegraphs as a means to facilitate the movement of information and commodities, Heaton believed they should serve "the needs of modern life," which was expressed in the vast mobility of people all across the globe.[187] In Heaton's theory, global communication was seen in a true enlightened manner as a "natural right" that should demolish boundaries not only of geographical distance and time, but also of social class and national identity.[188] In Anderson's reality, global communication solely nurtured the trade of a wealthy few and was essential to a modern materialism, which was also mirrored in the routes the cables took. It went without saying, for Anderson, that the cables were laid where the commerce was.[189] These conflicting worldviews show perfectly the competing forces at work in imaginaries of *Weltcommunication* and also the connected ideas of modernity itself.

As a consequence of the cable actors' strong emphasis on revenues, the new *Weltcommunication* was, from its beginning in 1866, based on geopolitical structures of economic and political interests, and from a structural point of view, it remained so until the twentieth century. In the nineteenth century, nobody would have invested in a telegraph cable from Great Britain to India to foster intercultural understanding.[190] However, there were repeated attempts to challenge such a materiality of global communication, first in the form of news agents in the 1880s and, later on, in Heaton's campaign for a social tariff. News agents' great interest in the technology of submarine telegraphy demonstrated that the entanglements of the Euro-American world soon extended beyond mere trading relations. People in Europe and the United States were interested in the fate of David Livingstone, who was "lost" in the "dark continent" of Africa. In particular, with his two independent editions of the *Herald*, one in New York and the other in Paris, James Gordon Bennett nourished the development of an Anglo-American yellow press and journalistic sensationalism that encompassed the entire globe in its forms of news distribution, news gathering, and news production. The advent of the newsmaker also marked a systemic shift in the organization of the global communication system. A market advantage was no longer obtained via control over vertical business networks that integrated the manufacture, laying, and operating of cables, but rather through the concerted organization of the cables and their content. This characterized a development from initial point-to-point communication to the inclusion of Euro-America's "hinterlands" and the combined working of ocean and land lines. Bennett's fight against Jay Gould and Western Union for the independence of the AP, in contrast, illustrates the influence of local systems and forces on global systems, as well as the interrelatedness of the business of communication and news. The conflict between the Commercial Cable Company and the Atlantic pool was as much a cable war as it was a news war.

The twentieth century witnessed a new attempt to challenge the conventions of a cable communicational materiality. Henniker Heaton died in 1914, coincidently the same year as the start of World War I. His crusade for a universal penny press prior to the outbreak of the war benefited from an increasing nationalist and social activist movement. This made it easy for him to merge these two aspects and find alliances among nationalists within the cable reform movement, who neither took notice of nor challenged his internationalism. Heaton made himself the spokesman of the 90 percent of the globe unconnected to global communication, and his treatment in this chapter argues for a broader perspective on those nonusers of the global media network. When it came to cable charges, little changed on the Atlantic market. Between 1888 and 1923, the tariff remained steady at one shilling a word.

In consequence, the submarine telegraphic system led to the phenomenon that while the boundaries of time and distance were challenged, other boundaries, such as class, were reinforced. Moreover, although the system's operation restricted global communication to a specialty service, it allowed newsmakers a global reach for producing and selling news. In their use, the ocean telegraphs were highly ambiguous. From a systemic perspective, the ocean telegraphs provided one of the key mechanisms that simultaneously nurtured processes of both global integration and global fragmentation, whereas distance and proximity were not only created by space but also by time. Concerning global communication, two sets of people were facing the same geographical distance between London and New York, which by means of news seemed to be constantly decreasing, yet they were moving in entirely different spheres of time, which determined whether New York's distance from London was one of minutes or, in particular concerning social news (i.e., personal communication between individuals), weeks. For the scholars of today, the geographical maps of ocean cables are highly deceptive if we take them as yardsticks of global communication. If we look at them from the perspective of global news, however, as exemplified in the work of Gordon Bennett, we must assume that the scope of experiencing "globality" was far larger than the cable maps suggest, even if predominantly only for Euro-America.

5

THE PROFESSIONALIZATION OF THE TELEGRAPH ENGINEER

Will any of "Ours" kindly inform me how I can obtain a situation as telegraphist in the Australian Colonies?

"Queries," *The Telegraphist*, April 2, 1884.

TELEGRAPH MAGAZINES, such as *The Telegraphist*, received requests like the epigraph in large numbers. Indeed, it was part of the attraction of the job as telegraph operator, engineer, or electrician that a young man could "try [his] fortune in other countries."[1] A massive worldwide dissemination of a predominantly British workforce accompanied the emergence of a global cable network. While local firms usually handled and staffed the landlines, the great majority of submarine cable stations were manned by the British companies' own staff. Both the Electra House Group as well as the Atlantic pool companies not only manufactured their cable equipment in Great Britain, but also recruited British staff. The companies' management claimed that not everyone could be entrusted with the job, as submarine cables were a highly complex and difficult technology bound to absolute secrecy. Men of predominantly English, Welsh, or Scottish and sometimes Irish origin operated almost 80 percent of the world's ocean cables. This translated into a globe-spanning network of British specialists who moved back and forth between the various stations and the companies' headquarters in London.

Professionally, these telegraph operators and engineers organized themselves into the Society for Telegraph Engineers (STE). Over and above the level of interactions between individuals, the STE provided the link between the predominantly British and continental European telegraph expatriates manning the cable stations around the globe. It established a common identity, while each company created an individual company man, and furthered the exchange, generation, and homogenization of telegraphic knowledge around the world.[2] Famous members of the Class of 1866, such as William Siemens or William Thomson, helped the STE flourish and also used it as their scientific and political platform. At the time, scientific societies often took on a "gatekeeping role" by defining eligibility. They not only sorted out "those whose views [did] not fit in," but also defined professional membership.[3] Through their presence, the eminent cable engineers managed to shape and influence the image and archive of their own profession. In the end, however, the STE as a "cosmopolitan institution" represented another failing vision of an electric globe in union. Toward the turn of the century, the cable engineers had to witness the transition of their society from the STE, an international, but largely British, society to the national Institution of Electrical Engineers.

TELEGRAPHIC KNOWLEDGE AS COSMOPOLITAN SCIENCE

The profession of submarine telegraphy emerged into a world where science and engineering were expanding enormously. New and diverse fields of science, such as archaeology and cell biology, were established, and others, such as chemistry and geology, matured.[4] With constant new inventions and innovations in the fields of steam engines, railways, and electricity, the number of civil engineers rose considerably.[5] In contrast to the eighteenth century, when people who performed the acts of *engineering* had been a loose assortment of mechanics, stonemasons, millwrights, or instrument makers, engineering practitioners had by the nineteenth century begun to acquire the characteristics of a professional group.[6] Moreover, Great Britain represented the leading industrial nation, as well as the uncontested forerunner in the fields of science, engineering, and invention. In 1870, the country produced nearly a third of the world's manufactured goods. In comparison, the United States produced less than a quarter and Germany only 13 percent.[7] The Great Exhibition in 1851 in London's Crystal Palace had displayed the achievements of the British Industrial Revolution to local and foreign visitors.

It simultaneously became the symbol for the "heroic age of British engineering," a period from roughly 1800 to 1870.[8]

The cornerstones of scientific exchange at the time were a selected number of learned societies, such as the Royal Society or the Society of Civil Engineers. They had been established in the late eighteenth and early nineteenth centuries. Most of them were located in London.[9] Similar to other emerging professions at the time, telegraph engineers came to organize themselves professionally. In 1871, a group of eight, then less famous, cable engineers, among them Wildman Whitehouse, Robert Sabine, and Ludwig (Louis) Löffler, founded the Society of Telegraph Engineers (STE) as a platform for professional exchange.[10] Within weeks, the more eminent Great Atlantic Cable engineers, such as William Siemens, Samuel Canning, Latimer Clark, Willoughby Smith, Cromwell Varley, and William Thomson, took over the undertaking. As they occupied important offices and positions within the society, they attracted other telegraph professionals from around the world. Soon the society represented the focal point for all (ocean) telegraph engineers and operators throughout the world.[11]

Formally, the STE was established in May 1871. Its membership body consisted of "Members, Associates, Students, Foreign Members and Honorary Members."[12] Drawing from the example of the Society of Civil Engineers, the telegraphers established a hierarchy with two principal grades, members and associates, who had to pay an annual subscription of two guineas and one guinea, respectively. In principle, the STE was open "to all persons . . . interested in Telegraphy without being necessarily Telegraph Engineers by profession."[13] The founders conceptualized the society to be as inclusive as possible. They reached out to the practical men of science as well as talented "amateurs" and, finally, telegraph operators and administrators of state-owned organizations, such as British Post Office officials. They engaged with landline and submarine telegraphers, engineers, operators, scientists, and practitioners. This inclusive approach was mirrored in the constellation of the institution's first president and its two vice presidents. Charles William Siemens, electrical manufacturer, engineer, and head of Siemens Brothers in London, represented as president the "practical applications of electricity," whereas vice president Lord Lindsay, a British astronomer and twenty-sixth Earl of Crawford with an extensive personal library and a private observatory, stood for the group of "gentlemanly specialists."[14] Finally, Frank Ives

Scudamore, head of the Postal Telegraph Department and himself not a telegraph engineer, represented the administrative and political aspects of the telegraph business. The STE's initial membership body consisted of 70 persons, but it quickly expanded to 278 members in 1872 and more than tripled this number by 1881 to 981 members.[15] In the words of William Thomson, the society had been growing "with telegraphic speed."[16]

The STE's proclaimed goal was "the general advancement of Electrical and Telegraphic Science, and more particularly . . . facilitating the exchange of information and ideas among its Members."[17] According to its founders, this exchange of knowledge had to be organized according to the rules of cosmopolitanism and along the lines of the eighteenth-century Republic of Letters. Already in his inaugural speech of 1872, STE president William Siemens made clear that it was "necessary for a Society of Telegraph Engineers to be a cosmopolitan institution."[18] His programmatic statement underlined the institution's heritage from cross-boundary science, as represented in the eighteenth-century Republic of Letters; the present condition of distinct cable transnationalism, drawing from the inherent international character of ocean cabling; and the STE's future claim to "universal" leadership. Siemens weaved all these points together in his address as the first president of the telegraphers' institutional representation. In his notion of "cosmopolitan," Siemens was influenced by German scientific discourses on Kant's *Metaphysical Foundations of Natural Science* and post-Kantian German *Naturphilosophie*. He derived his notion from Kant's ideal of a worldwide community of human beings or, in this case, telegraph engineers and operators.[19] Importantly, however, this form of cosmopolitanism did not oppose absolutist statism, the empire, or the nation-state. Rather, these structures seemed to be recognized as useful but relatively unimportant in shaping the telegraph agents' identity. For Siemens and other cable engineers, this cosmopolitanism defined the relationship between state structures and the inherently transnational ocean cable business in a nutshell.[20] Siemens also played on a then still typically British self-image. British intellectuals and politicians at the time argued that imperial Britain was inherently cosmopolitan and not as infected by nationalist thinking as continental Europe.[21] Cosmopolitanism was an integral part of British identity. Thus, the STE as a cosmopolitan institution simultaneously served British imperial self-perception as well as the transboundary nature of submarine telegraphic engineering.

In his conceptualization of submarine telegraphy as cosmopolitan science, one group of members received particular emphasis in Siemens's speech: the foreign members. This category was applied to "all members (English or Foreign) residing permanently abroad" and signified the key elements of the newly established society.[22] With the installment of foreign members, Siemens expressed not only the society's aspirations to cosmopolitanism, but also its claim to international leadership. This orientation toward an international membership body demonstrated the strong influence of submarine telegraphic thinking. Landline agents and their rather "national" topics were considered to be of lesser importance. Finally, the concept of foreign members indirectly weakened the argument of those opposed to a society of telegraph engineers; an institution might be small when conceptualized in the geographical realms of Great Britain and its empire, but exponentially larger and more significant when conceived within the realm of a submarine telegraphic globe. This strong emphasis on foreign members also set the STE apart from other learned institutions in Britain at the time, which were solely nationally organized. In the end, Siemens's strategy of "think big" not only was a means to strengthen a scientific institution based on one single product, but also stemmed from the notion that transnationalism was inherent to cable work.

Indeed, submarine telegraphy had already in its formation been, in the words of Charles Bright, "the work of many hands." In many respects, it represented the prime example of transnational science, albeit solely within a Euro-American setting.[23] As early as the 1840s, various attempts to take the electric telegraph *submarine*, or under water, had been carried out independently by Charles Wheatstone in Swansea Bay in 1844, by Samuel F. B. Morse across New York Harbor in 1842, and by Ezra Cornell through the Hudson River in 1845. Other schemes are attributed to Charles West, who had in 1846 obtained permission from the British government to connect Dover and Calais and laid a cable through the Portsmouth Harbor; a Mr. Armstrong who experimented in the Hudson River with an insulated cable in 1848; and, in the same year, Werner Siemens who used a wire to fire underwater mines in Kiel.[24] Although research and development were predominantly done in Great Britain, submarine telegraphers gained an astute awareness of research and experiments conducted elsewhere.[25] In his 1867 monograph *The Electric Telegraph*, Robert Sabine, telegraph engineer, author, and consulting partner to Atlantic cable engineer Samuel Canning, revealed that he was widely read. He was equally well versed with the experiments of the Germans Soemering and Schilling, as with those of the British Cooke and Wheatstone and the American Samuel F. B. Morse.[26] An even more striking example is Dionysius Lardner's 1855 book on electric telegraphy; its content was similar to Sabine's and aimed "to render intelligible to all who can read . . . the various forms of telegraph in actual operation *in different parts of the world*."[27] This high interconnectivity and flow of knowledge across the borders of Europe and America now nurtured the understanding of a scientific transnationalism among the engineers that was connected to an assumed Western universality of knowledge when it came to submarine telegraphy. For the primarily Euro-American discourse, technical works from Great Britain, France, Germany, and the United States sufficed for such "universality." Although highly Eurocentric in its structures, in mid-century, there was a high degree of scientific exchange, translations, and foreign literacy of which Siemens, himself a German scientific émigré to Britain, was keenly aware.[28]

In his inaugural address in 1872, William Siemens explained the concept behind foreign membership. Its justification lay in the extension of the "great network of international telegraphy to every portion of the civilised and semi-civilised world," traversing "deserts and mountain chains," passing over "the deep plateau of the Atlantic and over the more dangerous bottom of tropical seas." Hence it was "necessary" for the STE to be a "cosmopolitan institution." Its goal was "to combine the knowledge of these diverse circumstances, and of the diverse practice resulting therefrom, . . . to be a focus into which the thoughts and observations of all countries flow, in order to be again radiated in every direction for the general advancement of this important branch of applied science."[29]

In order to ensure such a result, Siemens continued, "the Council have agreed to the creation of another class of members,—the 'Foreign Members.'"[30] Siemens himself recruited the first cohort at the international telegraph conference in Rome organized by the International Telegraph Union (ITU) in late 1871. Siemens had participated as representative not only of the Indo-European Telegraph Company but also of the newly established society. The council had authorized him "to invite the representatives of the telegraph administrations of the world" at the gathering, and in his address, he read out the cordial responses he had received to his appeal.[31] The new foreign members included General Lüders,

Director-General of the Russian Imperial Telegraphs; Signor D'Amico, Director-General of Telegraphs in Italy; Signore Salvatori, Inspector General of Telegraphs in Italy; Monsieur Ailhaud, the representative of France to the ITU; Monsieur Vinchant, the representative of Belgium; R. de St. Martial, Secretary of the International Bureau; and, of course, Dr. Werner Siemens, his older brother. Furthermore, the representatives of the Netherlands, Spain, and Switzerland as well as Professor Capanena, Director-General of Brazilian Telegraphs, had consented to join the STE. Samuel F. B. Morse was at that point the only American member. The director-generals and chief engineers of the telegraph administrations of Great Britain and India represented British telegraph engineers abroad.[32]

The creation of the category of foreign members represented one of the shrewdest moves in the conceptualization of the STE. It was equally clever to tap the international network of the ITU, which had been established in 1865 and in which all important telegraph nations were represented. Soon, the STE represented all major telegraph networks and associated engineers, electricians, administrators, and operators of the various globally active companies, as well as the ITU. Although this might have delayed the creation of similar learned institutions on telegraphy outside of Great Britain, it allowed the existence of an almost strange parallelism between an institutional transnationalism as represented in the STE and an institutional internationalism as represented in the ITU. In fact, the parallel existence of the ITU and the STE demonstrates the simultaneity of concepts of cosmopolitanism and internationalism in the late nineteenth century.

In the following decade, the class of foreign members grew from 17 in 1872 to 152 in 1881, at that point representing 15 percent of the entire membership.[33] From the outset, the STE saw to it that the foreign members were prominently represented; one example was its multilingual publishing policy.[34] At a time when Latin had ceased to be the lingua franca, this seemed to be the sole means of expression for a scientific universalism. Many foreign engineers were well versed in the English language; others were not. Their application to the STE could hardly be expected "unless [they] offered them [their] proceedings either in their own language, or at least in another language besides English." English, French, and German were the languages that "it may safely be assumed . . . every educated person throughout the civilized world speaks."[35] As a result, articles were printed in all three languages.[36]

The increase of foreign members prompted the STE's council to create another tool for promoting its global reach. In 1873, it decided to establish "in each country an honorary appointment," thereby also strengthening its influence abroad. The appointment was called Local Honorary Secretary. The main objective was "to advance the Society by obtaining an increase in members, and to act . . . between the Council and the various foreign members resident in the same country as the Local Honorary Secretary."[37] Starting out with five foreign, that is, non-English, branches in India, Japan, Italy, Norway, and Belgium, the number had increased to thirteen by 1877. In addition to branches in Europe (in France, the Netherlands, Denmark, and Germany), the council expanded the society's network to almost all other continents. Aside from branches in India and Japan, new foreign branches were established in the Argentine Republic, New South Wales (Australia), Chile, and North America.[38] The setup of local honorary secretaries followed the major submarine cable routes as well as the ITU membership lists. It resembled the geopolitical structure of the time as it came to overlap as well as expand on the structure of the British Empire.

In most cases, the honorary secretaries of the extra-European foreign branches were British engineers or scientists residing abroad. Professor William Ayrton, for example, who later became president of the STE, occupied the post of honorary secretary while he was employed as professor of natural philosophy at the Imperial College in Tokyo. He was one of the approximately 3,000 foreign scientists and engineers Meiji Japan had hired to initiate its technological and industrial revolution.[39] The list of foreign local honorary secretaries of 1874 only contains one person without an obvious Anglo-American name among the secretaries from extra-European regions: Don Ramon Pias, director-general of the Chilean Telegraph in Santiago, Chile.[40] Despite the institution's claim to "combine the knowledge of [the] diverse circumstances" of the globe where submarine telegraphy found its application, this did not extend to the inclusion of knowledge generated by those other than British or Euro-American telegraph operators.[41]

Exceptions in this constellation were the engineers from Japan, who shared certain social and professional characteristics with their Euro-American peers. Usually middle or upper class, they were the technological elite of a rapidly industrializing country. Most important, however, they were trained according to European standards. Between 1868 and

1912, Japan underwent a phase of social, political, and economic transformation and industrialization. After Japan's confrontation with Europe and the revolution of 1867–1868, Japan's new leaders believed that in order to avoid becoming a colony, their country had to become not only militarily powerful but also technologically sophisticated. Starting in the 1870s and "[w]aving the flag of techno-nationalism," the Japanese government sparked an industrial revolution. They invested heavily in sectors that they considered important for economic development, such as mining, railroads, electric power, and communication.[42] As Japan was determined to "catch up with the West" militarily and industrially, its engineers were met on equal terms.[43]

Through its alleged cosmopolitan character, the STE pursued a policy of global outreach and inclusion and also of global control. The STE shaped the picture not only of the telegraph operator nationally as well as internationally, but also of his knowledge set, drawing from a sense of (Euro-American) epistemic universalism. Despite the all-inclusive approach toward membership for everyone with an interest in telegraphy, the institution did not set out as a network for anybody interested in telegraphs. As gatekeeper of telegraphic engineering, the society defined eligibility. When the first transatlantic cables failed in 1858, many people came forward with suggestions for improvements. As this massive response indicates, submarine telegraphy was very much a technology of public, and scientific, interest. Many from the scientifically interested public, such as the Prince of Wales himself, took part in the discourse on best submarine telegraphic practice. As engineer Charles T. Bright reported in a letter to Field in 1858, they had "more than twenty machines for the like purpose (paying out) brought to our office."[44] As his son, Charles Bright, put it, the very project of an Atlantic cable "appeared . . . to stimulate and excite the brains of many a sanguine inventor," and it was "both amusing and sad to think of some of the ideas put forward."[45] And yet, only the ideas of the "professionals" were considered.

The statutes of the STE show that it certainly was not meant to be all-inclusive, but rather to serve as an exclusive club of learned telegraph practitioners. To become a member, one had to fulfill one of the following conditions: to have been "educated as a Telegraph Engineer" with subsequent employment for "at least five years in responsible positions"; to "have practiced on his own account" in the profession for at least two years and have "acquired a degree of eminence in the same"; or to "be so intimately associated with the science of Electricity or the progress of Telegraphy that the Council consider his admission to Membership would conduce to the interests of the Society."[46] Knowledge and education were markers for inclusion, making the STE a middle- and upper-class representation. Admission to the STE was by proposal and seconding, whereby every member first had to go through the status of associate before becoming a full member. George Spratt, second superintendent at the Porthcurno telegraph station in Cornwall, vividly recorded in his diary strong indignation that his superior, Bull, had been suggested as member (with the company paying his membership fee), while he was not.[47] Sources from Valentia and Heart's Content further suggest that only senior clerks were members.[48] In its early decades, the STE managed to exert tight control over the telegraph professionals it represented. With its global outreach and great mass of foreign members, its gatekeeping function became even more prominent, and it globalized a distinct image of a telegraph agent and telegraphic knowledge.

ALL THE GLOBE IS A LABORATORY: ON THE CONSTRUCTION OF TELEGRAPHIC KNOWLEDGE ACROSS SPACE

No other profession in the nineteenth century so fully incorporated the entire globe in its representation, in its practical exertion, and also in the generation of its profession's knowledge as submarine telegraph engineers. Euro-American operators, engineers, and electricians constructed, generated, and dispersed the world's telegraphic knowledge within their global network of submarine telegraphs. Visions of form and content of this knowledge, however, were very diverse and entailed a strong tension between *local* and *global* on top of a broader debate on the role of research and development. Although the STE played an important role as focal point of all knowledge on telegraphy, the importance of telegraph stations and telegraph ships as sites of knowledge should not be underestimated. Yet, the relationship between those two sites and so between men of science, such as William Thomson and William Siemens, and men of practice, such as James Graves, was complicated and often tense. In the 1870s and 1880s, the STE, which came to symbolize the archive of global telegraphic knowledge, provided the battleground for these negotiation processes.

As the early history of submarine telegraphy with its frequent cable failures illustrates, the profession was one in which the practical application of its technology seemed to dominate over scientific theorizing. Cable manufacturers, such as Richard Glass, built and operated the earliest cable submarine "without much reference to precision measurement." Instead, their work appeared to obey the principle of trial and error.[49] Although submarine telegraphy essentially drew from scientific discoveries and inventions, such as electricity and magnetism, as well as from the mathematical calculations of William Thomson, much of its practical knowledge originated from actual work on the cables. The engineers and operators tested and developed their instruments and apparatuses while paying out or repairing cables on board a ship or while operating them in transmitting messages. Werner Siemens, for instance, developed his *Legungstheorie* (theory of laying submarine cables) of 1857 and devised an apparatus to regulate the strain on the cable while assisting the laying of a cable in the Mediterranean.[50] Similarly, the electricians and engineers on board the *Great Eastern* adapted and perfected the paying out and propelling machinery during each of the four Atlantic expeditions.[51] In this fashion, they gathered practical knowledge in the 1850s and 1860s from various cable expeditions in the English Channel, the Mediterranean, the Red Sea, and the Atlantic involving engineers who were predominantly from the Class of 1866.

The telegraph companies also applied the method of learning-by-doing in the training of the telegraph operators. Telegraphy was not necessarily something you learned at a general school near the London headquarters, but while working with the cables within the company's realm at the respective telegraph stations—the actual spaces of knowledge. The companies themselves took care of the training of their operators, continuing the master-apprentice model typical for Great Britain at the time. The apprenticeship, however, followed rigorous preselection. The telegraph profession was not something one could simply join or acquire; a good sense of spelling was the minimum expected, and a middle- or upper-class upbringing was the standard. In addition, the Anglo-American Telegraph Company, for instance, asked questions concerning the background, education, and former employment of the applicant. Their application form was found in the appendix to the 1880 *General Orders, Rules and Regulations to Be Observed by the Officers, Clerks and Servants of the Company*, and thus was probably designed to be passed on to family members and friends of the active staff.[52] In the end, the technology was just as exclusive in its telegraphing usage as it was in its telegramming usage.

The Pender-Siemens controversy in the 1870s influenced this initial setup of submarine telegraphy as learning-by-doing significantly. As a result of Pender's victory against Siemens, submarine telegraphy became based on a rather service-oriented business model with few funds for research and development. The industry tended to be rather "conservative" because it preserved techniques and procedures over a long time. Most of the techniques used for manufacturing, laying, and repairing the sea cables had developed during the early period of the 1850s and 1860s. Thereafter, little changed, and the design of the cables remained unaltered for almost a century.[53] Until the turn of the century, the cable companies, in particular Pender's Eastern and Associated Companies conglomerate, saw outside inventors as the principle source of technical change with regard to the telegraphic instruments. Between 1872 and 1929, the company entered into 21 patent licensing agreements, such as William Thomson's syphon recorder of 1870 (which produced a permanent record of the signals), Alexander Muirhead's duplex apparatus of 1876 (which enabled simultaneous transmission and reception), and Sidney Brown's drum relay of 1899 (which improved the signal and speed with which signals were automatically passed on to recording instruments or another cable). The annual royalties for these inventions often reached several thousand pounds but allowed the company to spread the risk of technological development to outside experts.[54]

Although this conservative approach toward research and development with only small changes to the actual cables made submarine telegraphy appear to be a "stagnant technology," this does not mean that knowledge was only preserved and not generated.[55] Although the scale of research and development at the Eastern and Associated Companies was small in comparison to the American companies Western Union, Bell, and Western Electric, it was by no means insignificant.[56] In addition to a small in-house research staff, the individual cable stations were important generators of knowledge. Scientists and engineers could hardly reproduce the behavior of long deep-sea telegraph cables within the realms of a scientific laboratory. As a result, the cable stations served not only as spaces of translation and transmission, but also as spaces of research and testing, sometimes joined together as one globe-spanning research network. According to the stations' letter books and diaries, the engineers often brought new inventions out to the stations to develop and test them. Both James Graves, superintendent at Valentia, Ireland, between 1866 and 1909, and Ezra Weedon, superintendent at Heart's Content between 1866/1867 and 1884, reported that the American J. B. Stearns, who is credited with the breakthrough in duplex telegraphy, came out to the stations in the 1870s with his instruments for testing purposes.[57] During these testing visits, the engineers conducted research in close conjunction with the telegraph operators, who then had to send reports about the results to them and to the companies' London headquarters.

In the 1890s, for instance, F. Perry, Weedon's successor as superintendent at Heart's Content, and Alexander Muirhead, inventor of a patent to use duplex on submarine cables, corresponded intensively on the duplex system on the Atlantic cables in the 1890s.[58] Sometimes, as Graves remarked in his diary, the instruments would even arrive without their respective inventor. Stearns, for example, "did not come at first to Valentia but sent his apparatus and a copy of his specifications with written instructions by the aid of which it was started and worked at this Station."[59] Similarly, in 1871 and again in 1877, William Thomson submitted different prototypes of his automatic curb sender to Spratt at Porthcurno and Graves at Valentia "for trial on the cables."[60] Cable stations served as an outpost of the scientist's home laboratory, where the scientist, in this case William Thomson, would stop by briefly on a tour with his yacht *Lalla Rookh*, gather data, and return home to his desk.[61] Practical and theoretical additions to the pool of telegraphic knowledge were generated in conjunction with, albeit relatively independent of, each other as they were separated by geographical distance.

However, research and testing did not just happen in interactions between the scientist in London, Berlin (Siemens), Glasgow (Thomson), or Boston (Stearns) and his personal laboratory outpost, cable station, or cable ships. The plurality of the cable stations in their global embedding also played an important part in research designs. The generation of knowledge was globally imagined and executed. At times, the cables were turned into gigantic test tubes encircling the globe, as during the "thimble experiment" in 1866. One of the first theories discussed by telegraph engineers was that due to the unbroken length of submarine cables, a battery with high voltage was required for them to function correctly.

This dispute on high and low voltage ran in particular between Wildman Whitehouse and William Thomson during and after the first failed Atlantic cable of 1858. Wildman Whitehouse worked as electrician on the Great Atlantic Cable until he was dismissed by the board of directors in late 1858. Contemporaries claim that it was his fault that the 1858 cable failed. Attempting to receive a stronger signal, he had sent extremely high voltage through the cable, causing the jacket to melt.[62] For the thimble experiment, as Graves reported, on September 12, 1866, both Atlantic cables were joined at Newfoundland making a loop circuit of 3,748 nautical miles. Valentia, the operating end of the cable, used a silver thimble and a piece of zinc as a battery.[63] The success of this experiment across the Atlantic and back proved William Thomson's theory that a long ocean cable could be operated "by a current generated in a lady's thimble."[64]

Regular tests encompassed a similar geographical range and ran between the various stations at opposite shore ends throughout the companies' entire network and from cable ship to shore. Such maintenance testing was part of the daily cable routine.[65] With every "Can you read?," operators and engineers traversed distance at unprecedented speed and generated new knowledge. In their research, they integrated the globe in its geographical premises and thus shrank intermediary distance to a negligible size. Different continents were brought within simultaneous instants. In this manner, one of the first telegrams William Thomson sent through the Atlantic cable in 1858 was: "Where are the keys of glass cases and drawers in the apparatus room?"[66] This message was sent over more than 2,000 miles as though the addressee of the request were in the other room, creating a feeling of a door-to-door working atmosphere despite its long-distance nature.

In a similar fashion, a larger Euro-American scientific community used the cable stations as laboratory outposts: ocean and land surveyors and astronomers journeyed to the stations to make use of the cables. During October and November 1866, for example, Benjamin A. Gould of the U.S. Coast Survey used the Atlantic cables to determine the difference in longitude between Greenwich, United Kingdom; Valentia, Ireland; and Heart's Content, Newfoundland.[67] In 1874, Captain Browne of the Royal Observatory conducted similar time experiments on the lines between Porthcurno and Alexandria, Egypt.[68] Aside from measurements of the earth, scientists considered the cable stations to be optimal places for experiments in thermodynamics. Various institutions, such as the London Board of Trade and the London Meteorological Institution, as well as individual experimenters, used the cable stations as weather stations.[69] Usually the clerk in chief or the superintendent sent daily weather reports to various recipients and thus earned himself some extra money. In September 1860, for instance, London's Board of Trade established a meteorological station at Jersey for which they employed James Graves. Every morning, Graves forwarded a report to the Board of Trade, which was then published in the *Times*, the *Shipping Gazette*, and the *Globe*.[70] These attempts at measuring and mapping the world by means of telegraphy were part of a broader construction of systematic knowledge developed since the European Renaissance in a massive attempt to measure, map, and categorize the world. It all contributed to Euro-America's appropriation of the world through science.[71]

In the world of submarine telegraphy, this appropriation of the world through science was soon entangled in a fierce dispute between science and technology or, more explicitly, mathematical theorizing and practical applications. The STE provided the battleground for these negotiation processes. Already upon the society's launching, William Siemens had seen its purpose as "to combine the knowledge of these diverse circumstances, and of the diverse practice . . . in order to be again radiated in every direction."[72] Two years after Siemens, William Thomson devised another strategy to make use of the cosmopolitan character of the society for research. Foreign members were key figures in his master plan for global research, which he described in great detail in his inaugural address in 1874. According to Thomson, the advancement of electric science was not just to be pursued in the "scientific laboratories of Europe," and he was "looking forward to the benefits which science may derive from its practical applications in telegraph engineering."[73] The STE was to function as a channel "through which these benefits may flow back to science," while it simultaneously supplied "the counter-channels by which pure science may exercise its perennially beneficial influence on practice."[74]

The scientific problem Thomson had in mind was one most urgently puzzling telegraph practitioners: the magnetic storm. At the time of Thomson's speech, engineers believed that magnetic storms caused the needle of the telegraph apparatus to "fly . . . as much as two or three degrees. . . from its proper position." The magnetic storm was always associated with a visible phenomenon: the aurora borealis.[75] Today, scientists define a magnetic storm as a temporary disturbance of the earth's magnetosphere caused by changing environmental conditions in near-Earth space. In the second half of the nineteenth century, it was unclear what caused magnetic storms and the aurora borealis. The aurora was a phenomenon that had for centuries fascinated scientists, poets, and the common man alike. Both science and superstition struggled to find an explanation for it, and in the "olden days," the aurora was believed "to portend death or other calamity and disaster."[76] For the telegraphers, it was not the aurora borealis that was troubling but its accompanying magnetic effects, as these impeded their work. For the duration of magnetic storms, operators were unable to send or receive messages. As James Graves pointed out, these natural phenomena were "much disliked by the telegraphers on account of the general interruption to business which ensue[d]."[77]

The first noted experiments on the aurora borealis date as far back as 1744, when scientists attempted to reproduce the luminous phenomena that are characteristic of the aurora borealis in the laboratory.[78] During the 1830s, loose networks of magnetic observatories were established throughout the world, and it was discovered that the magnetic disturbances associated with the occurrence of the aurora could be measured on a global scale.[79] The experience, or rather the perception, of simultaneity expanded with the growing telegraph network. For the first time, the occurrence of the aurora and its "spread almost simultaneously over the telegraph world" could also be measured and experienced outside the scientific observatories because it was almost concurrently putting telegraph stations all over the globe out of order.[80] After the occurrence of a magnetic storm in February 1872, George Drapner, from the British Indian Submarine Company, informed the *Times* that "the brilliant aurora which [had been] visible in London last night [had] also [been] visible in Bombay, Suez, and Malta."[81]

Quite naturally, these phenomena were discussed before the STE, and in 1872, several articles from various parts of the telegraph world were published in the *Journal of the Society of Telegraph Engineers* that focused on the aurora of February 4th. Apart from George Drapner's note, its editors also published a lengthy article by James Graves, superintendent at Valentia. Graves provided the reader with scientific data from five different Atlantic cable stations in Europe and North America that he had collected during the occurrence of the magnetic storm.[82] In a second article, a telegraph operator stationed in the Persian Gulf offered further information.[83] It might have been this uncoordinated and spontaneous compilation of knowledge on the aurora of 1872 that intrigued William Thomson in his inaugural address of 1874 to suggest a coordinated approach by the practical telegraphers. He turned the entire telegraph world connected via the STE into one big laboratory. Thomson was convinced that if they could have "simultaneous observations," they should have "a mass of evidence from which . . . [they] ought to be able to conclude an answer more or less definite to the question."[84] Thomson dedicated a considerable part of his presidential address to detailed research instructions for the telegraph operators. For him the value lay not in the single operator's data but in the entire telegraph network's connection into one big experiment. Concerning his curiosity about atmospheric electricity, Thomson did not even feel hesitant about sending the operators climbing the "peaks" surrounding their telegraph stations with electrical equipment in order to make adequate measurements in their spare time.[85] It seems as if Thomson perceived the telegraph operators, represented in the STE, as an army of research assistants to be used on a global scale. Distance and geographical and national borders were so irrelevant as to remain unconsidered by Thomson—the ocean cables transgressed it all.

Although little is known as to what came of Thomson's joint scientific operations, it is probable that the telegraph operators were not quite as keen on the knowledge exchange and their job as personal research assistants for William Thomson as Thomson himself. Underlying this lay a conflict between scientists and practitioners—between theoretical treatise, mathematical analysis, and experimentation on the one side and routine (i.e., industry-based) tasks such as instrumental testing, technical development, patent protection, standardization, and quality control on the other. It also points to a more general discourse on the nature of industrial research.

With submarine telegraphy, this conflict centered on the fact that the cable stations were not only places of testing and knowledge gathering but also places of genuine invention. The telegraph operators adapted their instruments to their surroundings and their daily work. Several entries of the cable station books relate to the manufacture, adaption, or invention of telegraph equipment. Similarly, the anonymous diary of a Telcon employee and James Graves's technical biography are full of technical notes and sketches.[86] In January 1859, for instance, Graves invented an alarm bell "for the purpose of giving the clerk notice when his machine required winding up—and thereby avoiding the inconvenience of the machine stopping in the middle of a dispatch."[87] In most adaptations, the characteristics of the respective locality, such as climate, temperature, and humidity, played a decisive role; telegraph operators adapted their inventions to the characteristics of their surroundings at cable stations throughout the globe.

The vast material on "inventions" made by the telegraph operators suggests that they considered themselves as inventors and true experts in their field. Between 1872 and 1887, Graves alone produced sixteen papers to be read before the STE on as broad a range of topics including earth currents and earthquakes, conductors and resistance, and the construction of cable keys.[88] Furthermore, the operators filed for patents, whereby jealousies not infrequently arose between the two stations at each end of a cable centering on the usefulness of the other's novelty.[89] Also, they were not shy in advising the management at the headquarters in London about how best to proceed with practical as well as economic-strategic cable matters.[90] But the telegraph operators also took on the role of active scientists in other fields of science, such as astronomy. Upon the occurrence of a "most 'unastronomical [sic] sky,'" which translated into a "brilliant meteoric shower," Graves organized his staff "dividing the heavens into sections for each one to watch" the meteors for the remaining night.[91] Over and above their claim to producing genuine knowledge, the respective cable stations' superintendents claimed to be teachers. In the fashion of scientific Britain, they gave lectures for their staff and locals on "electricity," "magnetism," or "submarine telegraphy."[92] The operators' basis for their inventions and their practices as teachers was their practical experience. They saw scientific knowledge produced in laboratories and through mathematical calculations as on a par with practical knowledge and honored the proverb "experientia docet."[93]

Yet, the operators' self-perception as producers of universal knowledge was seen critically in London. Neither the companies' headquarters nor the STE fully recognized their work. The underlying conflict was one of discursive hegemony, authority, and power. Such tensions between a perceived periphery and center of telegraphic knowledge erupted, for instance, over the automatic curb sender Thomson left Graves for trial.

Graves published his findings in a paper titled "On Curbed Signals for Long Cables," which was read before the STE.[94] The engineers and scientists present hotly debated and contested his article. Some argued that Graves was "not acquainted with what was the true nature of 'curbed' signals" as his paper only dealt with "results obtained in actual experiment with instruments supplied by the inventors and submitted for trial."[95] What ended in a debate between theorists and practitioners was a demonstration of power. A similar encounter occurred between Weedon and the Anglo-American Telegraph Company's general manager Henry Weaver in London. Responding to a letter from Weedon commenting on Stearns's setup of duplex instruments, Weaver's response was more than harsh: "The idea of your *protesting* against the form of the artificial line is absurd. *You* are setting up as the inventor against Stearns."[96] In the end, as science led the way to modernity, the scientific community became even more defined. The operators served increasingly as assistants, clerks, and collectors of data and were attributed little genuine reasoning. On February 15, 1876, Spratt commented in his diary that "Sir James [Anderson was] tired of expmts! [experiments]" at the cable stations.[97] Spratt gives no explanation for Anderson's reasoning. Certainly, stations and station managers were easier to handle from London if the staff was merely doing the work and not acting independently and carrying the habitus of inventors who claimed patent rights and respective remuneration. In the years to come, London slowly gave up on its global research network at a large scale and thus halted a further integration of the telegraphic globe on the base of knowledge.

Although there were still disputes among telegraph professionals on "telegraphic knowledge," there existed a complete gap between these professionals and local people. Although submarine telegraphers were dispersed all over the globe, local or "native" knowledge, that is, non-Western knowledge, found no entrance into their pool of telegraphic knowledge. Someone in India, for instance, who, as a colonial subject, was affected by telegraphy, as it facilitated colonial rule and trade, had no real relationship to the telegraphs as apparatuses and no influence on their operation.[98] The demarcation line was not so much between the telegraph operators and the respective indigenous population, as the latter possessed no knowledge of relevance for the improvement of submarine telegraphy. In its Eurocentric universalism, submarine telegraphy was no exception from the rule when compared to other fields of science.

In the nineteenth century, generally only very few non-Western concepts were incorporated into a canon of acclaimed universal knowledge.[99]

The only exception was the exploitation and production of gutta percha, a gum used for the cables' insulation, which strongly depended on native methods as well as native knowledge about where this resource could be found. Gutta percha, which only grew in the Dutch, British, and French colonies of the Far East, first came to the attention of the Europeans in 1656, when an English traveler brought samples back to London. For a long time, it remained a mere curiosity, and it was only in 1832 that a Scottish physician learned of the extraordinary qualities of the gum from a Malay worker: gutta percha became pliable in hot water and hardened as it cooled and was thus fashioned into canes and a variety of tools in the Malay world. In 1847, Werner Siemens first employed the substance as an insulator for an electric telegraph cable. The first Atlantic cable was composed of seven three-eighth-inch copper wires twisted tightly together, each wire individually coated with three layers of refined gutta percha; minute inspections guaranteed that the insulation and protection from seawater were as perfect as possible.

Although gutta percha came to be the sine qua non for the success of the submarine cables, its usage rested on a contemporary paradox. According to the standards of the time, submarine telegraphy was a high-tech industry, but it depended on the most "primitive" extraction industry for the gutta percha, which had not changed since the days before it was "discovered" by Europeans. To extract gutta percha, a group of native woodsmen, equipped with axe-like *biliongs* and machete-like *parangs*, would enter the jungle in search of a grove of *Isonandra* trees. The latex, which runs in black lines throughout the heartwood, would then be drained from the trunk into bamboo bowls or coconut shells, washed, and folded into blocks.[100] Occasionally, as the *Birmingham Daily Post* put it, Euro-American workers would be startled "by the sight of a grim looking Indian idol, or a dragon, or a lizard, or a gigantic butterfly, or a distorted elephant, or some other grotesque object" into which natives had formed the gutta percha.[101] In their cable production, Euro-American industries were entirely dependent on this native branch. In fact, huge amounts of gutta percha were necessary for just one cable. The 1857 Atlantic cable, for instance, weighed 2,000 tons, 250 tons of which were gutta percha. By the 1890s, the cable industry was consuming about four million pounds of gutta percha every year, a demand that was unsustainable. In the 1890s, two French cable companies even withdrew from competing for a cable to Africa, owing to the scarcity and poor quality of the material available. As tree extinction became a greater problem and because commercial plantations only originated in the twentieth century, native knowledge on where to find and how to exploit these trees was crucial.[102]

Yet, in the end, it remained a Eurocentric discourse and debate on what the form and content of "universal" telegraphic knowledge should contain. The mere existence of the telegraph technology had already set the "West" and "the rest" apart and established a system of inherent difference. The disputes within the telegraphy community over technical knowledge must be seen as an intra-Western discourse within a white system of knowledge in which only their distance from London granted telegraph operators and station managers an autonomy to experiment and innovate that they would never have had closer to home. In the end, telegraphic knowledge was defined in and through scientific terms as well as via a strong London-centrism stemming from the companies' directives as well as the STE. Practical knowledge, which still played a tremendous role on the local scale, was increasingly neglected.

THE INTERNATIONALIZATION OF SCIENCE AND THE WORLD OF TELEGRAPHY

In the late nineteenth century, the organization of science changed significantly. Almost immediately upon the STE's foundation, its engineers were enveloped in accelerating processes of global technological and scientific developments, paralleled by an increasing nationalization and internationalization of science. Entangled in both, the STE failed in its cosmopolitan character and became one of many nationally organized, learned societies within an international framework. By the 1880s, the STE was no longer gathering as the global representation of all telegraph engineers around the world, but as the British association of all electrical engineering within the British Empire. Simultaneously, William Siemens lost out against John Pender with his economic and technological vision of submarine telegraphy. Pender's Direct United States Cable Company (DUSC) takeover was symptomatic of the institutionalization of submarine engineering as well as its loss of significance for individual engineers. Submarine telegraphy was no longer in the hands of the engineering giants but in the hands of the "huge octopus," the representatives of the Eastern and Associated Companies.[103]

The literature analyzed this radical shift in the program of the STE as the transition of telegraphy from *new* to *old* technology framed by Britain's general *relative* industrial "decline."[104] Set in a global context, the downfall of the submarine engineers shows their struggle with the very forces they produced: an increasing integration and acceleration of the world. In fact, the institution was hit by globalization. Simultaneously, it had to give way to the nationalization and internationalism of science, as represented, for instance, by the ITU or by the legal idea of *l'esprit d'internationalité* brought forward by David Dudley Field and Louis Renault. The STE's decline was not only a matter of scientific change, but also a failure of its cosmopolitan model for world governance.

Even before the STE had been established, its purpose was a point of ample discussion. Already in his inaugural speech of 1872, William Siemens took umbrage at a critique directed at the institution, namely "whether there was need and scope for a Society of Telegraph Engineers" that would after its foundation only "degenerate. . . into 'specialists,' or what may be called 'fractional quantities of scientific men.'"[105] During the inaugural meeting in 1872, Atlantic telegraph engineer Cromwell Varley confessed his doubts whether "a society [could] truly be kept on its legs simply by telegraphy."[106] Questions had arisen: was not "telegraph engineering . . . a branch of civil engineering," and would not all the society's proceedings fall "within the legitimate sphere of action of the Institution of Civil Engineers?" Furthermore, was not "the Royal Society or . . . the British Association open for [them]" to discuss "difficult questions regarding physical or mathematical science?"[107]

Siemens's stand against this reproach underscored the professionals' perception of their importance. He was convinced that the STE was "necessary for the more rapid development of a new and important branch of applied science . . ., desirable in order to afford Telegraph Engineers frequent opportunities of meeting . . . and of impressing them with the conviction that their united actions will be advantageous to the material interests of all."[108] An "occasional paper" discussing matters concerning submarine telegraphy before the already established societies of learning was "quite inadequate to constitute a record of the progress of a branch of Engineering which gives daily proof of its public importance, which is distinguished for its rapid development, and which comprises within itself a wide range of scientific enquiry."[109]

Siemens's vanity comes into relief when compared to the setup of the Institution of Mechanical Engineers in 1846. As the railway age superseded the canal age, railway engineers felt misrepresented by the Society of Civil Engineers, which was dominated by canal professionals, and launched their own society to represent them professionally and serve as a platform for exchange.[110] Despite the clear preponderance of railway people within the institution, they still designed their society according to the type of engineering (i.e., mechanical engineering), and not according to one single product of their profession's work, the railways. The situation was entirely different with the telegraph engineers. In the aftermath of the success of 1866, Samuel Canning, Latimer Clark, Willoughby Smith, Cromwell Varley, and William Thomson had been heralded not only as "benefactors of their country" but as benefactors of their "race" and of "mankind" in general.[111] Almost unparalleled by any other scientific achievement of the day, these public perceptions and aspirations had nurtured a profession's self-esteem that asked for its reward: its own society.

However, despite Siemens's convincing presidential address, there remained the uneasy feeling that telegraph engineering was still too narrow a basis and that unless the STE "could range a good deal further over electrical science than testing the joints of insulated wire, it would not flourish."[112] The skeptics seemed to be proven correct when the institution resorted only a decade after its establishment to a pervasive process of reorientation, paying tribute to the continuous emergence of new applications of electricity such as the telephone, electric lighting, and the storage of electricity. In 1884, the STE altered its name to the Society of Telegraph Engineers and Electricians and, in 1889, to the Institution of Electrical Engineers. In effect, the founding of the STE in 1871 marked a snapshot of the time when public excitement correlated with the expansion of a technology that was considered to be promising but, in retrospect, had actually already seen its prime.[113]

Since the 1830s and the "discovery" of electricity, telegraph engineers had virtually been the sole representatives of the profession of electrical engineering because no other technical application of electricity came anywhere near the telegraph in importance. This situation changed dramatically in the 1870s and 1880s when electrical engineering underwent enormous expansion, and new applications such as the telephone, the electric tramway, the electrodynamometer, and electric lighting excited engineers, entrepreneurs, and the public alike. These new innovations had their base not primarily in Great Britain but in the United States and Germany. Based on submarine telegraphy's manufacture, business, and scientific organization on the banks of the Thames, London's preeminence in electrical engineering now started to end.[114] In contrast, in continental Europe and North America, the years after mid-century (for the United States, particularly after the 1870s) mark the beginning of a period of great inventive power and immense industrial progress. During the *Gründerjahre*, or the time of the Second Industrial Revolution, these countries were "closing the gap" with Great Britain.[115]

Within the STE, the loud voices demanding a change in its structure in the face of the diversification of electrical engineering soon came to be heard. The main advocate for change was the eminent (Atlantic) telegraph engineer Latimer Clark. In 1875 and again in 1876, he suggested adding the words "and of Electricians" to the society's name to broaden its appeal.[116] In 1878, he read a paper titled "On a Standard Voltaic Battery" whose content was geared toward the need of electricians.[117] Although the subheading of the *Journal of the Society of Telegraph Engineers* was "Original Communications on Telegraphy *and* Electrical Science," Clark considered the STE to rest upon too narrow a basis.[118] He was convinced that focusing solely on telegraphy would alienate those who they ought to attract, namely, the electricians. In a letter from 1880 to Edward Graves, engineer-in-chief of the Post Office, Clark argued that as the science of electricity was advancing fast, the men occupying themselves with it, such as William Thomson, Silvanus P. Thomson, and John Hopkinson, could not "afford to stand still merely because we prefer to remain a Society of Telegraphers & do not care to invite them into our ranks." There was no alternative: an "electrical society must be formed or else existing societies must supply the need."[119] Although the journal's tables of contents suggest by a fair selection of noncable papers that it was to be a predominantly symbolic act, Clark's pleading proved successful. At a general meeting in December 1880, the membership decided to "very appropriately" alter the society's title to the Society of Telegraph Engineers and of Electricians, which it carried from 1881 to 1889.[120]

Just a few years after these initial changes, new discontent found its way into the open. Some members complained that most of the papers and discussions held were by members of the council and telegraph engineers, to the exclusion of the rank and file. Moreover, they remarked, some of these papers were "mere compilations, histories and essays" and were not sufficiently related to original work.[121] In the words of Colonel R. E. B. Crompton, one of the major pioneers of Britain's electrical industry, there was a widespread opinion that the STE did not "adequately represent the present body of Electrical Engineers." Crompton warned that "this feeling may lead to the formation of a new and rival Society" by the power engineers, those working on electric lighting or the storage of electric power.[122] Because of these reproaches, the STE moved to alter its name again. Edward Graves, president in 1888, explained the move in his inaugural address. Although the telegraph engineers were "still the most numerous units in our body," they could no longer be said, "from the force of circumstances," i.e., the course of technological developments, "to be the special representatives of its character." In Graves's eyes, "The Institution of 'Electrical Engineers'" was a title "comprehensive enough" to pay tribute to these "circumstances" and "to include all devotees of the science." Moreover, this way, "no class of works [would be] singled out for undue prominence, no class [would] by implication [be] excluded."[123]

In February 1888, the secretary of the STE reported that a circular had been issued to the members asking for their opinion on the proposed change in the title. By October 1888, 857 replies had been received; a great majority assented to the proposed motion. On January 1, 1889, the Institution of Electrical Engineers was incorporated.[124] As indicated, most members welcomed the STE's change in name to one that "the catholicity [i.e., the diversity] of its programme [sic] seems to justify," wrote George Spratt.[125] Rising membership numbers justified these measures. In 1890, membership accounts had gone up to 2,100 and in 1910 to 6,218, making the Institution of Electrical Engineers, along with the Institution of Civil Engineers and the Institution of Mechanical Engineers, one of the big three of learned societies in Great Britain.[126]

The change in program and name not only paid tribute to the developments and diversification of global science, but also symbolized changes concerning the importance of singular engineers in general and the Class of 1866 in particular. The success of the Atlantic cable made its protagonists, such as Charles T. Bright, William Thomson, and Willoughby Smith, engineering royalty. As a consequence, these individuals came to be *the* face of ocean telegraphy, just as Isambard Kingdom Brunel or

Richard Stephenson had been the giants of the steam and railway age two decades earlier. Those engineering giants translated their fame into money, as they came to work as independent advisors or consulting engineers and electricians to the various submarine companies throughout the globe as well as to governments that might require their services. Essentially, they were in the position to pick and choose. Over the next few decades, telegraph professionalism changed. With the greater experience and efficiency of the contractors' staffs and the gradual extension of the work generally, the scope for independently working consulting engineers became increasingly limited. Ocean cabling was no longer the work of one singular engineer but of any engineer who worked with a particular company.[127] By the 1880s, the profession of submarine telegraphy had seen the end of its age of giants, many of whom also passed away in the 1880s. William Siemens and Cromwell Varley died in 1883; Fleeming Jenkin, who had worked for the French Cable Company as well as the Commercial Cable Company and had been in a patent-pooling agreement with William Thomson and Cromwell Varley, died in 1885; and Charles T. Bright died in 1888. In the same year, Willoughby Smith, former chief electrician of Telcon, retired from his management post.[128] Although some, such as William Thomson, Latimer Clark, and Werner Siemens, were still alive, their elitist circle was severely decimated; in addition, each of the three had moved on to occupy themselves with *newer* applications of electricity. The names that had once upheld the STE were no longer there or willing to keep it running.

The change from the Society of Telegraph Engineers to the Institution of Electrical Engineers had to do not only with technological and generational developments in engineering and science but also with the rapidity of these changes. Reflecting upon their times, contemporaries often highlighted the speed with which social, economic, or political changes were taking place. With its ability to "annihilate time and distance," albeit only in relative and not in absolute terms, the telegraph was one of the greatest accelerators of globalization processes. For the STE, it is one of history's greatest ironies that the telegraph engineers were steamrolled by the rapidity of global change. Even among those who had their fingers on the pulse of time, there existed great insecurity and incomprehension of the acceleration of globalization.

Indeed, the telegraph engineers constantly grappled with this rapidity with which science and technology developed. Repeatedly, they debated whether submarine telegraphy was a new or old technology. Already in 1872, William Siemens pointed to the fact that "so rapid" had been the progress of their branch of science, that, while he was "obliged to speak of these men [Oerstead, Ampère, Faraday, Weber, Steinheil, Schilling, Ronalds, Wheatstone, Cooke, and Morse] as belonging to [their] early history," they were still "almost without exception, living amongst [them] in full enjoyment of their faculties."[129] Electric telegraphy's "invention" in only 1837 gave the impression that it was a fairly new application, whereas the changes in electrical engineering in the 1870s and 1880s made it appear old and outdated.

This entanglement of old and new caused constant puzzlement among the engineers and remained seemingly irresolvable within the society. Willoughby Smith's inaugural address of 1883 or William Preece's lecture "On Practical Applications of Telegraphy" of 1884 both unsuccessfully attempted to tackle this question.[130] More than a decade after Siemens, they still ruminated on whether their profession was old or new. Theirs was a discourse on the relations of past and present in an ever accelerating world. They captured, in the words of Senteney Shami, "the in-betweenness of a world always on the brinks of newness."[131] In 1883, STE president Willoughby Smith set the stage by arguing for the novelty of their science: in comparison with "many of the other branches of the same tree" of science, theirs was only a "very young shoot." Telegraph engineers and electricians had "no great masters of antiquity to imitate or revere" as most of their great men were "of the present age."[132] This thought was expanded and juxtaposed by Preece. He claimed that telegraphy was "the oldest and the first of these practical applications." And though it was "the oldest, and the first, nevertheless it [was] very young, for it dates its birth only from the year 1837."[133] Both men expressed the idea of a merger of past and present, of old and new, and thus an odd parallelism that was typical for the globalizing nineteenth century. Despite the rapid expansion of the railways, the use of mules for transport remained a common sight. Similarly, the telegram did not render the letter unnecessary and rather increased the gap between those who participated in the rapidity of modern progress and those who did not. The late nineteenth century was characterized by a simultaneity of tradition and modernity and their respective discourses, but also by progress and "standstill"—to return to Latimer Clark's remark on the fast development of the field of electrical science. These antitheses could find expression in the mule and the railroad just as well as in the telegraph and the telephone or electric lighting.

In this narrative, standstill and decline refer to questions of global significance and the STE's originally inherent character of cosmopolitanism. Although the STE began as a cosmopolitan institution, which came to represent all telegraph engineers of the world despite its London-centrism, in the 1880s and 1890s, the institution was reduced to a British representation only. The telegraph engineers missed out not only on the transition in science but also on the transformation of the organization of science and ocean telegraphy in an increasingly nationalizing world. Siemens's Kantian cosmopolitanism was not adequate to meet the (inter)nationalization of science or the international regulation of ocean telegraphy through the ITU. His approach to science was not the right tool to keep the STE the leading learned institution on electrical engineering.

It was not only in Meiji Japan that science and technology played an important part for the national or imperial project. In Europe, beginning in mid-century, the various scientific institutions, rituals, procedures, and performance measurements of the Western industrial system set up "a discourse of competition."[134] Science contributed to "the national welfare," and scientists played an important role in creating "high culture." Contemporaries celebrated them at the world's fairs, which they saw as technological and economic battlefields.[135] In addition, the progress of schools and university measured that of nationalism. In fact, over time, as places of learning and education, schools and especially universities became nationalism's "most conscious champions."[136] Despite the initial transboundary character of scientific exchange, reminiscent of the eighteenth-century Republic of Letters, nineteenth-century learned societies were spaces of increasing nationalization. The STE was no exception. In particular, discussion of the new applications of electricity displayed an increasingly nationalistic vocabulary. In 1877, Latimer Clark declared after hearing the American Alexander Graham Bell speak before the STE on "Electric Telephony" that "two or three of the most important recent electrical inventions have come to us from the other side of the Atlantic." For him, America and other nations, such as Germany, were looming on the horizon, while Great Britain was "falling behind."[137]

The international comparison of each nation's engineering progress became increasingly common and increasingly important. In his position as a post office official, William Preece visited the United States three times to "inspect . . . and examine . . . the telegraph, telephone, and other electrical industries, including railways and electric lighting."[138] He returned reassured and concluded his report with praise for British preeminence. According to Preece, it was "satisfactory to point out that invention ha[d] not progressed in America as much as it ha[d] in England." Quite the contrary, Great Britain showed "greater signs of progress," and their apparatus, their mode of working, and the general transactions of their business compared "most favourably with those in America." He concluded that Great Britain was still ahead of the United States concerning "the rapidity of transmission, cheapness of telegraphy, and expansion of knowledge."[139] In the national contest, there was no reason to worry. The STE responded to these national challenges by engaging in its usual cosmopolitan manner: it invited the new leading figures of science, such as the American Alexander Graham Bell and the Russian Pavel Yablochkov, an electrical engineer and inventor of the Yablochkov candle (an electric carbon arc lamp), to be members. Yet, aside from incorporating new foreign members and their knowledge, the STE had little to offer for countering the increasing organization, codification, and regulation of science in an age of national comparison.

Originally, William Siemens and other telegraph engineers had envisioned their society as a universal network in which to combine all the knowledge on ocean telegraphy and then develop practical devices thereof.[140] The regulation and codification of (ocean) telegraphic knowledge would work via the STE. They disregarded the fact that simultaneous to the foundation of their institution, a state-based model to regulate ocean telegraphy found its practical application, the International Telegraph Union (ITU). With the founding of the ITU in 1865, which was before the establishment of the STE, practical matters of landline telegraphy, such as universal tariffs, telegraph codes, and cable routes were already dealt with on the international level. From 1865 on, a conference on international communication was held every five years in one of the capitals of the countries represented and rules and regulations discussed and resolved.[141] In 1871, the year of the STE's founding, the ITU began to hold conferences to regulate the submarine telegraphy business, and the ocean cable companies were officially "invited" to participate. The ITU codified the forms of telegraphs; defined what a word was, telegraphically speaking; and regulated the use of codes and cipher and state taxes. Although the ITU primarily dealt with telegraphic form and transmission costs, it encroached upon the claims staked out by the STE.

The clash between the ITU and the STE came over the issue of electrical standards. Since the creation in the 1860s of the Joint Committee to Inquire into the Construction of Submarine Telegraph Cables and the Committee on Electrical Standards, this question had been closely connected to the profession of submarine telegraphy. Both committees were set up in response to the early cable failures of the Atlantic and the Red Sea cables. The work of the Committee on Electrical Standards lasted eight years and involved many well-known cable figures, such as Charles T. Bright, Latimer Clark, and William Thomson. It resulted in a system of electromagnetic absolute units, from which were derived the ohm, ampere, farad, volt, and coulomb, which are universally used today.[142] Nevertheless, the STE could not claim this as their success. Although this system of global units had been developed by some of the most eminent members of the STE and had emerged from their work with the ocean cables, it was proclaimed by an international body. In 1881, an international congress on electrical standards confirmed the global validity of the telegraphers' system.[143]

This International Congress of Electricians, hosted during the world exhibition in Paris in 1881, symbolized the first turning point and rebuff for the STE's claim as a global representative of the world's telegraphers and their knowledge. The Congress had been called together not only in response to the new applications of electricity, but also to settle matters on the exchange of standards. Previously, the STE had laid claim to offer such a forum to discuss and resolve pressing scientific questions. Their debates on the aurora borealis are only one example of this. Now the STE only reported on decisions taken elsewhere. These international congresses on electricity continued at regular intervals. In Chicago in 1893, their concerns about electrical standards led to the definition of the *international ohm*. At the fifth congress in 1904, the decision was taken to establish a permanent body, the International Electrotechnical Commission (1906), to organize regular international meetings of electricians.[144] The standardization and homogenization of the (scientific) worldview had been started as a response of the British government and the Atlantic Cable Company to submarine cable failures; now it was discussed internationally.

In addition to this shift away from the STE's forum to the international level, the telegraph engineers were facing further challenges. Simultaneous with their society's realignment as the Institute of Electrical Engineers, several other societies emerged: the *Elektrotechnischer Verein* in Berlin (1879), the *Société Internationale des Electriciens* in Paris (1883), and the American Institute of Electrical Engineers (AIEE) in New York (1884).[145] These institutions' interest lay in electrical science per se and not explicitly in telegraphy. In fact, the AIEE had been established in response to the need for national representation during the International Electrical Exhibition in Philadelphia. Because so many "famous foreign electrical savants, engineers and manufacturers" would be visiting, it would be "a lasting disgrace to American electricians if no American National Electrical Society were in existence to receive them with the honors due from their co-laborers of the United States."[146]

Although some of the societies were modeled after the association in London—the *Elektrotechnischer Verein*, for example, probably through Werner Siemens, took up William Siemens's idea of foreign membership and multilingualism—they were set up as equals and rivals on the international scale.[147] The STE no longer filled its self-proclaimed position of global leadership. During the period of submarine telegraphy, everyone had turned to London; now in the era of power engineering, this stopped. Consequently, the number of foreign members dropped. By 1891 the increase in foreign members had slowed down remarkably; there had been 152 foreign members in 1881, but only 177 in 1891. From 15 percent, foreign members now only accounted for 9 percent of the total membership.[148] Over time, the designation "foreign member" gradually fell into disuse until, in 1911, the society resolved that the 103 foreign members that remained should be transferred to the class of members and that they should continue to pay £1 per annum as long as they resided abroad.[149]

However, international exchange between the societies and their members was continued, promoted, and, considering the establishment of regular international electrical exhibitions from 1881 on, even expanded. In telegraph science, internationalism and nationalism coexisted.[150] Yet, with regard to Siemens's cosmopolitan STE, the character of these international exchanges was altered. Instead of Local Honorary Secretaries arranging meetings, meetings and visits were arranged between *national* delegations, such as the combined meeting of the AIEE and the Institute of Electrical Engineers in Paris in 1900, a visit of another foreign institute's delegation to Germany in 1901, or a visit of electrical engineers to Italy in 1903.[151] Instead of global outreach arranged by the society in London, exchange was organized as one between nations meeting on an equal level. The failure of the STE in its ambition for global leadership was also a failure to regulate and codify the telegraphic world as a whole through Siemens's cosmopolitan model of governance.

In the end, the STE's history is emblematic not only for the history of the Class of 1866 and the development of submarine telegraphy as science and technology, but also for broader developments of the nineteenth century. Siemens's idea of a "cosmopolitan institution" mirrored the Atlantic cable engineers' self-proclaimed importance as well as their deep entanglement in notions of submarine telegraphy as transboundary science. They constructed their society to represent the entire, albeit small, Euro-American community of telegraph engineers and so claimed "universality." After all, the experience of simultaneity within a global network of cables was essential to their research. The STE's members failed to recognize the transitions in science and its organization and to respond accordingly to these global challenges. Science and technology diversified and moved from being centered in London to being multicentered and global. This reflected the rapidity of global changes and the difficulties STE members had in coming to terms with underlying systemic changes.

6

CABLE DIPLOMACY AND IMPERIAL CONTROL

> *Whether we call ourselves Englishmen, Americans, Australians, or what not, we shall prosper none the less, . . . because our neighbours—because the whole world in fact—are "moving." Therefore, by all means let every nation that wishes and is able to develop its own cable systems and trains up its own army of telegraph engineers and electricians, do so with our very best wishes. All that it is our business to see to is, that we at least don't lag behind. And if any ringfence [sic] (of preferential rates or other privileges) is to be established, let us make sure that those admitted within it are also those who by kinship, community of language, or historical association, can be expected to get on well and harmoniously together both with ourselves and with each other.*
>
> Charles Bright, *Submarine Telegraphs: Their History, Construction and Working* (London: C. Lockwood, 1898), p. 173.

IN THE face of growing nationalist sentiments and international competition toward the turn of the century, British telegraph engineer and writer, Charles Bright, still argued for the compatibility of national and economic competition and universal peace with regard to ocean cables. Writing this part of his treatise on submarine telegraphy only months before the outbreak of the Spanish-American War and the Fashoda Incident in 1898, during which imperial interests of Euro-American powers clashed, he explicitly invoked what he believed to be the friendly nature of imperial nations and the principle of cooperation. Only a couple of weeks later with the Spanish-American War raging on the other side of the Atlantic and only days before the publication of his book, Charles Bright was singing a different tune. In an interview conducted by the *Pall Mall Gazette*, he voiced his conviction that in this conflict "cables would be cut right and left" notwithstanding existing international agreements and a commonly cherished belief of the cables' safety in times of peace and war. The only solution for Great Britain to avoid communicational isolation was an "all-British cable" around the world only landing at secret places on imperial shores.[1] The realities of the military conflict between the United States and Spain had induced Charles Bright to drastically change his rhetoric about the peaceful nature of economic and imperial competition.

Indeed, ideas of an all-British, all-French, or all-German cable dominated the respective government debates on global communication. Strategically relevant ocean cables were cut almost immediately upon the commencement of the Spanish-American War and the First World War. At the same time, international cooperation on tariffs, rules, and regulations as well as the cable companies' transnational business model survived almost unperturbed through times of military conflicts and upheaval. The "politics of the world's electric nerves" depended on a complex system in which the relationship between governments and cable companies was continuously redefined according to changing priorities of economic or military control over global communication.[2] For Charles Bright, one of the key turning points was the year 1898. It marked, however, not an end to international cooperation and transnational business conduct, but a change in rhetoric on the cable business side. Various cable actors, such as John W. Mackay or Henniker Heaton, from this point on increasingly used nationalist rhetoric to please imperial governments and to disguise nonnationalist structures and goals. This strategic nationalism, which had characterized the cable business to varying degrees from its very beginning, was essential not only to balance, but also to combine, imperial and economic interests in the world of submarine telegraphy.

THE ROLE OF NATION AND EMPIRE IN THE WORLD OF OCEAN TELEGRAPHY

Submarine telegraphy conceptually represented a transboundary entity through its transgression of limits of historical territoriality. Rivers, seas, and oceans, similarly to mountain ranges, had from earliest times served as natural border lines. Although in the nineteenth century, borders—foremost in the United States and Africa—were made on the drawing table according to degrees of latitude, water lines remained popular characteristics of defined territoriality and belonging. Therefore, it was in the very nature of this kind of telegraphy to expand beyond its respective national territorialities by going *submarine*.

From early on there existed a general and international consensus that submarine telegraphy should be a private enterprise.[3] Unresolved questions concerning landing rights and national jurisdiction at the extending end of the cable called for a neutral intermediary, the cable companies, to provide for transnational communication. Already during a congressional debate on the Ocean Cable Bill of 1857, allowing for the landing of the first Atlantic cable, Congressman William Smith of Virginia asked the question of how far the jurisdiction and constitutional power of a government could stretch: could it extend across oceans and thus beyond the borders of historical territoriality? He questioned the "constitutional power" of the United States "to establish a telegraphic line through foreign parts and beyond the jurisdiction of [their] Government."[4] In 1857, Smith's question remained unresolved. The dimension of international law only took to the subject of submarine telegraphy in the 1870s and 1880s. Yet, because it was generally assumed that a cable owned by a government would necessarily be restricted to that government's territorial possessions, it was concluded that submarine telegraphs could only be private undertakings. Governments could grant certain necessary rights to the cable business, such as the purchase of land upon its territory by a private enterprise, but not by a foreign government, which would either have to abide by foreign jurisdiction or a claim for extraterritoriality; both were unlikely options.[5] Moreover, because states were worried about serious international complications that might arise in the case of war, "no Government would permit cables belonging to a foreign Government to land on its shores."[6] As a consequence, in 1898, of the total of over 165,000 nautical miles, almost 90 percent of the long distance lines had been provided by private enterprise.[7]

Although submarine companies were predominantly private enterprises, this did not deter nation states from attempting to influence, control, or take interest in them. Governments provided financial subsidies, ships, and data from ocean soundings, as well as personnel to the cable business.[8] In 1859, Great Britain's Board of Trade and the Atlantic Telegraph Company established the Joint Cable Committee in 1859 to investigate best practices, and during the Great Atlantic Cable project, the British Royal Navy and the American Navy supplied the undertaking with the British ship HMS *Agamemnon* and the USS *Niagara* and took on the costs for the necessary refurbishing of the two ships.[9] Prior to 1866, government support was essential. Many of the early undertakings would not have materialized without the support of the British government in particular.[10] The Great Atlantic Cable project, for instance, received an annual sum of £14,000 from the British and American governments for twenty-five years.[11] After 1866, although the various Euro-American governments followed different models of state involvement, from British and American noninterventionism to French state entrepreneurship, government subsidies generally became less important and common, as the cable companies managed to find ample private investors and had their own cable ships and personnel.[12] Government subsidies only again gained prominence around the turn of the century, in particular in the case of France and Germany. As both nations attempted to attain independence from the "English" cable system, they poured large amounts of money in national cable undertakings. The Deutsch-Atlantische Telegraphengesellschaft received 1,300,000 Reichsmark annually without which, as a member of the German *Reichspostamt* (General Post Office) confidentially reported to the German Foreign Office, "it could not exist."[13]

Although government subsidies waned after 1866, no cable enterprise was ever entirely free of government support or manipulation.[14] By means of international agreements on cable tariffs, routes, or regulations as well as landing rights, the respective nation-states retained an important stake in the cable business. In fact, no ocean cable could ever be landed without the respective government's accordance. While the entrepreneurs strove for state guaranties, seeking to connect landing rights with a twenty- to thirty-year monopoly concession, governments in return would ask for preference of way and reduced rates or impose special regulations, such as a ban on the use of code and cipher, concerning the handling of telegraphic traffic.[15] In the case of the Great Atlantic Cable, in 1854, the government of Newfoundland had granted a 50-year landing monopoly to the New York, Newfoundland and London Telegraph Company.[16] This landing monopoly represented an important cornerstone in the ongoing success story of the Class of 1866 and the Atlantic pool. It put all rival Atlantic cable companies in an inferior position because they could not land their cables on the Newfoundland shore but had to extend their lines to the North American mainland, adding another hundred miles of cable. Because cable length was strongly related to transmission speed, rivaling companies, such as the Direct United States Cable Company (DUSC), time and again attempted to convince the Newfoundland government to suspend this monopoly, or they tried to land their cables on the shore of St. Pierre, the French-owned island next to Newfoundland.[17] Immediately upon the expiration of the Anglo-American Telegraph Company's monopoly, which was legal successor of the New York, Newfoundland and London Telegraph Company, all other Atlantic companies redirected their cables and moved up to Newfoundland. By 1905, the Newfoundland government started exploiting its favorable position by introducing an annual tax of £822 for each cable landed.[18]

Questions of landing rights were particularly complicated because these concerned not only state–cable company relations but also relationships between the states where the cable was to be laid. Landing rights were issues not only of international business but also of international diplomacy, and they could easily grow into diplomatic disputes as the nation-states at each end of the cable called for equal treatment. The first of these incidents stretched from 1867 to 1869 and concerned the French Atlantic cable, establishing telegraphic communication between France and the United States. Initially pleased with the chance of lowering transmission rates to Europe, American officials changed their position radically when informed that the French landing concessions excluded potential American businesses from landing cables on the shore of France. Due to the "political furor" this provoked, the United States established the principle of reciprocity—in contrast to issuing monopoly-like landing concessions—which the U.S. government acted on from then on. The principle of reciprocity meant that landing rights were only granted if an American company had the equal right to land a cable on the shore at the other end. By refusing to grant sole rights, the United States formed an exception on the cable market.[19]

U.S. protests were allegedly initiated by Cyrus W. Field, who had a personal business interest in the failure of the French Atlantic Cable Company, but he evoked nationalist feelings when rallying against the unfair treatment of American businesses through the French landing concession.[20] According to James A. Scrymser, Field's American rival in the business of submarine telegraphy between the North and South Americas, Cyrus W. Field had been the one to "induce" General Ulysses Grant, then president of the United States, "to officially notify the Emperor of France that the French cable could not be landed on American shores unless reciprocal rights were granted to an American company for French territory."[21] Hamilton Fish, Secretary of State under Grant, even threatened "to send a naval vessel to Duxbury, Mass., with order to tear up the French cable, if landed without his permission."[22] In December 1869, four months after the cable had been landed at Duxbury, President Grant addressed the U.S. Senate and House of Representatives in a speech on the state of the nation. Those unresolved landing rights were one of the "grave questions the United States has with any foreign nation." The issue at stake was that landing rights in France had only been obtained "with the very objectionable feature of subjecting all messages conveyed thereby to the scrutiny and control of the French government." Furthermore, the concession excluded "the capital and the citizens of the United States from competition upon the shores of France" because the French government had not given consent to a possible American cable to be landed upon their shores. As a consequence, the Committee on Foreign Affairs was requested to inquire into the matter.[23]

After the French government agreed to the principle of reciprocity as demanded by the U.S. government, Scrymser claimed that Field lobbied the Massachusetts legislature to withdraw its consent for landing the cable on their shores. Referring to Baron d'Erlanger's service for the Confederacy during the American Civil War, Field was again stirring nationalist sentiments.[24] Allegedly, it had been Scrymser that thwarted Field's plan and made the landing of the cable at Duxbury possible.[25] Although there were personal animosities at play between Scrymser and Field, Scrymser's narrative shows how shrewdly Field called upon nationalist sentiments and managed to influence the American government to his way of thinking. Throughout the decades, the cable agents knew how to play their cards so that the respective governments would act accordingly. Interestingly, in 1907, it was Scrymser who applied a strategy very similar to Field's when he saw his rights as the owner of an "American company," the Central and South American Telegraph Company, infringed upon by his German rival, the Deutsch-Atlantische Telegraphengesellschaft, concerning traffic to South America. Several diplomatic notes, telegrams, and letters had to be exchanged between the American and the German foreign offices to smooth the debate. Unfortunately for Scrymser, this diplomatic outburst of communication between the American and German foreign offices and the German cable company remained the only result to his complaint.[26]

Alongside the issue of landing rights, nation-states represented an important player by framing the cable business through international regulations. In 1865, the International Telegraph Union (ITU) was founded in Paris, representing the culmination of various previously established bi- and multilateral agreements on transboundary telegraphic traffic between diverse European governments. The governments' main objective was "the promotion of a uniform system of traffic exchange."[27] Bern was selected as the organization's headquarters, and international telegraph conferences were to be held once every five years at a different capital of the various member countries. The member nation-states were represented by one or more delegates, some of the British colonies such as Australia and Canada had their own representative, and from the Rome conference in 1871 on, representatives from the various submarine telegraph companies were also invited. However, these representatives only had the right to speak, and not to vote.[28]

Generally, the work of the ITU was geared to the collection and distribution of relevant information and statistics, the publication of a monthly journal, and the preparation of conferences.[29] It was at these conferences that details concerning international rates, technological standards, the distribution of revenues, citizens' rights to privacy, and the prerogative of states to censorship were debated and added as amendments to the original International Telegraph Convention of 1865.[30] By 1873, thirty countries from Europe, Asia, and Africa were members of the ITU, lending its decisions a truly global character.[31] Because only government representatives were allowed to vote at the conferences, the cable industry's representatives seemed to be only onlookers unable to control any of the fundamental decisions. Yet, although deterred from voting at ITU conferences, the companies' representatives still had "every facility . . . to state their views and to take part in the discussions."[32] In fact, behind closed doors, submarine telegraph employees exerted great influence over state officials. Emma Pender's papers reveal how frequently her husband, John Pender, had gone to see Herbert, referring to the important figure Sir Robert George Wyndham Herbert, undersecretary of the British Colonial Office. At a banquet celebrating the opening of the second cable line to Australia in 1876, Herbert was reported to have offered "to attend personally to any business that Pender might have with the Colonial Office about cables."[33]

Formal and informal visits and correspondence were also key means in the cable agents' strategy to influence their government representative to vote in their interests at the international conferences of the ITU. Although Great Britain only joined the ITU in 1872 upon the privatization of its domestic telegraph network, five years later, it had already become customary that C. H. B. Patey, third secretary to the British Post Office, or some other representative of the Post Office corresponded with representatives of the submarine cable companies prior to the international conferences in order to communicate a joint strategy.[34] In 1879, however, it was still unclear if the cable companies came as suppliants or as those dictating conditions. This was soon to change. The 1870s mark the period of a trial of strength between the British Post Office and the submarine company's political machinery in London. It was a contest of power between two institutions, the General Post Office (GPO) and the submarine cable business, which had both grown immensely in the previous decade. After the introduction of the uniform penny postal service in 1840, the GPO greatly expanded, and by 1860, its income had increased by 35 percent with a constant upward tendency. Between 1860 and 1868, its revenues again went up by an additional 25 percent. Simultaneously, the GPO introduced a series of reforms that made it more efficient; in the same time period, service costs had only increased by 13 percent. By the mid-1860s, the GPO had developed a nationwide system of communication and a reputation for a financially efficient provision of service.[35] When, at the same time, discussions on the nationalization of the British telegraph landlines commenced, the GPO "capitalized" on public opinion and followed suit. In 1869, the Telegraph Act was passed, and landlines in Great Britain were officially nationalized on January 31, 1870.[36] Although this nationalization was an important prerequisite to free money for ocean cable investment, it also created a powerful institution controlling the entirety of British national telegraph communication as well as some of the shorter submarine lines across the Irish Sea and the English Channel. The Society of Telegraph Engineers paid tribute to this power when they elected Frank Ives Scudamore, head of the Postal Telegraph Department and himself not a telegraph engineer, as second vice president in 1872.[37]

The showdown between these two heavyweights of communication took place in 1877–1878, prior to the international ITU conference to be held in London in June 1879, and culminated at the conference itself. The incentive was the question of cable tariffs. At the end of 1877, C. H. B. Patey from the GPO, who was the designated president of the ITU conference, communicated to the submarine companies that he intended to suggest a radical tariff reform. Originally, international tariffs were calculated by the telegram's place of origin and destination and not by the length of the route taken. There were, for example, five different connections between Great Britain and Greece, each amounting to the same tariff despite their varying lengths and routes. Patey's idea was to change this and charge by distance, introduce the word rate for the inter-European traffic, and generally induce a reduction in tariffs. He was stern that he could not have "the dividends of submarine companies . . . stand in the way of a reduction of tariff."[38]

The response of the Eastern and Associated Companies manager, James Anderson, to Patey's "startling proposition" was more than a simple argumentative rebuff and plea to let them obtain a tariff "at which [they] could live." It was a demonstration of power. Aside from the fact that Patey's plans would mean that thirteen of the adhering states (with the exception of Great Britain) would "lose an aggregate of £579,382 yearly" in tax revenue for the state, Anderson made it clear that a reduction of tariffs was not in his or his cable peers' interest. Anderson's lengthy counterargument, which he wrote "recollecting [Patey's argument that they] could not expect [him] to abstain from lowering tariffs in order that cable companies should make dividends," concluded with a clear warning, emphasizing the great importance of the submarine cables within British economy and society. Whereas Patey was merely dealing with "the money of a nation which can bear taxation," the cable companies were "dealing with twenty two millions of British capital in submarine cables, subscribed by thirty thousand shareholders." They constituted "a very important and essential department of the telegraph system of this country."[39]

Anderson implicitly argued that he was not merely protecting his company's capitalist interest in dividends and making profit, but also protecting the investment of thirty thousand British shareholders worth several million pounds. Not only did this make up an essential part of the British gross income, but also the shareholders were an important sector of the voting public. Anderson turned attention away from the question of financial output to set up the companies as the true representatives of the people.[40] Later on, Anderson became even more explicit as to the importance of submarine companies for Great Britain: these were "a credit to the nation—not excelled in either respect by any Administration in the World—equally by few, if any."[41]

This is probably the key quote for understanding the cable agents' self-image. Not only did Anderson name private undertakings alongside government administration, but he clearly assigned them even greater importance for the progress of the nation and the world. In his attempt to bully the GPO official into concessions, Anderson plainly suggested that the submarine cable machinery with its worldwide reach was superior to the national network of the British Post Office, because it provided for global growth and progress and administered the wealth and well-being of millions of Euro-American citizens. He was not only making a strong and certainly exaggerated statement of power, but also signifying the radical change that had been taking place concerning the carriers of global "progress" and authority. Implicitly, Anderson proved that Manchester Liberalism, which was at the same time abandoned in the course of the Pender-Siemens controversy over a monopolistic market structure, could yet again be used as a strategy to argue against state involvement. At a time when states attempted to control their local communication system, either by nationalizing the lines, as was the case in Europe, or by proposing antimonopoly laws, such as the National Telegraph Act of 1866 and the Butler Amendment of 1879 in the United States, the ocean cable companies fought and won to keep the role of the state at a minimum.[42] Almost in a friendly voice, Anderson closed the letter expressing his hopes of being "allied" with the GPO and that they "might depend upon [its] support and vote at the conference" for this "invaluable deep-sea department of England's telegraphy system."[43]

Although Patey's response is unrecorded, we can presume that Anderson's argumentation, in addition to the adoption of a joint threatening position of most of the ocean cable companies, was effective. In a subsequent letter written a month later by representatives of five of the largest submarine cable companies, the Eastern Telegraph Company, the Indo-European Telegraph Company, the Direct-Spanish Telegraph Company, the Anglo-American Telegraph Company, and the German Union Telegraph Company, Anderson's point was further enforced.[44] During the next meeting between the GPO and the submarine cable companies in February 1879, Patey not only refrained from his initial proposition, but informed the gentlemen present that he would now even support an *increase* in tariffs, but only on extra-European routes (which were the routes of importance to the ocean cable companies).[45]

The London conference of 1879, which lasted several weeks, was another demonstration of the cable companies' and, in particular, the Pender-Anderson consortium's power. Although they had no voting rights, the sheer presence of the cable representatives at the conference was almost oppressive and illustrated how fervently they attempted to steer ITU decisions. Of the sixty-eight delegates, thirty-five were government officials, three came directly from the ITU, and thirty came from the cable companies.[46] The important social events, during which proposed legislation could be discussed in a private manner and which were stages of backdoor diplomacy, were primarily organized by the ocean cable companies headquartered in London. Indeed, they were "busied," as the *Daily News* pointed out, "in organizing fêtes of various kinds to make the sojourn of the delegates as pleasant as possible."[47] After the conference, Emma Pender reported in a letter to her daughter that she was exhausted from her duties of hospitality: "[The conference] has sat nearly six weeks & been feted to its heart's content!"[48]

One of the social highlights was a banquet at the Freemason's Tavern given by the Joint Reception Committee of the GPO and the cable companies. The chair of this banquet was taken by John Pender, and not the secretary of the GPO, who was the president of the conference. Pender used this position to bring the other delegates in line with his point of view. In his opening address, he reminded the delegates of their duties and their responsibility toward the ocean cable companies. The conference should, given that the submarine connections were so "vitally essential to telegraphs, well consider the position which the companies occupy." This particularly concerned conditions "which affect them most materially," namely telegraph rates. Confirming Anderson's argument on the companies' indebtedness to its shareholders, Pender made it clear that tariff reductions, which had been urged upon the delegates by London's Lord Mayor only a few days earlier, were out of the question.[49] In the end, the conference delegates found an ostensible compromise. No reduction of tariffs had been decided upon, but concessions as to the form of the telegram had been made. These confirmed, for example, the word rate and the abandonment of a twenty-word minimum, because the cable companies had already introduced the word rate in 1872. The various decisions made about the form of the message certainly impacted the ocean companies, but only to a negligible degree. They had won their battle. Seeing the obvious, the *Daily News* concluded that this international conference's result was "the destruction of any hope of cheap and popular international telegraphy."[50] Over the years this power constellation within the ITU hardly changed. ITU decisions on tariffs or codes generally favored the ocean cable companies, which continued to extend their power through their sheer presence at the conferences.[51]

The relationship between the submarine cable companies and the respective governmental institutions was highly complex and interdependent. Through questions concerning landing rights and international regulations, the respective nation-states did play a decisive role in the system of submarine telegraphy, despite the fact that after 1866 governments predominantly did not own or finance submarine companies. These were private companies. Negotiations took place almost on equal footing as the cable agents marked their high (and according to Anderson and Pender, even superior) standing, referring to the benefits for national as well as global progress that they had made possible. Indeed, from the very beginning, the cable agents knew how to agitate governments, as well as the public, and to sell them the cables as national necessities.

THE POLITICS OF CABLE DIPLOMACY

Internationalism, in the form of regulations, organizations, and interstate relations, shaped the system of ocean telegraphy. In the communicational linkage between two nation-states, the private cable companies represented the "neutral" party, equally providing both nations with the advantages of a global network without interfering in matters of jurisdiction or government. From the cable agents' perspective, cable neutrality and guaranteed friendly relations between the two cabling nations were essential for their businesses' success. Friendly relations, or as others called it "universal peace," were, in the eyes of the cable entrepreneurs, a prerequisite for the maintenance and development of commerce and global telegraphic traffic. Consequently, the great men of ocean cabling acted as *cable diplomats* to restore or maintain friendly feelings among the world's few capitalist nations. They used their excellent connections to the global business and public elite, which came with the nature of their cosmopolitan social and business lifestyle. Also those employed at the cable stations were cable diplomats. The relationship between foreign locals and the cable stations and their staff all around the globe was equally important for the stations' well-being as for business per se. These two aspects form the politics of *cable diplomacy*.[52]

The Class of 1866 formed a particularly important group of cable diplomats, composed of the companies' managers and directors and all those that could establish personal ties across the world's oceans by means of frequent transatlantic traveling. They were, according to James Anderson, "something of an ambassador, seeking to carry feelings of good fellowship."[53] The most vivid example was Cyrus W. Field. Through the success of the first Atlantic cable business, he had taken on such an outstanding position within the Anglo-American world that he was frequently asked to mediate conflicts between the old world and the new.

From the beginning of Atlantic cabling in the 1850s, the American businessman Cyrus W. Field had to negotiate between the British and the American side. To Field's surprise, he met considerable resistance and skepticism among the Americans with regard to the cable project.[54] Instead of rejoicing at the possibilities of foreign markets for American agricultural products, national representatives were skeptical about such a close connection with its former mother country. During the congressional debates of 1857 on the Ocean Telegraph Bill, Congressman William Smith of Virginia exclaimed in his vigorous speech against the undertaking that "[e]very consideration, whether [one] regard[ed] it in a business light, or as a question of power, or as a question of dignity, ought to restrain [the U.S. Government] from the passage of this bill."[55] Opponents of the cable saw its greatest danger in the fact that both its termini were on British territory and that the British "would have control of it, and by it could control the trade of this country, and aid in speculation."[56]

The background of such openly expressed distrust in Great Britain lay in the strained relationship between former colonial power and colony. Anglophobic rhetoric and actions were a common characteristic of the U.S. American antebellum, or rather *postcolonial*, period. Instead of growing closer to Great Britain, its former colony sought to break free. Anglophobic rhetoric in the United States ranged from notions of

British cultural hegemony to trade imperialism. Congressman Smith's objections drew from the notion that Great Britain could not only injure the United States but also threaten "to hold back, thwart, or dispute the establishment of America's political, economic, or cultural manifest destiny."[57] Former wartime and present territorial tensions, such as the dispute over Oregon Country in the 1840s and 1850s, fed American Anglophobia. Early land telegraph schemes along the ocean coast were also discussed as an early warning system against foreign, i.e., British, attack.[58] The debate on the ocean cable bill was permeated with such rhetoric. Even ardent supporters of the cable, such as Congressman Lewis D. Campbell of Ohio, spoke of England as the "ancient enemy."[59] In the final days before James Buchanan succeeded President Pierce in office, the U.S. Congress finally passed the bill, granting U.S. governmental support for the undertaking.[60] The fierce debates preceding it had given Field a taste of what was to come. When in 1866 the Atlantic cable was finally laid, American enthusiasm was nowhere near the festivities of 1858. There existed a general skepticism that the cable was only "the product of British work and capital."[61] The general tone in public statements was reminiscent of Senator James C. Jones's statements in Congress, namely that they "didn't want anything to do with England or the Englishmen."[62]

Field, however, believed in Anglo-American kinship. He repeatedly took on the role as mediator in the development of Anglo-American special relations, that is, the exceptionally close political, cultural, and economic relations between the United States and Great Britain. His excellent relations with the British and the American establishment helped him in his performance. Field enjoyed a close relationship with British Prime Minister William Gladstone, as well as several members of the British Parliament, such as Lord Clarendon, and James Wilson, Secretary to the Treasury in the 1850s.[63] Field's diary is filled with appointments during his time in London in 1868. One day he had breakfast with Captain Galton of the Royal Engineers; second breakfast with Mr. Collett, traffic manager of the Anglo-American Telegraph Company; and lunched with Lord and Lady Russell. In the evening, he visited the House of Commons and heard Shaw Lefevre, Lord Stanley, W. E. Forster, and others speak on the Alabama Claims case, a particularly bitter legal dispute between the United States and Britain growing out of the Civil War. Field subsequently had the speeches cabled to the United States in full.[64] It was in between these engagements that Field was "attending to telegraph business and the ladies to shopping."[65] In the United States, Field was comfortably situated at Gramercy Park, which symbolized his membership in New York's elite. In Washington, one of his closest allies was Henry Seward, the American Secretary of State.

Field knew of his favorable position within Anglo-American networks and employed it accordingly, in particular during the period of the Alabama Claims negotiations in the early 1870s. These were a series of demands put forward by the American government against the British Empire given its aid to the Confederacy during the American Civil War. The British had sold warships such as the CSS *Alabama* to the Confederacy and so aided the buildup of the Confederate navy. The Treaty of Washington of May 8, 1871, between Great Britain, Canada, and the United States, established the basis for a settlement for the Alabama Claims, as well as allegations made by the British concerning Fenian raids from the United States into Canada. In 1872, the International Tribunal of Arbitration at Geneva endorsed the position of the United States. They were awarded $15,500,000 in 1872 dollars pursuant to the terms of the treaty, and the British officially apologized.[66]

Upon the signing of the Treaty of Washington, Cyrus W. Field held a dinner in New York for the English commissioners who had been negotiating for the British side. Although Field officially only represented a private citizen of the city of New York, the banquet was perceived as highly political. Justin McCarthy, an Irishman, edited the proceedings of the banquet. In his introduction, he stated that although the banquet "was indeed private in its origin," its object of celebration and the character of the guests removed the dinner "out of the category of mere private entertainments." Rather, it was "thoroughly representative in its character" and had "so profound an international interest."[67] McCarthy unwittingly exemplified the role that the transatlantic telegraph agents had taken on as informal mediators between Great Britain, the United States, and Canada, and their achievement of a truly hyphenated Anglo-American character. Finally, it was no coincidence that Field had picked McCarthy, novelist and former editor of the London *Star*, to edit the proceedings for the British audience. Already in 1869, Field and Peter Cooper had invited Justin McCarthy to speak on "England and the *Alabama*" before the New York public at Cooper Union, a college founded by Peter Cooper and dedicated to the advancement of science and art, thereby marking their continuous commitment to international understanding.[68]

McCarthy's talk in 1869 was ill-received by an American public disinterested in an Englishman's opinion on the Alabama Claims.[69] Negative public reactions illuminated the difficulties faced by Field, Cooper, and others in their mission to translate between the countries. Often they failed and misread their own countrymen. In the 1880s, Field got into trouble, overstepping the boundaries of American patriotism by erecting a monument to an English "spy" from the Revolutionary War. Major John André was a British army officer who had been captured behind American lines on his mission to aid the American general (and traitor to the revolutionary cause) Benedict Arnold's attempt to surrender West Point to the British. He was tried and executed as a spy. In 1821, his remains were removed to Westminster Abbey. The Dean of Westminster Abbey, Arthur Penrhyn Stanley, when visiting his friend in Tarrytown, New York, called Field's attention to the very spot where André was caught. Field consented that if Dean Stanley would write the inscription, he would buy the property and erect a monument.[70] The inscription reveals that both men understood their act as an expression of Anglo-American special relations. It was a sign "not to perpetuate the record of strife, but in token of those better feelings which have since united two nations, one in race, in language and in religion, in the hope that the friendly understanding will never be broken."[71] To some Americans, however, it was an affront against their patriotism and a betrayal of their history and their nation's very foundations. The nationalistic Washington Heights Century Club condemned Field as Anglophile and utterly unpatriotic. In their eyes, the act of erecting a monument to a spy "who was aiding in trying to defeat the struggle for American Liberty" was only an attempt "to please the English nobility" and a great "insult" to the memory of the founding fathers. Field had made himself "ridiculous by toadying to the British aristocracy" and was setting an example that could not be "too strongly denounced."[72]

Three times, people attempted to destroy the monument.[73] "Mr. Field's Pet Memorial," as the *New York Times* called it, came to signify his utter estrangement from his fellow countrymen. In a letter, Alexander Hamilton, grandson of the same-named hero of the Revolutionary War, called the monument "a sham and a fraud" and the inscription "sentimental nonsense."[74] Field's attempts to restore the monument time and again and to protest against its destruction through "indignation meetings" in Tappan were at best laughed at.[75] Hamilton was convinced that there was "not another citizen with sufficient audacity to outrage a just national sentiment as Mr. Field [had] done in this instance."[76] Even residents of Tappan, most of them close friends of Field, realized that "the monument would never be allowed to remain whole in Tappan."[77] Field's inability to comprehend this fact relates to his social frame of reference. Due to his frequent Atlantic crossings, he was a truly hybrid figure.[78] Most Americans, however, were anchored in their local, regional, and increasingly national frame of social and historic reference and not in Field's global one. Even Field's plans to finance a monument for the Revolutionary War hero Nathan Hale or to build a Washington Park around the spot of André's execution could not appease the American majority.[79] In the end, the André monument remained in ruins.

From the American Cyrus W. Field, who represented a rather unique and elitist example of cable diplomacy, let's now turn to the actual cable engineers and electricians paying out the ocean cables as well as the cable workers at the various stations throughout the globe. Over the years, the procedure of cable laying was increasingly standardized. This concerned not only a standardization in technology, but also the development of rituals and sequences making up a distinct cable-laying protocol obeyed in all parts of the world. It involved festivities and dinners, which had to be given and received, as well as a routine of calling on distinct people.[80] The cable festivities in particular played a decisive role in the process of interconnecting the various parts of the globe via ocean cables. Such ritualization of social relations not only expressed a sense of the recently achieved telegraph union, but was also part of peace negotiations. The rituals were, according to Charles Bright, a firm "feature in cable expeditions" pointing to the "necessity of ensuring friendly relations with those to whom the cable ha[d] been taken."[81] Over the years, banquets celebrating the laying of new cables or remembering the laying of old ones became an important and distinct part within global ocean cable politics. At Heart's Content, the successful laying of the Atlantic cable was celebrated anew every year, renewing also the friendly alliance established with the Newfoundland public (elites).[82] Similarly, the frequent meetings between cable and public officials, usually upon the opening of a cable station, followed a distinct routine of visit and countervisits.[83]

Once the cables were laid, their stations remained a source of influence on the local surroundings. This influence was not so much economic or political, as the stations did not represent a great source for employment and its staff was advised to stay out of local politics, but rather cultural. The cable companies exported a large number of British exiles with their cultural traits and global imaginaries all across the earth, thereby aiding British imperialism, yet also furthering the peaceful constellations necessary for the success of ocean cabling.[84] Whereas Cyrus W. Field was the man of the unofficial banquets negotiating on the international level, these cable operators were the men on the ground negotiating on the local level; both actions were taken unofficially, not directed or initiated by a nation-state, and it is this very nongovernmental character that characterizes cable diplomacy. For the success of the global undertaking of ocean telegraphy, ambassadorial characters such as Cyrus W. Field and James Anderson were as important as the diplomatic cable operators, such as Ezra Weedon and James Graves, at their remote rural stations, such as Heart's Content in Newfoundland or Valentia in Ireland.

Almost all submarine cable stations represented British microcosms dispersed all around the globe. Until the late 1880s, the cable companies brought their British staff with them, manning even the "French" and "German" sea cables. Locals were only gradually employed and, in the beginning, solely on the landlines. At Heart's Content, for instance, for almost twenty years after the opening of the station in Newfoundland in 1866, "men from England were brought to work the cable while Newfoundlanders and Nova Scotians were found on the land lines."[85] This seemed to have been as much a security issue as one of education on the working of the cables. Only in the late 1880s did this distinction "between 'natives' and men coming from England" slowly start to dissolve.[86]

But it was not only the British staff who were exported; with them, they brought British manufacture, clothing, games, and money. London architects were sent out across the Atlantic to make changes to the houses. Similarly, furniture was shipped out from Manchester. Superintendent Weedon complained that "[t]hese people [the Newfoundlanders] kn[e]w nothing about Earth closets," that is, lavatories in which dry earth is used to cover excreta.[87] For the staff's well-being, even the station's doctor was sent out from England.[88] In their free time, the men engaged in cricket and curling, and throughout the nineteenth-century tenure of the station, they received their payments in British sterling even though the local currency was the Canadian dollar.[89] Establishing themselves as the new elite in Newfoundland, the predominantly British staff also distanced themselves from the local "others." In the beginning, they mixed little with the locals; the Anglo-American Telegraph Company provided them with their own church as well as their own school.[90] Over the years, they increasingly opened up to the locals, employing them in their service and allowing them to send their children to the station's school.[91]

Yet, even in the initial phase of minimal interaction between natives and staff, the company followed a diplomatic strategy. They realized that for the station's well-being, their men had to be perceived as civilized and friendly and had to stay out of trouble with the native population. Although in Newfoundland a company's employee did not necessarily run the risk of being killed by the natives, as happened to Edward Graves, younger brother of Valentia's superintendent James Graves, while inspecting the lines of the Indo-European Telegraph Company in Persian Baluchistan, establishing friendly relations with the locals was vital.[92]

In the mid-nineteenth century, Newfoundland had a population of barely 100,000; most of the people were English or Irish or descendants of these immigrant groups apart from the inhabitants of the so-called French shore at the island's west coast, where French settlers originating from St. Pierre and French North America dominated. The great majority of people depended on marine resources for their living, and fishing represented the island's only industry.[93] One of the central themes of mid- to late nineteenth-century Newfoundland was an ongoing struggle for the poor island's progress, with varying governments trying to introduce new technologies such as the telegraph or the railroad or, alternately, trying to keep them out and concentrating on the fisheries. The majority of Newfoundlanders appeared skeptical at best toward the new *revolutionary* technology that had found its terminating point at their shores. When in 1857 the landlines to connect the Atlantic cable with the commercial hub of New York were laid, they protested, occasionally even by cutting down telegraph poles. They opposed the Newfoundland government's enclosure of woods for the telegraph company as part of their landing contract. Whereas the government saw it as "waste land," the fishers, who had always used it for materials for the fishery, saw their livelihood encroached upon.[94] To win local sympathy, the station's superintendent Ezra Weedon extended the company's own development projects, such as running water, waste management, or the local school, to the "outsiders," the immediate local population surrounding at Heart's Content.[95] The most important of these measures was allowing their doctor an "outside practice" among those who were not company employees. In a place where during some days in the winter "no sum of money could get a doctor" to the area, this action meant the difference between life and death.[96]

Apart from these development projects, a strict code of moral behavior was enforced at the stations. According to the Anglo-American Telegraph Company's book, *General Orders, Rules and Regulations,* this code was to be observed by all officers, clerks, and servants in the company's service, regardless of where they were stationed around the globe. For the Anglo-American Telegraph Company, it was utterly important that its employees were civil and served as role models. Apart from the proper sending and receiving of telegrams, the guidebook also discussed the social behavior expected: gambling or betting of any kind was "absolutely prohibited" and so was alcohol in the staff's rooms. Rule 31 read that "*[a]ll persons* employed by the Company [were] required to *conduct themselves with civility;* incivility or rudeness [would] never be overlooked."[97] When in the 1880s a telegraph operator at Heart's Content took to drink and spousal abuse, superintendent Weedon threatened him with instant dismissal. He would "not have the Company's houses disgraced by any such proceedings."[98] "Wife beating coupled with turning wife and child out of doors" was bad enough among "'navvies' 'costermongers' 'coalheavers,'" but such practice "amongst a class whose education should teach them better [was] simply unpardonable." Moreover, the disgrace attached to it extended "to the whole of [their] little community."[99] It was only due to the wife's incessant pleading that this particular employee did not get turned out of service. The telegraph stations were neither tools of the British Empire enforcing British law and order, nor an official element within Euro-America's civilizing mission, and yet they indirectly and at points unwittingly served both ends through their policy of cable diplomacy. By importing not only the staff but also British culture to the remote stations, they were perceived as British exports and, as such, representatives of the British Empire. By aiding the development of their immediate outside and the company's emphasis on its staff's devotion to civility, they were indirectly part of a civilizational discourse and so agents of Euro-America's civilizing mission.

However, superintendent Weedon's attempts at winning the hearts and minds of the locals by development projects, medical aid, and civility did not show early success. As he reported to London in 1877, more than ten years after the cable station had been set up, the Newfoundland public and local officials still had a rather negative and, as he was convinced, absolutely faulty impression of them. Moreover, he claimed that the public was purposely kept in the dark on "how much this island has benefitted by the cable's coming." It was high time that they came up with a strategy so that the "Anglo Co should be shown in its true colours to the public of Newfoundland."[100]

Much of this public dislike derived from the skepticism mentioned earlier toward a modernization that did not seem to be beneficial for the local fishermen and the tokenism of the topic in local politics.[101] Unwillingly, the cable station was dragged into it as Newfoundland politicians either claimed credit for or verbally abused the cable line at a time of fierce political disputes in the 1870s and 1880s on how to best develop the poor colony. How decidedly the entire colony was against the cable company and its station in Newfoundland was best documented during the preemption scandal of 1873, which happened shortly before the financial panic of 1873. Upon establishing the landlines in the 1850s, the New York, Newfoundland and London Telegraph Company had in its charter conceded the right of preemption to the Newfoundland government, meaning that after twenty years the government could purchase the landlines at their original cost. Now in 1873 with the deadline drawing near, stock market speculators and cable opponents in London, such as Henry Labouchere of the DUSC, successfully convinced the liberal Newfoundland government of Charles Fox Bennett to help manipulate the price of shares. According to contemporary historian Prowse, the government sent a telegram "at the moment that it suited the stock-jobbers" saying that they would not wave their right of preemption.[102] On the stock market, the effect was "immediate and disastrous" as the value of the company's shares went down from £20 to £14, which amounted to a loss of £1,000,000.[103] According to Prowse, "nearly all the influential men in the Colony [were] drawn into this clever, unscrupulous game" as they were monetarily compensated for "working out this great financial dodge." Additionally, also the entire "unpaid and disinterested" Newfoundland public worked against the company and aided "a great stock-jobbing trick."[104] Superintendent Ezra Weedon repeatedly reported on these local developments in his correspondence with the London headquarters. If he had already been quite enervated by the scam of 1873, in 1877 he was fervently put out by the conservative politician William Whiteway who "in addressing his constituents at Trinity sa[id] '*I have given you this line.*'"[105] Weedon was right in his indignation, because Whiteway, who was to succeed Sir Frederick Carter as Premier of Newfoundland in 1878, had little influence on the cable laying, but Whiteway was entangled in his own political controversy on Newfoundland's development.

The first to voice the idea of developing Newfoundland by means of telegraphy was the local head of the Roman Catholic Church. In a letter to the St. John's *Courier*, Bishop Mullock suggested in 1850 that St. John's should be the landing place for a possible transatlantic telegraph, thereby envisioning bringing his "neglected island" into the beneficial "track of communication between Europe and America."[106] The bishop calculated that such a cable would bring steamers, news, and commerce, none of which occurred when it finally landed on Newfoundland's shore.[107] Quite similar to the Bishop's theory of progress through technology, a group of young politicians emerged in the 1870s and 1880s who believed, contrary to the government of Bennett that had carried out the preemption scandal, that "landward industrialization would make Newfoundland a neighbor of consequence to Canada and the United States, and one that Britain would have to treat better."[108] They were facing an opposition that thought that Newfoundland's future lay in fishing and that its government should rather see to securing sole rights for its island's fishing grounds.

During those turbulent times, despite Weedon's conviction that they should "do far better by leaving politics alone," the station was caught up in the political struggle.[109] Directives to the cable station usually came via London from St. John's where Alexander Mackay resided as overall superintendent for the land and ocean lines connecting at Heart's Content.[110] During the election campaigns of 1873 and 1877, he repeatedly gave orders to Weedon to support the candidates telegraphically by letting them use the lines free of charge. The Anglo-American Telegraph Company must have even sponsored some of Whiteway's running mates because superintendent Alexander Mackay gave all candidates for the opposition "full power . . . to send . . . election reports free by wire," much to the dislike of station manager Weedon. Nevertheless, Weedon vowed to be "civil" to the candidates of the opposition, "as [he was] to the other side," but certainly he would have "nothing to say about politics."[111] Sources are scarce on how far the Anglo-American Telegraph Company's engagement actually went beyond providing free communication and to what degree they were successful. Yet, it is undisputable that the company influenced the local elections in Newfoundland, attempting to help into office a government that would be more favorably disposed to them so that the debacle of 1873 should not be repeated.

While the company jumped in on the fight between conservatives (pro-modernization) and liberals (pro-fishery), it attempted, if at all possible, to stay out of another hot issue—Newfoundland's *Irish question*. A common theme in Canadian history is that of transatlantic Ireland, which encompasses a story of religion as much as one of mass migration. Newfoundland was the site of the first large-scale permanent Irish-Catholic settlement in North America in the early modern period, and Irish migrants remained the largest group of immigrants to the island until 1830. At that time, the collapse of the provision trade for fishing products between Newfoundland and the British Isles nearly entirely halted Irish migration. However, by that time, some 30,000 Irish, most of whom were Roman Catholic, had permanently moved to Newfoundland. They primarily settled along the shores of Conception Bay, one of the cable landing places, and at the capital, St. Johns.[112]

Religious denominations played an important part in Newfoundland's local politics. Bishop Mullock, in particular, exerted great influence on political issues, demonstrating the great power of the Catholic Church. In 1867, Newfoundland declined to join the Canadian Confederation for fear that it would mean domination by the Protestant Irish (immigrants or descendants of immigrants from Northern Ireland) of Ontario.[113] By the 1880s, ethnicity had become "a significant 'competitive strategy' for negotiating power and access to resources" in Newfoundland. And as the vast majority of the Protestant population of Newfoundland had its roots in the west of England and the Catholic population its roots in the south of Ireland, "religion became inextricably linked with articulations of ethnicity."[114] While the Catholic Church attempted to establish authority over the Irish-Catholic population, previously mixed living areas became increasingly homogenous. In addition, settlements in the Conception Bay area of Newfoundland, one of the cable landing places, became increasingly ethnically segregated. This ethnic dichotomy between English Protestants and Irish Catholics fueled a conflict that was yet to come.[115] The station was stuck between the Old World Irish Nationalism pleading for Irish-Catholic independence and Orange Unionism protecting Irish-Protestant interests as it played itself out in the locale of New World Newfoundland. While Field was negotiating on the international level between Great Britain and the United States on the Alabama

Claims and Fenian raids, Weedon and his staff were asked to handle a quite similarly situated problem on the local level.

First, let's return to the Newfoundland politician, William Whiteway, the cable station's desired candidate. After Whiteway took office, he turned, as had been hoped, away from the question of fisheries and its regulation to interior development by means of telegraphs and railroads.[116] Despite fierce opposition from the mercantile community, the Newfoundland Railway Company began construction in 1881. The opposition feared that the expenses of the railway project might bring the colony into bankruptcy and ultimately raise the taxes on their goods to finance the scheme. Opposition soon aggregated behind the New Party of Walter Baine Grieve and James J. Rogerson, which brought another element into Newfoundland politics, namely religious denomination. Although Whiteway was a Protestant, he won on the railway issue due to the strong support of the Roman Catholic Church. Thanks to Bishop Mullock, the Roman Catholic Church had always played an important role in local politics. Countering not only the railway issue but also the strong Roman Catholic support for Whiteway, the New Party's new leader aligned himself with the forces of Orangeism.[117] As the political lines shifted from one of railway versus fishery to Catholics versus the Orange Order, thereby reproducing the Irish question in colonial Newfoundland, the situation grew alarming for the telegraph station at Heart's Content.

Superintendent Weedon was convinced that staying out of politics was the best way of handling things in Newfoundland, in particular when it came to questions of religion. In 1877, the community at Heart's Content received its own little church, which was open to everybody living at Heart's Content and the surrounding area.[118] Members of the Anglican Church, Roman Catholics, and Wesleyans indiscriminately used the building for services.[119] In 1881, however, Weedon and the station found themselves in the midst of the seething conflict between Catholics and Protestants, when one of Weedon's cable hands who had been transferred from Valentia, Ireland, got into trouble with the locals over the Irish cause. Alarmed, Weedon reported to the office in London that since "the orange feeling [t]here (among the natives) [was] so very strong" and this particular cable hand was "a bigoted Roman Catholic far on the road to Fenianism" who had already "collided with the Orangemen," he would "never have any years" at the station.[120] Instantly, Weedon asked the London office to transfer the said operator either "to one of [their] other stations or . . . to some other company."[121] Although Weedon showed clemency in other cases of fights, drunkenness, or even domestic abuse, this case was different: he was determined that this Irish operator was not to remain at Heart's Content and get the cable station mixed up in the turmoil of local politics. In 1882, Weedon reported to London that he would prefer "not to have any more VA [Valentia] men" from Ireland. The last consignment had already been "a little too much for us & it's only by a lucky chance we are getting clear of them."[122]

Weedon proved far-sighted in immediately sending his cable hand away from Heart's Content. On Boxing Day in 1883, members of the Orange Order marched through a Catholic neighborhood in Harbor Grace, a town only some twenty miles away, where the station received most of their supplies. A conflict followed that ended in an open street riot leaving three Orangemen and one Catholic dead. The subsequent trial investigating the deaths of the Orangemen called on Whiteway as well as the Grand Master of the Orange Order as prosecutors and popular Catholic politicians as defenders; it was a grand political show. For fear of additional riots, a British naval ship stood by at St. John's, the capital of Newfoundland, where the trial took place.[123]

In conclusion, cable diplomacy predominantly entailed various means and methods of conflict resolution, be it Cyrus W. Field's banquets or superintendent Weedon's development projects. On the international and local levels, the cable agents attempted to keep relations running smoothly because good relations were vital for the businesses' success. The 1873 preemption scandal demonstrated the result of local opposition. As the support for Whiteway showed, partisanship was helpful to a certain degree, yet not overemphasized. Moreover, as the conflict over the Irish cause exemplifies, clear political or denominational stances on the side of the staff or the entire station were prohibited. Until the beginning of the twentieth century and the commencement of the First World War, it all came down to one aspect: neutrality.

CABLE NATIONALISTS AND STRATEGIC NATIONALISM

Loyal primarily to their profession, the cable agents worked with, around, and at times also against the respective national representatives. The cable agents appealed to national sentiments or to those of transnational

Euro-American kinship, depending on whether the obstacle in the field was another rival, government intervention, or international tensions. Generally, the relationship between a cable company and local, national, and international representation was highly complex and changeable. Yet, all in all, those working in the cable industry, such as John Pender, James Anderson, and Cyrus W. Field, employed various strategies to keep those relations smooth and cooperative. Their interests were grounded in their conviction that peace was beneficial to commerce, and hence to their own business, and that ocean cables were means of commerce and not of war. Until the 1890s, the negotiating parameters of the relationship between the international, transnational, and national dimensions of the submarine cable business remained relatively stable. This equilibrium was disturbed at the turn of the century as several things happened simultaneously: the expansion of the world's economy on a global scale, a growing pluralism in the submarine cable system, and a new period of territorial expansion of the Western countries. All contributed to a growing discourse on "cable nationalism."

One significant factor in this shift was the development of the world economy. After the period of the Long Depression between 1873 and 1879 or, for Great Britain, of the Great Depression of 1873–1896, the economy had again started to grow rapidly. Moreover, the global economy had geographically broadened its industrial basis, as countries like Sweden, Russia, and the Netherlands, as well as North America and to some extent Japan, were undergoing an industrial revolution. In particular, the economic power of Germany and the United States seemed to be bypassing Great Britain: "[t]he Age of Empire was no longer monocentric."[124] There was also a growing pluralism in the submarine cable business, which was no longer solely bound to the banks of the Thames. The advantage of cable firms registered in Britain had so far been that they had had the field all to themselves. Now Germany and France had become "fully competitive in manufacture of electric equipment, cables, and cable ships."[125] Aside from the French manufacturer Société Industrielle des Téléphones, it was the German cable manufacturer Felten & Guilleaume that came to play a central part.[126] The last external factor for the changing discourses of cable nationalism lay in the territorial expansion of the Western countries. The economic and military supremacy of capitalist countries was now systematically translated into formal conquest, annexation, and administration. In the period between 1880 and 1914, which is often called the phase of new imperialism, the world outside Europe and the Americas was "formally partitioned into territories under the formal rule or informal political dominance of one or other of a handful of states."[127] These were Great Britain, France, Germany, Italy, the Netherlands, Belgium, the United States, and Japan.[128] It was in this period, characterized by a growing pluralism of both the economy and the submarine cable business, in addition to enormous imperial expansion, that a discourse targeted at the cables as symbols of national progress shifted to a discourse of national necessity and imperial defense.

At the turn of the century, it was hardly a novelty that cable undertakings and their personnel were embedded in stories of nationhood. From the beginning of submarine cabling, its actors had had to face and react to national sentiments connected to their achievements. Most of the time, these dealt with questions of how these cables were represented as national achievements and were debated in the respective national public press. With their success, the ships, the cables, and their engineers were claimed for, and woven into, narratives of national progress and distinctiveness. The cable ship *Great Eastern*, for example, was considered to be a "national property," which was to England with its "monumental structure. . . what a pyramid was to Egypt—a practical trophy of art and power, a grand illustration of science and wealth, a prove [sic] that [they were] still moving in advance of every nation."[129] All over Euro-America, technological artifacts, such as steam engines, bridges, buildings, and cables, played an important part in narratives of technological nationalism. They assumed the nature of national icons and were consumed by the urban masses.[130] Although in the case of submarine telegraphy, nationality markers were not as clear-cut. Because the cables established connections between two countries, the public, the press, and the government representatives usually settled the issue by assuming that, because the various cable companies were registered in Great Britain, they had to represent Britain. Thus, contemporary public debates not only shaped an entire historiography, but also influenced later historians to focus strongly on the cable business's administrational London-centrism. In these recurring disputes on the respective cable's nationality, the cable agents themselves were usually strangely absent, concealing the multinational character of their companies while letting themselves be claimed equally for various discourses. They knew that a certain alignment to a national government, in particular the British, was extremely helpful. As a consequence, the submarine cables throughout the world carried a distinct national marker from the beginning, making each one a British, French, or American cable, without much disturbance to the daily business of submarine telegraphy. Matters changed toward the turn of the century when the nationality of the cable companies became essential beyond the narratives of nationhood.[131]

Two wars with colonial motives initiated the shift from a discourse on submarine cables as symbols of national progress to a means of national security. They exposed the cable actors' efforts at peacemaking as failures and the pacifist rhetoric as hollow. The years of 1898 and 1899 marked not only the Spanish-American War as well as the start of the Second Boer War, but also the end of cable neutrality and cable protection in times of war, which had originally been secured under the 1884 International Telegraph Convention. During the Spanish-American War, the U.S. government "panicked" upon finding that it had no U.S.-owned cable to Cuba. Using a loophole the ITU convention of 1884 offered with its statement that the convention would not "in any way restrict the freedom of belligerents," the U.S. military had cut cables in the Philippines—unfortunately the wrong ones. Soon after it had erroneously cut the British Eastern and Extension Company's cable, the U.S. government asked the company to repair the line. However, the company refused, obeying the terms of their license from the Spanish government that its cables could not be used against Spain. Only after the military and political situation had changed in August 1898 in favor of the United States was the line to Manila reopened.[132]

U.S. actions during the Spanish-American War with regard to cables revealed that the previously cherished belief that privately owned cables would be neutral in times of war was obsolete. As a consequence, access to cables as well as the ability to cut them "became a military imperative and the avoidance of such action became a strategic necessity."[133] Whereas the British governing elite considered the London-centric system of submarine cables a means of imperial defense, other national governments considered it a monopolistic tool that allowed Great Britain "control of information, of propaganda and censorship."[134] In his 1901 article "The Influence of Submarine Cables upon Military and Naval Supremacy," the American George Squier warned against British control of nearly four-fifths of the world's cables, which were "woven like a spider's web" and in which all other nations were caught up.[135] As a consequence, those national governments that had beforehand been content with using "British" cables increasingly sought to influence their ocean cable companies and simultaneously attempted to break free from the Eastern and Associated Companies, or, as they perceived it, the British monopoly. From the 1880s on, Heinrich von Stephan, Postmaster General of the German Reich, pushed for a German transatlantic cable via the Azores, mirroring the attempts undertaken by the French government along the same route.[136] German and French competition with the British was helped by the growing pluralism of submarine cable manufacture and the fact that landing concessions and contracts, such as between the German government and the Anglo-American Telegraph Company concerning German Atlantic traffic, were coming to a close or needed renewal.[137] In 1891, the French government invited tenders for a cable from Marseille to Oran and Tunis and, for the first time in history, excluded British contractors altogether.[138]

The British response to these changes was the all-British or All Red cable system, a network scheme of submarine cables touching only upon British territory. In the period prior to the First World War, this scheme of "imperial telegraphic communication" busied many British cable strategists and reformers, such as Charles Bright, Sandford Fleming, and Henniker Heaton.[139] As a consequence, longer distance lines with less economic value were laid for political reasons.[140] Upon the outbreak of the Second Boer War in 1899, for instance, the British government had a cable laid from Great Britain to the Cape of Good Hope via Porthcurno in Cornwall, Carcavelos in Portugal, Madeira, Cape Verde, Ascension, and St. Helena.[141] The "epitome of strategic cables," however, bound into the scheme of an All Red route, was the Pacific connection. In the late 1890s, after decades of unsuccessful attempts to attract government interest, a Pacific cable was finally taking shape.[142] For 45 years after the telegraphic breakthrough on the Atlantic, the Pacific Ocean was still "as innocent of cables as the pond of a country village."[143] But now with the British and American governments discovering their imperial and economic interests in the Pacific Rim, two cable schemes were brought to completion in 1902 and 1904. One connected Canada with its sister British colonies, Australia and New Zealand, and the other connected the United States with its newly acquired possessions in the Pacific and further on to the coast of Asia. Both are excellent examples of the underlying strains of the imperial discourses.

After the success of the Atlantic cable, contemporaries around the world were relishing the idea of a Pacific cable. In 1870, Cyrus W. Field and his American Atlantic cable business partners incorporated the Pacific Submarine Telegraph Company and obtained a landing rights bill from the U.S. Congress that granted naval support in laying and ocean sounding, as well as a payment of $100,000 in government bonds.[144] However, Field's Pacific plans were just as fruitless as those of the European, American, and Asiatic Telegraph Company, incorporated in February 1871 by an Anglo-American consortium, or the scheme brought forward in December 1871 by a group of U.S. politicians and entrepreneurs, including John F. Miller, U.S. Senator from California, Bradley Barlow, U.S. Representative from Vermont, and railway magnate William G. Fargo. The U.S. Congress had given each one of them only a year to find suitable investors.[145] During the Forty-Third Congress, 1873–1875, the Committee on Foreign Affairs dismissed three additional bills on ocean cables in the Pacific.[146] Without proper ocean soundings, costs for a submarine cable could have easily skyrocketed; later estimates speak of costs around $3,000,000—much greater than those of the Atlantic route.[147] From the American government's view, at that time, there seemed to be little reason to undertake such a commercial risk, and interest did not revive until the turn of the century.

At the end of the 1870s, Sandford Fleming, a Scottish émigré to Canada and railroad engineer, moved to the foreground of these initiatives with his advocacy for a Pacific cable from the west coast of Canada to Japan, China, Hong Kong, Australia, and New Zealand.[148] From the Canadian point of view, there were two advantages to a Pacific cable: Pacific trade and the promotion of a closer union among the "sister colonies" of Canada, Australia, and New Zealand, and thus an interimperial federation.[149] Both of these goals would aid Canada in ensuring for itself greater economic and political influence in the Pacific region. For the British metropole, there initially seemed to be little benefit in a Pacific connection. For more than twenty years, it kept thwarting Fleming's plans. The reasons were straightforward enough. The telegraphic shortcut via the Pacific would not only bring Australia and Canada in direct communication with each other, but would do that by omitting London. Consequently, the cable would not only promote a new, if small, center of trade and communication in the Pacific region, but also tremendously change the entire global communicational network structure. So far, London was the uncontested center of communication; now there would be a girdle round the world allowing messages to go in all directions, possibly without a detour through Europe. This would leave British merchants bereft of their centralized position and its associated benefits. In a letter, John Lamb, director of the General Post Office's telegraphic department, reveals that British government officials were particularly concerned about the advantage of the Pacific cable for American commercial centers such as Chicago, San Francisco, and New York. Without it, all market-sensitive information "must reach England before it [could] be sent to America"; however, if the Pacific cable scheme were to be carried out, "England would . . . lose the advantageous position it . . . occupie[d]."[150] This reinforced the General Post Office's view that "London should remain the center of world communication."[151] Despite various scenarios of a French threat, the Pacific still seemed to be strategically irrelevant to the British government, and they had no interest in a new center of power within their empire, especially one that lay in the Pacific and was led by Canada.[152] For years, the admiralty refused to undertake necessary ocean soundings.[153]

Finally, the British government's lack of support for Fleming's endeavor was motivated by its support for John Pender. Playing the national card, Pender managed to convince the government that any competition on the connection to East Asia would be detrimental to his cable consortium and, ultimately, the British people. In a memo from 1897, Treasury Official Hamilton explained that since Pender's cable companies "represent[ed] a real British interest, and one entitled to great consideration from the Imperial Government," it was "a matter of no small importance that the Government should continue to maintain friendly relations with them." Such consideration "could hardly be expected if the Government were to take an active part in the establishment of a cable in direct competition with the Eastern Company's system," meaning a Pacific cable.[154] Whenever possible, John Pender tried to thwart Canadian Pacific cable schemes, discouraging them openly or acting behind the scenes.[155] The highlight of Britain's policy of cable protectionism was a secret agreement with John Pender. As the *New York Times* reported in 1899, "a bombshell was hurled . . . at the promoters of the Pacific cable" when they learned of a private agreement made in 1893 between Great Britain and the Eastern and Associated Telegraph Companies guaranteeing a monopoly to the Eastern and binding the government neither to lay nor to assist any one to lay, and to not permit anyone else to lay, a cable to Hong Kong or Singapore.[156] With this agreement, Canadian schemes for a Pacific cable seemed to be utterly defeated.

Four years later, the Canadian Pacific cable scheme gained needed British governmental support, but only when Great Britain started to seriously worry about its state of imperial defense in the face of fluctuating alliances in Europe and mounting challenges overseas.[157] Fleming immediately recognized the writing on the wall and emphasized even more strongly the Pacific cable's importance for imperial defense schemes. He promoted the cable as the "missing link" in Britain's All Red cable, which would sustain communication with the east should the cables passing through the dangerous waters of the Mediterranean be interrupted.[158] Fleming benefited from the fact that his most serious opponent, John Pender, had died in 1896 and passed on his cable empire to the hands of his son John Denison-Pender. Now, his argument hit a nerve. The Mediterranean cable route from London to India and on to Southeast Asia, Australia, and New Zealand was, from a military point of view, one of the most sensitive connections. Not only were its landing points often on non-British territory, but also the shallow waters along most of the route made British cables easy to spot and to cut in the case of war. Already in 1888, at a meeting convened by the Pacific Telegraph Company, the Earl of Winchilsea warned that in times of war the existing lines would be "absolutely defenceless."[159] Potential enemies made similar statements that reinforced these fears. In 1898, the Russian journal *Novoe Vremya* concluded that in case of an armed conflict with England "[their] first task would be to block England's communications with India and Australia."[160] The Pacific route offered "an 'All-Red-line' operated by British clerks, which would lie in deep waters, far from potential enemies, and touch on foreign soil (if at all) only at the Sandwich islands [Hawaii, a U.S. possession]."[161] Because the British government had at last decided on the imperial necessity of a Pacific cable, John Denison-Pender, who had continued his father's battle, changed his strategy: if there was to be a Pacific cable, it was to be his. Within two weeks, he announced that a Pacific cable was already being manufactured and that, if the colonies and the government agreed, it could be laid within two years.[162] On December 7, 1902, a cable was opened connecting Canada to Australia and New Zealand. Sandford Fleming had the honor of sending the first telegram around the world.[163]

In the United States, interests in an ocean cable across the Pacific reawakened in the early 1890s with the first ocean soundings in 1892.[164] American interests were strongly connected to Hawaii's 1893 coup d'état when a group of antimonarchists, composed largely of American citizens and aided by the U.S. Navy, overthrew the queen. In 1895, the new Hawaiian government granted landing concessions to an American consortium and debates were, albeit unsuccessfully, resumed in the U.S. Congress.[165] In 1898, the *New York Times* pushed the issue by stating that, "[a]nnexation [of Hawaii] having been accomplished, the next important consideration is the cable to the United States."[166] In 1899, newly elected U.S. President McKinley took up the issue. Like German and French government officials, the American government also grew suspicious of the monopoly of British cable companies in the context of mounting nationalism and imperial rivalry. As a consequence, President McKinley announced in 1899 that the United States should no longer rely on foreign cables. One of his greatest concerns was that access to Asia would only be gained at the expense of relinquishing control to foreign interests, as the telegraph lines in Asia were controlled by the British-based Eastern and Associated Companies and the Danish-based Great Northern Telegraph Company. For the United States, a Pacific cable would facilitate a prolonged westward expansion as part of their policy of colonization and protectionism, especially toward the Philippines, Guam, and Hawaii. At this point, the correspondence between the old Atlantic friends, John Pender and the American Abram Hewitt, offers great insights into the growing importance of "jingo theory," or jingoism, a form of militaristic nationalism, in the politics of the United States.[167] Naturally, John Pender was of the opinion that although a Pacific cable was not wanted, "if insisted on, [*he* had] the people to do it."[168] During the 1890s, he was in frequent telegraphic communication with Abram Hewitt, who kept Pender updated on the latest congressional debates. Already in 1895, Hewitt warned Pender of the "hostility to England and English interests" and advised him to wait and see.[169] In the "present state of feeling," it would be unwise to submit an offer because "a prohibition against any contract with foreigners" was most likely. He would, however, in case the president should be empowered to make a contract "without limit as to nationality . . . communicate [his] proposition . . . and secure a fair hearing."[170]

As the Pacific cable turned into a "national necessity," it became equally essential that its manufacture, laying, and operating was "wholly under the control of the United States."[171] Many senators argued in favor of this kind of model. The initial cable bill sought the cable to be "American made, laid by American ships, and managed by the American Government."[172] Officially, Pender was out of the picture, leaving the field to American-based competitors such as John W. Mackay of the Commercial Cable Company and James Scrymser of the South and Central American Cable systems. Both American competitors "sought to outdo the other by wrapping themselves in the American flag and extolling their corporate virtues and national purity."[173] Scrymser even negated his own business interests and extended arguments in favor of government ownership of the cable "citing incidents of the Spanish-American war showing the importance of Governmental control of the cable."[174] In all likelihood, this was part of his strategy to show off as the better American, as he knew that this was Mackay's weak spot. The nationality question became a trying issue for John W. Mackay, as it was extended from the character of the cable company to the contractor himself. Some congressmen questioned his loyalty considering his status as an "ex-patriated American."[175] Although Mackay spent most of his time in the United States, his wife and son had long taken up residency in Europe. In addition, he was associated by cable business with James Gordon Bennett Jr., who represented that class of American expatriates par excellence.

Despite Scrymser's opposition, in the end, John W. Mackay and his Pacific Commercial Company laid the trans-Pacific cable, starting work in 1902 and reaching the Philippines in 1904. Two years later, cables were laid from Japan and China to the Philippines. Upon drafting the landing concession for the Pacific Commercial Cable Company in 1902, President Roosevelt had made "a stipulation that the cable company should be an all-American line, and should make no connections with any other companies except American companies."[176] This condition posed difficulties regarding not only interior Chinese telegraph connections, but also the questions of landing rights at the Southeast Asian coastline. Both the Danish-based Great Northern Telegraph Company and the London-based Eastern and Associated Telegraph Company held, singly and jointly, landing monopolies in Southeast Asia that forbade a cable landing of the American Pacific Commercial Cable Company. A solution was found behind closed doors. All along Mackay's company was mostly owned by foreign companies. Therefore the ostensible problem in Southeast Asia was easily, and somewhat mysteriously, solved. Although German telegraph engineers and government officials suspected as much as early as 1901, only in 1921 did the American government officially learn that Mackay's allegedly American company had been 75 percent owned by foreign companies: 50 percent by John Pender's Eastern company and 25 percent by the Danish Great Northern company. All but one copy of the contract was destroyed. The remaining copy was kept in London in a box with six locks that was never to be opened without court order or the written orders of all involved companies.[177] Such pompous secrecy contradicts the communality of knowledge, at least among administrators and engineers, concerning the workings of submarine telegraphy. For telegraph insiders such as Carl Wilhelm Guillaume of the German telegraph company Felten and Guillaume and board member of the German Atlantic Telegraph Company and Reinhold von Sydow, Secretary of the German Ministry of Postal Services [*Reichspostamt*], it was obvious that questions of landing right monopolies had to be attended to one way or another. In a letter to Sydow dating August 1901, Guillaume reveals how both suspected some sort of working agreement between the companies involved. [178] Details on the Commercial's ownership were kept top secret. There were no public statements indicating that the two rivals, the Eastern Companies and the Great Northern, both held substantial amounts of shares. Still, at the time of the cable construction, so it seems, no American government official *wanted* to suspect the Commercial of being anything else but an American undertaking. Believing in a nationalist mission concerning communications, they turned a blind eye to the transnational working modes and mechanisms so natural to the technology of submarine telegraphy.[179]

The Pacific cable undertaking was not the only one where nationalism had become a strategy. Neither the government's nor the public's determination on all-British, all-American, all-French, or all-German cables could be challenged, nor could the transboundary logic of the submarine cable business be altered. Similar to the Pacific cable case, the cooperation between German and Dutch entrepreneurs in the form of the Deutsch-Niederländische Telegraphengesellschaft (German-Dutch Cable Company) to lay cables in the East Indies or between German and French entrepreneurs to lay a cable across the South Atlantic showed that purely national enterprises were not feasible. To circumvent the restrictiveness of landing monopolies, companies of different "nationalities" were forced to enter into agreements with each other.[180] Hence, the cable agents settled on a strategy of changing national identity on an as-needed basis. In 1910, for instance, the American-based Commercial Cable Company used its German allegiance, as it owned a stake in the German Atlantic Telegraph Company (DAT), to put the American government under pressure. The U.S. government had previously forced the Commercial Cable Company as well as the DAT to relay their cables away from New York harbor to make room for expansion. Now George Ward, managing director of the Commercial Cable Company as well as honorary director of DAT, landed a coup. Using O. Moll, the German DAT director as his spokesman, he asked the German Foreign Office to intervene with the U.S. government on behalf of the two companies because genuinely German interests were at stake.[181] Diplomatic notes were sent back and forth between the German imperial representatives in Washington and Berlin debating how to proceed before, in the end, they declined to act. Yet, as Duke von Bernstoff, German ambassador to the United States in Washington, admitted, it was exceedingly clever and certainly comfortable for Ward to bring the German diplomats into action.[182]

This strategic nationalism applied as much to the companies as to the agents themselves. Following a similar philosophy of changing national alliances, former Commercial Cable and now Siemens Brothers manager Georg von Chauvin switched nationality as late as November 1914. He simply renounced his German citizenship, dropped the particle from his name, and utterly undisturbed, continued working for the Siemens Brothers in London until 1925.[183] Chauvin was not the only one among the cable agents to switch nationalities to meet working place requirements or retain personal benefits. Other cases include William Siemens and Ludwig Löffler, another Siemens Brothers engineer.[184] Once the cable agents had to give up their claim on cable neutrality, they opted for nationalism as a strategy to disguise their working structure. This was not necessarily done out of cosmopolitan conviction, for some of the agents were fervent patriots, but for business necessity stemming from the indisputable landing right concessions and monopolies of other countries, and their companies, in certain regions.

The relation between governmental representatives and cable agents over the means of global communication within local, national, and international settings was highly complex and changed over time. To protect their business interests from rival companies, as well as from government control, the actors switched between the strategies and rhetoric of cable diplomacy and cable nationalism. Generally, these strategic interventions can be divided into two periods, with 1898 and the Spanish-American War and the Fashoda Crisis as the decisive dividers. From the beginning of ocean cabling, national governments played an important role in the system of global communication. They gave out landing rights concessions and financial and material support. Through the ITU, they decided on cable rates and transmission policies. Nevertheless, their influence on John Pender's cable empire in that period was minimal. This was based on the economic principles of Manchester Liberalism and its focus on minimal state involvement, the fact that cable concessions were a one-time issue, and a general belief in cable "neutrality," which furthered ocean telegraphy as a private business. This gave the cable agents space to act relatively independently from their position in maritime space. People like James Anderson successfully developed a strategy of playing on national or international sentiments to influence government officials to act in the companies' interest, and on their behalf, in situations when officially they had no say, such as during the ITU proceedings or landing rights' negotiations.

Their great independence and the transnational character of ocean cabling, on behalf of which its managers, engineers, and operators moved about the globe and built up a variety of social networks on a transnational scale, helped their engagement as cable diplomats. Grounded in their belief that peaceful relations on the international level enhanced world trade and thus their global business, various actors, such as Cyrus W. Field, embarked on the project to secure and develop friendly international relations. In the eyes of Charles Bright, they fulfilled that position so successfully that their opposition to all-British or all-American government cables was grounded in the fact that "[h]itherto, the companies [had] acted somewhat in the capacity of diplomats" and thus would lose this position.[185] However, the cable agents engaged in cable diplomacy on the international level as well as the local level. The primarily British operators brought their British culture with them to the remote cable stations and were certainly, if unwittingly, perceived as intermediaries of the British Empire. In addition, the cable stations' development projects, such as water and sewage disposal systems or schools, in which the natives from their immediate surrounding were included, and the companies' emphasis on its staffs' civility made them part of Euro-America's civilizing mission.

Matters changed with the Spanish-American War, which had exposed the cable agents' belief in cables solely as a means of commerce as largely an illusion and one they abandoned when the nation called for them to take sides; thus the war sounded the death knell for cable neutrality. The growing world economy and the development of industrial powers that could compete with Great Britain, a growing pluralism in the ocean cable system, and the period of new imperialism additionally influenced the new situation for the cable agents in the period prior to the First World War. From then on, submarine telegraphy was seen as a question of imperial control and national security, which resulted in various governments striving for all-national cables. It seemed as if the actors' business interests had to accept a subordinate role to these strategic imperatives. Their response, however, to this paradigmatic shift was the employment of nationalism as a strategy. Officially they tied themselves closer to one particular national discourse, thereby disguising the financial and working setup of their company, which was not congruent with principles of all-national submarine telegraphy. This strategic nationalism was only partially due to the agents' cosmopolitan conviction; more importantly, it was driven by the transboundary logic of the submarine cable business.

7

THE WIRING OF THE WORLD

To the promoters of the Atlantic Telegraph belongs the credit of having given the original impetus to [the wiring of the world]; and to you gentlemen, as inventors, as electricians, as manufacturers, as financiers, as journalists, as statesmen, is due . . . the great expansion of submarine telegraphy. . . . You have covered both earth and sea with a subtle and instantaneous means of intercommunication which is destined, in its far-reaching consequences to bind together all the branches of the great English-speaking family, and to play an important part in advancing the general well-being and progress of mankind.

Cyrus W. Field, Twenty-Seventh Anniversary of the First Atlantic Cable, August 5, 1885.

WITH THESE words, Cyrus W. Field asked his roughly 250 assembled guests at the Richmond Star and Garter Hotel near London to raise a toast in commemoration of the laying of the Great Atlantic Cable. Similar in spirit to the previous cable banquets and telegraph soirées as well as the much bigger world exhibitions of the era, this gathering in August 1885 served to celebrate the century's course of progress and mankind's assumed mastery over nature; most important of all, the event represented an occasion to celebrate the actors of the various telegraph enterprises in a well-staged form of self-aggrandizement. Admittedly, the men assembled had reason to be proud. Already in 1879, experts estimated that about 100,000 miles of ocean cables were in operation, "more than sufficient to make three entire circuits of the globe."[1] By 1900, this number had almost doubled to 190,000 miles of ocean cables, and by 1923, the worldwide cable network consisted of roughly 366,000 miles of cables.[2] This communication network facilitated world trade and politics and allowed for a greater group of people to experience a unifying globe. A newspaper-saturated Euro-American public lived through every stage of U.S. President James Garfield's death struggles in 1881 and every minute of the excitement of the America's Cup in 1895. In 1895, ocean cables and landlines carried some 15,000 messages a day around the globe, and ever new records of speed and "instantaneity" confirmed the notion of a global simultaneity.[3] The latter half of the nineteenth century was a period of changes that succeeded so rapidly upon each other that to contemporaries the "miracles of yesterday," such as the Great Atlantic Cable, soon became "the familiar facets of to-day."[4]

The prime movers of these changes in the field of global communication were Cyrus W. Field's listed actors of globalization: the inventors, electricians, manufacturers, financiers, journalists, and statesmen who had promoted the wiring of the world. In addition, there were numerous small shareholders who carried the system financially and many faceless telegraph operators and postal administrators who ensured its daily operating. Navigating in and benefiting from a world shaped by global coloniality, these Euro-American actors structured and constructed the global cable system. They functioned as mediators of exchange, transfer, and translation in the expansion of a world-spanning ocean telegraph network. Continuously, they battled over the means and ends of global communication. Various and, at points, conflicting visions of an electric globe in union were tested and negotiated in cable wars, disputes on the relationship of science and business, and almost constant, albeit unsuccessful, attempts to make it possible for larger populations to use the telegraph. Previous works have portrayed the wiring of the world as an issue of technology, an issue of imperialism, or as driven by transcontinental business networks. This book inserts these men and women as actors of globalization. Through the actor networks' perspective, this study links macro-structural processes of imperialism and capitalist expansion with micro-structural processes of transfer and translation with regard to global communication. In the end, social and cultural considerations alongside political and economic issues played an equally important role in understanding the course of the wiring of the world and ultimately globalization processes.

THE GREAT ATLANTIC CABLE, NETWORKS, AND GLOBALIZATION

Technology was king in the mid-nineteenth century. The time of the wiring of the Atlantic, 1854–1866, was also the time governed by the engineering giants. Isambard Kingdom Brunel, Joseph Lock, and George Stephenson, for instance, built bridges, railways, tunnels, and ships, such as the *Great Eastern*, of gigantic reach and premises. It was an era when technology strongly connected with notions of progress. Indeed, contemporaries' technological optimism was rampant. The Euro-American world celebrated accomplishments in technology and "civilization" with the Crystal Palace Exhibitions in London and New York in 1851 and 1853, respectively. Ever since the laying of the first commercial submarine cable across the English Channel in 1851, many a mind was employed in finding the telegraphic passage across the Atlantic. Engineers and entrepreneurs proposed routes hopping from island to island in the north or the south of the Atlantic, as well as the direct crossing along Matthew F. Maury's Atlantic plateau. It was the Class of 1866, an Anglo-American group of engineers, entrepreneurs, journalists, and financiers, that finally landed the coup. In 1866 they durably connected the Old World and the New by electric wire. Their success with the Great Atlantic Cable enabled them not only to work on many additional cable projects around the earth, but also to determine the course of telegraphic science. More generally, it provided their passport into the globe's ruling elite. As their scientific and business networks, such as the Society of Telegraph Engineers or the Atlantic pool companies, dominated the global cable system, they emerged as gatekeepers of the wiring of the world. Based on their notions of monopolistic business structures, professionalism, and a Eurocentric and gendered worldview, they determined who and what was part of the world of telegraphy.

Wiring the world was essential for globalization processes. The growing networks of the Class of 1866, the Eastern and Associated Telegraph Companies, the Society of Telegraph Engineers, or those codifying the cable business according to standards of international law were manifestations of worldwide integration and coordination. These various social, technological, economic, and political networks were part of a more general growth and densification of international networks and spaces of transnational interaction in the nineteenth century. Yet, what does the actor networks' perspective tell us about the nature and course of globalization processes in the nineteenth century?

First, the network analysis unveils the intricate connections between actors whose intertwined lives scholars have previously left unexplored. It demonstrates that particular social networks that functioned on a local level directed transnational movements. Neither the lone entrepreneur, like Cyrus W. Field, nor ingenious inventors or engineers, like Samuel F. B. Morse or Charles T. Bright, can be credited with directing globalization processes; their success depended on transnational networks that relied upon preexisting local connections. Personal networks stemming from business and neighborhood relations or friendship and family ties played a central role in the success story of the Great Atlantic Cable, the buildup of the Eastern and Associated Companies network, the Siemens Brothers, the Society of Telegraph Engineers, and the creation and dispersion of knowledge about electric telegraphy. Moreover, these networks also represented an important source to fall back on in times of difficulty. Without the advocacy of John Brett, Cyrus W. Field's ambitious plans for an Atlantic cable would have failed in the early 1850s. Brett's far-reaching network, which reached into the fields of politics, finance, science, and ocean telegraphy, and the support of the American expatriates in London helped Field secure much-needed financial and scientific support from Great Britain. Without these preexisting local networks, the cable would have suffered the same fate as its rival enterprises. The idea of a North Atlantic or a South Atlantic route, for instance, never got beyond the initial planning stage.[5] In the end, the cable established connections between local networks instead of singular nodes. In the words of contemporaries, the cable brought two *worlds* and not solely two cities, New York and London, into instantaneous communication with each other.

In the case of the Class of 1866, initial business or neighborhood relations were essential for their success. Moreover, the emotional connections that held together the small community of British submarine telegraphers and the enterprising Americans of New York's Gramercy Park resembled, in their own words, that of family ties. With this term, its members alluded not only to the prevalent concept of Anglo-American kinship but also to the then still dominant form of enterprises as family undertakings.[6] The joint experience of twelve strenuous years of fighting for a scheme that ranked in public opinion "only one degree in the scale of absurdity below that of raising a ladder to the moon" had formed the Class of 1866 into a close(d) group.[7] After 1866, the "old Atlantic friends" carefully guarded access to their enterprise and created a distinct and exclusive group narrative. In the 1880s, new competitors, such as the American entrepreneurs Jay Gould and Western Union or James Gordon Bennett and John W. Mackay and their Commercial Cable Company, successfully entered the Atlantic market. Yet, they did not become part of the community of the "cable kings" of their time. They were neither invited to one of the many commemorative banquets and telegraph soirées nor considered for a telegraph memorial. This rank remained exclusively reserved for those early cable pioneers, who excelled not only in mastering the Atlantic but also in mastering their own historiography. Indeed, one of the reasons for the enduring legacy of those "old Atlantic friends" was their success in steering public commemoration. They orchestrated acts of commemoration and renewal of their communality through dinners, anniversary celebrations, and objects of material culture, such as framed sections of the original Atlantic cable. To this day, this shapes the remembrance of the wiring of the Atlantic.

In a second step, the network analysis illuminates the social dimension of globalization processes as it reveals the social background of those involved in globalization processes and so regional differences within the Euro-American setting. Although crowned as "cable kings," these predominantly male actors of globalization were part of the emerging European middle classes. Similarly to other industrialists in Europe or Meiji Japan, these middle-class entrepreneurs and engineers gained commercial and political importance and thus helped to steer the course of the respective nations as well as the wider world. On the British side, those engaged in the wiring of the Atlantic did not initially belong to London's gentleman's clubs, but were outsiders, geographically as well as socially. They were not part of the land-owning gentry that formally ruled the British Empire, but the sons of chemists, merchants, or schoolmasters. Regionally, many of them, such as the merchant John Pender, the scientist William Thomson, and the politicians John Bright and Richard Cobden, originated from Britain's industrial north and the Manchester area, Scotland, or even Ireland.[8] Strongly influenced by the industrial setting of Britain's north, the philosophy of Manchester Liberalism, and the high level of Scottish education, they joined the ranks of many other Scotsmen moving the course of the British Empire at the time as merchants, missionaries, or soldiers.[9] Matters were similar in the case of the German actors. The Siemens brothers and the various German engineers connected to the Siemens brothers' cable undertakings, such as George von Chauvin or Ludwig Löffler, were important players in the British-dominated world of cable engineering. Moreover, similarly to their German peers, such as the Krupp or the Thyssen families, the Siemens brothers represented a new type of transnationally active engineer-entrepreneurs who economically and politically influenced the course of the German Empire in a global world.[10] From a European perspective, representatives of the middle classes, and not the land-owning gentry, were the movers of globalization projects. Their success furthered their upward mobility into the ranks of the globe's governing elite. After 1866, many members of the Atlantic cable undertaking enjoyed the benefits that came with carrying the title Sir or Baron, the distinction of the highest decoration in France, the *Légion d'honneur*, or the Order of Saint Jago da Espada from the Portuguese king.

The situation was slightly different for the American protagonists. They emerged from a national context in which a broad bourgeois consensus blurred social distinctions and seemingly allowed for greater social mobility. Certainly, Cyrus W. Field, Abram Hewitt, and especially Peter Cooper, who had been born into very poor circumstances, would have considered themselves part of a middle class and archetypes of the self-made man. Nevertheless, already before 1866, they represented the elite of American society. It was only in relation to Europe that they were not received on equal footing with the land-owning gentry. Furthermore, the exclusion of the Irish-American cable entrepreneurs John W. Mackay and James Gordon Bennett from New York's gentleman's clubs showed that American society could be even more restrictive than the old peerage systems of Europe. Still, the Class of 1866 exemplifies how the pioneers of globalization projects, such as the wiring of the world, were initially outsiders to Europe's ruling class but soon moved into its circles as that ruling class itself expanded.[11]

Finally, the actor network perspective reveals some of the logic of the mapping of globalization processes. It provides answers to the question of why these preexisting local networks went global and not national or regional and thus supports the thesis of early global modernities.[12] The case of the wiring of the Atlantic suggests a combination of economic, political, and cultural reasons interrelated with the fact that they already *were* connected: John Pender traded as a merchant in U.S. cotton, and Samuel F. B. Morse had studied art in London and spent time as an artist in Paris and Italy before he embarked on the project of transatlantic telegraphy. Although global entanglements attained an entirely new quantity and quality in the nineteenth century, they followed well-trodden paths.[13] In 1914, A. W. Holland argued that a cable between Greenland and South America would have been just as feasible as one between Great Britain and the United States. But most likely the cable would have remained unused and allowed to rot away because no common cultural, social, or political codes existed: "The Eskimos in Greenland would not want to know anything about the tall and strange and copper-coloured inhabitants of Tierra del Fuego, and even if they did, they would not be able to understand each other's signs."[14] Until the Pacific cables in 1902 and 1904, the ocean telegraphs were never a means of exploring new connections or markets, but only instruments to accelerate existing economic, political, and cultural connections. Globalization in the nineteenth century followed global maps of earlier times.

THE 1870S AND THE CONSTRUCTION OF A GLOBAL NETWORK

The story of the wiring of the world commences in the 1870s when a cable boom and rapid network expansion followed the Great Atlantic Cable; in this decade, ocean telegraphy established itself as a key technology for a unifying world. Moreover, in this period, important systemic decisions were made, and alternative concepts of how to govern the global communication system became institutionalized. Most of all, the Class of 1866 ruled the global communication system in the 1870s. In this period, the business consortium around Atlantic protagonists John Pender and James Anderson outran almost all competitors on the global market, and the Atlantic engineers, such as Charles T. Bright and Samuel Canning and scientist William Thomson, excelled as consulting engineers and presidents of the newly established Society of Telegraph Engineers. The "cable kings" of the time extended their influence and consolidated their telegraph empires.

The prime mover during this time was the former cotton merchant and cable financier John Pender. Together with James Anderson, former captain of the *Great Eastern*, as well as a group of other businessmen from the Class of 1866, he set up a series of companies that later merged as the Eastern and Associated Telegraph Companies. Operating from London, which came to be the center of ocean cabling, they facilitated communication with India, East Asia, Australia, and South Africa. By 1900, their business consortium controlled four-fifths of the world's ocean cable traffic. The remaining connections were under the control of a number of smaller rivals, such as the Great Northern Telegraph Company, or national governments. These connections predominantly covered shorter state cables, such as those across the English Channel and the Irish Sea, but also the Atlantic pool's North Atlantic connection, which was by that time leased to the American-based Western Union. In the end, after 1866, these actors had built up a monopoly, which had, "like a huge octopus, fastened its tentacles upon almost every part of the eastern and southern world."[15]

Two factors were essential for this success story of the Eastern and Associated Companies: the social network of the Class of 1866 and the vertical business network established on behalf of the Atlantic connection. The rapid expansion of ocean telegraphy in the 1870s was an undertaking carried out by some of the key players of the Atlantic connection. Multiple directorships within the Pender-Anderson business consortium demonstrated that the system was managed by a small group of former Atlantic entrepreneurs, such as John Pender, William Montague Hay, Daniel Gooch, and James Anderson. Besides their Atlantic fame, their success stemmed from the creation of a vertically integrated business network during the Great Atlantic Cable enterprise that controlled all processes of ocean telegraph communication: cable manufacture, laying, and operating.

The 1870s also witnessed the rise of the Pender-Anderson consortium's most important rival: William Siemens and the Siemens family's cable manufacturer Siemens Brothers. As a subcontractor for R.S. Newall, this Anglo-German engineer company had manufactured parts of the 1858 Atlantic cable and, between 1868 and 1870, established a rival connection to India via the land route. In the 1870s, Siemens recognized that the Class of 1866's main market advantage was based on their control over a vertical business network. The only way to beat the established players was to offer not an alternative provider of communication using cables laid by others, but an alternative network with its own capabilities to manufacture, lay, operate, and finance an ocean cable. For this purpose, Siemens Brothers had their own cable ship, the *Faraday*, constructed and fitted solely for the purpose of cable laying. Although the Siemens brothers' attempt to establish a rival cable company, the Direct United States Cable Company, failed, the success of future rivals on the Atlantic market relied upon their approach of business diversification. The Americans Jay Gould and John W. Mackay, James Gordon Bennett, and also French and German state enterprises employed alternative cable operating companies as well as manufacturers and cable layers. In this way, Siemens Brothers laid eight of the thirteen Atlantic cables that were in working order by 1900.[16]

The cable war between John Pender and William Siemens in the 1870s, however, did not just affect the business of submarine telegraphy and access to the Atlantic market; its outcome also defined the market and scientific structures as such. John Pender and William Siemens not only headed rival enterprises, but also stood for opposing market ideologies and different professional fields. Primarily, the cable war installed what the *Glasgow Herald* called a system of ruinous competition followed by amalgamation.[17] This pattern repeated with every new rival on the Atlantic market until the 1920s: through the employment of various tactics, such as price wars, buying up shares, or public denouncement of the rival, the Anglo-American Company forced all new competitors on the North Atlantic market into a joint working agreement, the Atlantic pool, and a monopoly. On a systemic level, this cable war represented a "drama of progress" between a system of monopoly aiming at capital accumulation, as pushed by John Pender, and a system of competition allowing for technological progress, as proposed by William Siemens.[18] The year 1877 marked the final break with the market liberalism of Atlantic cable supporters John Bright and Richard Cobden. After the financial crisis of 1873, market liberalism had seen its prime. The cable business was increasingly in "visible hands."[19]

This systemic change in entrepreneurial structure did not influence the cable agents' attitudes toward state involvement in the cable business. According to them, states could play a role in providing infrastructure through landing rights or ocean sounding data or coordinate technological and user standards through the International Telegraph Union (ITU), but they were not to regulate the ocean cable business through tariffs or taxes. The entrepreneurs remained against any kind of state involvement if it meant an infringement of their rights. When C. H. B. Patey, secretary of the British Post Office and designated president of the ITU's London conference, attempted to pursue an independent tariff policy in 1879, he met harsh opposition. James Anderson, in concert with other telegraph company managers, made every effort to deter Patey from enacting individual state policies and to minimize the Post Office's influence in general. In the end, they were successful. Patey decided to drop policies that antagonized the companies and instead mediated between them and the ITU. The 1870s marked a period of a trial of strength not only between different market systems but also between the British Post Office and the submarine machinery in London, and hence between the state and the market. Although, apart from the United States, most landline systems were run as state enterprise, the ocean cable business remained firmly in the hands of private entrepreneurs.

Additionally, when Pender took over the Siemens' Direct United States Cable Company in 1877, this victory not only installed a monopolistic market system but also disconnected the engineer from the enterprise of ocean telegraphy. In his war with Siemens, John Pender was stern in his opinion of the scientists; they were "all very well in their place, but their place was in the laboratory, or at any rate not in the directory of big business."[20] As a result, the importance of research and development dwindled in the ocean telegraph business. Soon after its breakthrough, submarine telegraphy became a stagnant technology, if only seen in relative terms.[21] But how much did the entrepreneurial decision for less research and development play into the British industrial decline? Commencing in the 1870s, British industry was overtaken by Germany and the United States, and the world became increasingly multicentered. *Prima facie*, the example of the Siemens-Pender controversy confirms the arguments of scholars in favor of the decline. It supports the notion of an "entrepreneurial failure" based on a British industrialist who did not innovate new technology with the alacrity of foreign competitors, in addition to British scientists and engineers who did not run an enterprise.[22] Although the technology of submarine telegraphy changed very little over time, the Eastern and Associated Companies was very successful as a result of its monopolistic structure. Moreover, Pender was quite innovative as an entrepreneur, as the launch of the Globe Trust and Telegraph Company in 1872 showed. The Globe was an investment trust company, which spread its shares over a number of companies and weakened the financial risks of cable laying. It was an important entry point for many small investors into the world of global ocean telegraphy. From a technological point of view, there was no need to innovate. Even rivals, such as the Commercial Cable Company, that employed the "latest" technology and provided faster cables did not manage to bypass the Anglo-American Telegraph Company. Market positions were negotiated via a monopolistic system of landing rights and not by deploying the latest technology. When ocean telegraphy became outdated after the First World War as the dominant means of global communication, it was not surpassed by a more innovative rival but by an entirely different technology, wireless, which allowed participants to disregard the question of landing rights.

At a time when the economic and political structuring of the global communication system was contested, its cultural and legal premises were also put to trial. The 1870s were a foundational period not just technologically and economically. The decade also witnessed the institutionalization of two alternative ideologies on how to govern the world as a whole: cosmopolitanism and internationalism. In the 1850s and 1860s, there were two dominant explanatory models for a unifying globe and its peaceful implications. Within the European context, contemporaries believed in the peaceful forces of Manchester Liberalism and a world ruled by international trade and dependencies. In the American setting, by contrast, the more cosmopolitan idea of a *Societas Christiana* prevailed. The decisive decades in terms of "governing" a global system of submarine telegraphs were the 1870s and 1880s. The ideologies of cosmopolitanism and internationalism were "institutionalized" in the Society of Telegraph Engineers, the codifications of international law, and the regulations of the ITU.

Almost simultaneously in the early 1870s, three institutions were established to conceptualize the electric union: in 1871, the Society of Telegraph Engineers, and in 1873, the Institute of International Law and the International Law Association. The Society of Telegraph Engineers was an association of all telegraph professionals around the world. Its founding in 1871 mirrored the acclaimed importance of ocean telegraphy at the time. Simultaneously, it advanced the (ocean) telegraphers' strategy of meeting the challenges connected with the transboundary movements of ocean telegraphy, as well as the threat posed by the monopolistic system of John Pender, which disconnected them from the business of submarine telegraphy. According to William Siemens, this organization could be nothing else but a "cosmopolitan institution." It was bound in its purpose to serve the great network of international telegraphy that extended "to every portion of the civilized and semi-civilized world."[23] The society drew from a transboundary understanding of telegraph engineering and its knowledge, following its own policy of globalization. The many foreign members and local honorary secretaries dispersed all over the world played an important role in establishing a global community of telegraph operators and engineers. The society's president, William Thomson, believed that this community of telegraph operators should be employed to conduct research on a global scale simultaneously. The network of the ocean cables was to be joined into one big research laboratory to explore, for example, the aurora borealis. From its London-centric position, this "cosmopolitan" community attempted to influence not only the image of the telegraph agent but also the generation, homogenization, and regulation of telegraphic knowledge and standards. The society's ultimate goal was to regulate and negotiate through its network all matters concerning telegraphy anywhere in the world, scientifically.

The two international law institutions, the Institute of International Law and the International Law Association, represented the opposite side. They had been established by a group of law reformers who believed that *l'esprit d'internationalité* would teach the global community of states that it was bound by common interests, which needed to be regulated by international institutions and, most importantly, international law. The establishment of these two organizations coincided with the launching of a large number of international organizations in the second half of the nineteenth century to regulate and structure the world as a whole. For the field of communications, the most important representatives within this group were David Dudley Field, brother of Cyrus W. Field, and the French law reformer Louis Renault, who were founding members of the International Law Association and the Institute of International Law, respectively. Renault and Dudley Field were key players in various attempts to codify the global media system and transform the world into an international society. The ITU, established initially in 1865 to coordinate the landline traffic, represented one of the key elements in their theory of an electric union, and Renault wrote many of its early contracts. The main focus of these law reformers was to regulate the protection of submarine cables in peace and especially in times of war; the world was not to be disconnected. Their major, but only, breakthrough was the International Telegraph Convention for the Protection of Submarine Cables of 1884.

Neither of these two global visions of internationalism and cosmopolitanism was truly universal for all practical purposes. Both conceptualized and enacted a global electric union from the position of Euro-American supremacy and its definition of "civilization." Even within the "cosmopolitan" Society of Telegraph Engineers, there were very few members, except for a select number of Japanese engineers, from outside of the Euro-American realm. In contrast to the cosmopolitanism proclaimed by the Society of Telegraph Engineers, the two law institutions based their model for structuring the world as a whole on the instruments and institutions that the nation-states provided. Nevertheless, although the 1870s and the outcome of the Siemens-Pender controversy marked a milestone in forging the global media system economically, the decade was still a relatively open period in terms of its ideological foundations marked by a parallelism of a Eurocentric cosmopolitanism and internationalism.

THE 1880S AND 1890S: THE SYSTEM'S SATURATION AND POPULARIZATION

In the 1880s, the system of global communication via ocean telegraphy became ever more defined. The cosmopolitanism of the Society of Telegraph Engineers failed as a model to structure the world, and the telegraph system as such matured and saturated. At a time when processes of global integration accelerated ever more quickly, the system, which provided the basis for these processes, became increasingly stagnant. No new cable routes of importance were exploited; only old ones were duplicated. Cable rates were fixed at a certain price and did not change until after the First World War. Yet, the 1880s were still a relatively turbulent time for the ocean cable system because far-reaching decisions were made regarding the networks' ideological structuration as well as a merger of content and technology. Decisive events were the cable war between the Atlantic pool and the Commercial Cable Company between 1884 and 1888 and the realignment of the Society of Telegraph Engineers beginning in 1884 and culminating in its renaming in 1889 as the Institute of Electrical Engineers. The outcome of the cable war not only set an end to tariff changes, but also established a duopoly. In the face of this Anglo-American multicenteredness on the Atlantic market, all future rivals had to take sides. Finally, the 1880s were a time when the global imaginary, as it was expressed and mediated through the ocean telegraphs, started to become popularized. Newsmakers, such as James Gordon Bennett, brought the world to their readers, and changes in the shareholders' structure brought the ocean telegraphs as a financial investment home to the ordinary public.

Ideologically, the failure of the cosmopolitanism of the Society of Telegraphy Engineers overshadowed the 1880s. In 1884, the year of the Convention for the Protection of Submarine Cables, the Society of Telegraph Engineers altered its program and name, and in 1889, it fully realigned itself as the Institute of Electrical Engineers. This move secured the survival of the society as a learned institution by broadening its membership body and thematic approach. But it also signified a slow demise in its claim to worldwide leadership in electric science. The number of foreign members decreased significantly. In 1911, the category was abandoned altogether. By the turn of the century, the Institute of Electrical Engineers was only one among many within an international system of science.

The Society of Telegraphy Engineers had failed as a global regulatory mechanism. Its cosmopolitan, yet London-centric, structures could meet neither the challenges of technological progress and development happening outside of the British Empire nor the increasing institutionalization and internationalization of science or the growing multicentric world. In the long run, the engineers lacked the substance to uphold an almost supranational standing. In the 1880s, the cable engineers' monopolistic position was weakened not only by their defeat in the Pender-Siemens controversy, but also by the deaths of some of their most important representatives, such as William Siemens and Charles T. Bright. Additionally, they were also overtaken by new inventions and innovations, such as electric lighting and the telephone, which weakened them more than rival cable companies did. Moreover, unlike the ocean cables, these new technologies were not inherently transnational, but found their initial applications and scientific organization in the local and national realms. With the demise of the society as a global key player, a cosmopolitan worldview gave way to internationalism as the dominant structuring ideology of the electric union. Given the vast expansion of international organizations to codify time and space and to regulate governments at that time, the internationalist worldview also structured the global system as a whole.

In the 1880s, entrepreneurs also made important decisions regarding markers of disconnection and connection. As networks function in their constitution through a clear distinction between the net and the in between, the actors always also defined the disconnected through their decisions with regard to tariffs, routes, and business policies.[24] From the providers' perspective, these decisions were taken in the 1870s and 1880s and remained, after this relatively turbulent period of systemic negotiations, almost entirely unchanged until 1914. The cable to South Africa in 1879 was the last major ocean cable for two decades, and during this time, the face of the ocean cable system did not truly change. The laying of submarine cables across the Pacific in 1902 and 1904 was the first new important route since 1879. Before the Pacific cable, old routes were merely duplicated or even laid three times over. By 1900, thirteen working cables crossed the North Atlantic, making it the densest ocean crossing in the world. This relative cessation in ocean cable expansion resulted from the entrepreneurs' decision not to lay cables to places for other than commercial reasons unless "Governments care[d] to subsidize them" and the governments' reluctance to do so until essentially 1898 and the rise of cable nationalism.[25] An important side effect was that the world's coastlines were almost entirely divided up between the different cable companies. The policy of giving out landing concessions to land a cable lasting several decades furthered a monopoly for a particular cable company.

The charges for telegrams were another means to steer connections and disconnections. The run of cable tariffs illustrated that submarine telegraphy never was a means of social or mass communication. Certainly an intrinsic quality of the medium, its form of brief and public cablegrams, did not allow for the same degree of personal intimacy and explanatory length as a letter. However, providers also placed restrictions on telegraphic usage. They followed, in the words of James Anderson, a theory of global communication that stated that en masse social interaction decreased in relation to distance. There was no global sociability that would make a low tariff profitable for the telegraph companies. Throughout the years, tariffs had only been lowered as a means to force competitors into bankruptcy and a joint working agreement, not to make telegraphy more affordable for larger numbers of people. Indeed, after

September 1, 1888, and the setup of a stable duopoly on the Atlantic market between the Atlantic pool and the Commercial Cable Company, tariffs remained stable until 1923 at one shilling per word. Translated into contemporary wages, this amounted to a small fortune for somebody from the working or lower middle classes. Even Henniker Heaton's crusade for a universal cable penny post in the 1890s and 1900s remained unsuccessful. He could neither convince the cable companies that the increase in traffic would even out their loss in revenues connected to a lowering of rates, nor persuade the governments that the benefits of global social communication via cable made the political project of buying out the cable companies and nationalizing the ocean lines worthwhile. Consequently, the providers' establishment of the submarine telegraphic system meant that although the ocean cables challenged temporal and spatial boundaries of time and distance, boundaries of social hierarchy were reinforced.

The wiring of the world further continued in the 1880s. Although with regard to ocean cables, decisions over connections and disconnections were made irreversibly early on, the network still expanded on land. New technological inventions, such as the telephone, created ever denser local and regional networks that became increasingly intertwined with the ocean cables. Atlantic cable newcomers, Jay Gould, John W. Mackay, and James Gordon Bennett Jr., had market advantages over the Atlantic pool: they also ran a landline system and controlled several newspapers in the United States. This way, they could create and direct their own customers to the networks that they controlled. Hence, once the oceans were spanned, network expansion happened between 1879 and 1902 predominantly in the local, national, and regional context. Both communication systems, however, must be seen as part of one system. The spread of terrestrial lines into ever smaller communities and the advent of the telephone, in addition to developments in international news and the cables' financing structure, allowed contemporaries outside of commercial and political centers, such as London, New York, or Buenos Aires, to also experience a global connectivity. They did not need to be directly connected to ocean cables.

Indeed, direct access to global telegraphing and the experience of a global communication network were two different things. Although the providers controlled the access to global communication, they could only partially influence, for example by means of press work and selective release of information, the sphere of an experience and understanding for the importance of global connectivity among those whom they actually disconnected. Certainly, the ocean telegraphs were not a means of mass global communication and not a Victorian version of the Internet. Yet those who could not afford to telegraph were not entirely disconnected. News and cable financing still popularized a global imaginary connected to the worldwide network of ocean telegraphy. Network structures were taken to extremes from the 1880s on when the combined use of developed social, business, and technical networks truly allowed for a worldwide reach. At a time when news became a manufactured good, James Gordon Bennett Jr. used his Euro-American newspaper system and his personal cable system, the Commercial Cable Company, to make global headlines.[26] He brought international news, such as the "discovery" of British missionary David Livingstone or the American explorer George W. DeLong's voyage to the North Pole, home to a Euro-American public. Bennett thereby also enjoyed the benefits of vertical control over a news-transmitting cable company and a news-making enterprise of several papers on two continents and an army of journalists he could order back and forth around the world thanks to the technical network of cables. Vertical control, however, no longer solely concerned the management of the cables but also the management of their content and thus the management of news.

However, with the study of news, scholars are left guessing as to who actually read their global newspapers. Here lies an advantage in studying financial connections and the shareholder lists, which reveal exactly who owned company stock. In 1872, John Pender launched the Globe Trust and Telegraph Company, an investment trust company that mitigated the financial risks of cable laying by spreading it over a number of companies. Thus, cable company shares became an option for the small and inexperienced investor of the emerging middle classes. Among Direct United States Cable Company shareholders, an increasing number of small shareholders with less than one hundred shares came from small to mid-size towns, often in agricultural areas, outside of Britain's industrial and trading centers. They were gardeners, art students, or shoemakers and certainly not among the primary customers of the ocean cable companies. Nevertheless, they owned part of the company. Toward the turn of the century, another trend emerged: more and more shareholders were female. Women had generally been excluded from the system of ocean telegraphy. Although female telegraphers were employed on the landlines, this was not the case on the submarine connections. Furthermore, women were also not very familiar with the art of ocean cabling, that is, with sending telegram messages. Nevertheless, women became one of the most important shareholder groups around the turn of the century. Striving to challenge a male-dominated society, they saw the stock market as a means for their new freedom and apparently took a particular liking to the ocean cables. In 1887, 25 percent of the shareholders were married women, widows, or spinsters; by 1909, this number had increased to 44 percent. By the end of the nineteenth century, the notion of global communication as a sound investment had spread beyond the capitalist classes directly engaged with, and communicating through, the medium. Barriers that had been created by the providers through tariffs, in combination with social structures, were undermined as society changed in the late nineteenth century. People consumed global products, such as news or cable shares, in different ways and experienced a global connectivity through a variety of channels.

Race remained the only barrier that was reinforced. Contemporaries saw telegraphs as an expression of Euro-American "superiority" and "civilization." In such a technologized understanding of civilization, as promoted by news gatherers like Gordon Bennett, the spread of news and information about faraway places through the telegraphs' instantaneity reinforced notions of the "dark continent" or "faraway" places that lay outside of the system.[27] As a consequence, those who did not belong to that "civilized world" were automatically deemed unfit to engage with submarine telegraphy in any way. This unfit status could be applied to the Newfoundland fishermen as well as the Indian colonial subjects. From the point of view of communication, the Newfoundlander remained utterly disconnected, and it was only in the late 1880s that "natives" were slowly employed to work on the "complicated" ocean lines. Nevertheless, those "unfit" Newfoundlanders were drawn into dependency on the global system, as the preemption scandal of 1873 demonstrated. In a joint effort, Newfoundland politicians, merchants, and the general public, including the ordinary fisherman, caused ocean cable shares to plummet on the London stock market in 1873 only months before the worldwide financial crisis. The successful manipulation of London stock from Newfoundland caused a great loss in profits for the Anglo-American Telegraph Company and large revenues for the Newfoundlanders. Globalization was entailed in active participation in networking and also had taken on the form of a perspective.

Finally, the popularity of jokes on "playing chess by cable" around 1900 illustrated another facet of how far (ocean) telegraphy had become popularized beyond its actual use.[28] The turn of the century was a time when official Anglo-American chess matches across the Atlantic marked the technology's ultimate breakthrough in terms of its incredible speed, its technological possibilities, and man's seemingly absolute mastery over time and distance. In 1858, the transmission of Queen Victoria's congratulatory telegram had taken sixteen hours; now move and reply were transmitted within two minutes over 3,483 miles.[29] These events vividly illustrated the vast extent of global integration since the mid-nineteenth century and the beginnings of a global communication system. Moreover, it also characterized a period when the ocean telegraphs had arrived as a technology-in-use in the midst of a Euro-American society. Long-distance telegraphy was popularized, if not entirely in the fields of social communication, then certainly in finance and news reports. Small investors from outside of industrial centers, as well as women, who were inspired by the ideals of the New Woman, expressed their understanding of the telegraphs' importance for global interrelations as well as their new sense of liberty by buying ocean cable shares. Served via cable by journalists, politicians, merchants, and telegraphers, contemporaries learned new conventions of simultaneity and a new sense of chronology and could thus grasp the notion of a global imaginary as suggested by the cable maps.

THE 1900S: CABLE NATIONALISM REVISITED

The global imagination of the electric union changed again after 1900. Spurred on by the Spanish-American War in 1898, Euro-America's industrialized, imperial nations started to conceive of the ocean telegraphs as a means of national security and imperial control on a large scale. Imperial rivalry started to grow, and mass migration made the ability to 'talk' globally exigent. Consequently, the governments of France, Germany, and the United States interpreted matters of ocean telegraphy as an integral part of their foreign policy. At the same time, Great Britain was caught in the debate over a Pacific cable as the centerpiece of an "All-Red route," a girdle of cables "round about the earth" that only touched upon British territory. Ocean cables had always been part of narratives of national progress, but now they turned into a "national necessity" from a government's perspective.[30] After almost three decades of private financing, governments returned to a policy of subsidizing cables that the companies considered commercially unattractive. This policy's most vivid expression was the Anglo-American scramble for a Pacific cable around 1900. In 1914, this cable nationalism brought to an end the international regulatory system of the electric union that had been established with the ITU and the Convention for the Protection of Submarine Cables in 1884. Immediately upon the outbreak of World War I, the "German" Atlantic cables, which actually operated in working agreement with the American-based Commercial Cable Company, were cut. This demonstrated that *l'esprit d'internationalité* could not prevent the entire world going to war with each other. The Convention of 1884 had only been an ostensible success: it had no power over the belligerents and the ocean cables in times of war.

Previous works have portrayed this period through a national and imperial framing.[31] Proponents of this approach argue that the development of submarine telegraphy was primarily driven by strategic purposes and that the cables functioned as tools of empire.[32] Approaching the matter from the actors' perspective, this study showed that this interpretation works from a government perspective but not from the cable actors' point of view. Given that the ocean cables, in contrast to landlines, were predominantly private enterprises, this is an important aspect. Entangled in the worldwide reach of the ocean cable system, engineers, entrepreneurs, and operators were thinking and acting beyond the nation-state. They employed nationalism as a strategy that sustained their system despite changing parameters and brought it through the First World War.

From the beginning of ocean telegraphy, the companies and their actors took on neutral or intermediary functions between the respective nation-states to be connected. They saw themselves as providers of a service that could be equally used from both ends. Their loyalty and their identity were primarily grounded in their profession, and they acted on the proposition that peace was beneficial to (their) commerce and their relatively nonterritorialized position in maritime space. Still, the entrepreneurs recognized nation-states as the fundamental governing structure, the building blocks of the globe, and acted accordingly. During the

Alabama Claims crisis and Newfoundland's elections, the cable engineers and operators even took on the function of "cable diplomats." Throughout the nineteenth century, there were several instances when they would actively seek state support. The entrepreneurs turned to government officials for help to influence the discussions of the ITU, where they had no voting rights, or the regulation of landing rights.

This rather loose and laissez-faire relationship between companies and national governments changed in the twentieth century. Especially after the Spanish-American War of 1898, companies enforced nationalist strategies. Responding to the notion of ocean telegraphs as a national necessity, they painted their international maritime business in the most national colors; their response to nationalism, however, looked different with every national context they attempted to serve with a cable connection. They found different local translations for a globally applied strategy: companies, single cables, managers, and operators, such as George (von) Chauvin, changed nationality as they saw fit; nationalism was a strategy they employed to sustain and protect their businesses, which functioned, due to landing rights questions, along entirely different lines. The restrictiveness of landing monopolies secured exclusive rights for the individual cable companies for decades. The Anglo-American's landing right in Newfoundland, for instance, expired in 1904, fifty years after the Newfoundland government had issued it. Consequently, entrepreneurs had to enter into joint agreements if they wanted to lay a cable at all. The most vivid example of such strategic nationalism was the Commercial Pacific Cable Company employed to lay the American Pacific cable. It was sold to the American Congress as a genuinely national enterprise in the prewar period. After the war, Congress learned that 75 percent of this allegedly American enterprise was foreign owned. Similarly, the cooperation between German and Dutch entrepreneurs to lay cables in the East Indies and between German and French entrepreneurs to lay a cable across the South Atlantic showed that purely national enterprises were not feasible.[33] Although the cable agents may have been privately patriotic, they were certainly not nationalistic.

The entrepreneurs' market advantage was maritime space. Laying and operating ocean cables, they were taking a middle-ground, or *maritime*, position, from which they recognized the international, but not the national, system as their frame of action on a global scale. It mattered little for their scope of action what kind of governmental regulatory apparatus they were facing: that of an empire such as Great Britain, a nation-state such as the United States, or a colonial government such as Newfoundland. Within this international system, the various nation-states represented other actors with which to act in concert or to act against. Strategic nationalism, especially after 1900, was their response to a shift in the authoritative discourse within that system, but it neither changed their working methods nor truly infringed upon their realm of action within that system.

The entrepreneurs were not the only ones to employ nationalism as a strategy. Many cable reformers in the 1890s and 1900s argued for the nationalization of ocean cables, but they did not necessarily do so to further nationalistic notions. Henniker Heaton, for instance, was truly cosmopolitan in his crusade for a universal cable penny post. In his view, ocean cables should first be nationalized to break the companies' monopoly. Then, as a second step, he wanted to eliminate national borders within the global media system and the various tariffs and taxes applied and regulated by the ITU. These were "[t]he chief obstacles" in his fight for affordable global communication for all.[34] Sandford Fleming employed nationalism in a similar matter. For more than two decades, the Canadians fought for a Pacific cable connecting the imperial sister colonies Canada, Australia, and New Zealand. While Fleming argued for interimperial union, Canada's main motive was to strengthen its position in the Pacific as a means to maintain its ground against U.S. westward expansion as well as to emerge as a British junior partner. For a long time, Fleming's most ardent opponent was cable magnate John Pender, who in turn convinced British government officials that such a cable certainly was not part of his or national interests. The cable nationalism of the 1890s and 1900s was a question of point of view and local embedding. In the case of the global media system, processes of globalization and nationalism were not opposed to each other or mutually exclusive. Rather, there existed an interdependency as one was employed to sustain the other.

The end of the ocean cables and the monopoly of the cable companies within the system of global communication came neither because of internal rivalry nor because of mounting nationalism and the divisive effects of the First World War, but essentially with the advent of a new technology: wireless telegraphy.[35] After the technology's development in the 1890s by Guglielmo Marconi and Ferdinand Braun, it was soon widely used. In 1902, Marconi sent his first transatlantic radio message.[36] In 1903, the German company Telefunken was established, and in 1907, the French Compagnie Générale de Radiotélégraphie was established.[37] As early as 1900, Guglielmo Marconi occupied an important market niche by offering wireless ship-to-ship and ship-to-coast communication, something the cables could not provide. In fact, this "niche of the sea" was wireless's biggest potential market.[38] Wireless entirely changed the construction of maritime space in interrelation with the global communication system. Simultaneously, it allowed for a denser structure of ocean space by control over ships in mid-ocean and yet made maritime space irrelevant.

This redefinition of maritime space particularly affected the agents' position. For more than half a century, the protagonists of ocean telegraphy had acted from their intermediary position between the continents and the nation-states in maritime space. The Class of 1866 lived off its fame as "conquerors" of the Atlantic and so controlled this maritime communicational space. With wireless, maritime space was not conquered, but transgressed. In a certain sense, maritime space became irrelevant. The messages were no longer bound to the routes of the cables on the ocean floor and their landing upon rocks in mid-ocean or the ocean shores; wireless communication was dematerialized and also undirected (hence *Rundfunk*, meaning broadcast). This also meant that the course of communication and the structuring of the system became less easy to control, as anyone, from the amateur radio telegrapher to government institutions, could tap the wireless or employ their own wireless system. Germany's initial interest in wireless had been to circumvent British cable routes. During the First World War, the German news agency Transocean worked with wireless transmission to successfully broadcast its news between Japan and the United States.[39] With these new technological developments, the telegraph companies' spheres of influence and the course of nationally defined communication were no longer regulated through landing right monopolies, cooperation among themselves, and the agents' position as neutral intermediaries or diplomats between two shorelines. In fact, from a systemic point of view, it was wireless that allowed for the nationalization of international news by returning the power over news back to the hands of the various nation-states. For the ocean cable system, in the end, wireless led to the (un)systematic change that they could not cope with.

However, it was not until the 1920s that the cable companies acknowledged radiotelegraphy as a serious competitor. But eventually they had to face the fact that their cables were getting old. In 1920, nine of the seventeen cables across the Atlantic were more than forty years old. Due to the fateful decision in 1877 to reestablish the Society of Telegraph Engineers as the Institute of Electrical Engineers, science had played only a minor role in the technology of submarine telegraphy. The makeup of the cables had changed little. In a frenetic attempt to keep up, the cable companies conducted new research targeted at making the cables faster without making them necessarily heavier. In 1926, when the Pacific cable was duplicated, it could handle 250 words per minute and was ten times faster than the old cable.[40] Nevertheless, especially after the successful implementation of short-wave radio technology, the cables could not compete with wireless, which was not only cheaper but also faster. By 1927, about half, and in some places even two-thirds, of the cable traffic had migrated to wireless transmission. As a consequence, the Eastern and Associated Cable Companies announced that it would sell its ocean cables and go out of business. To forestall the financial collapse of the cable companies, the British government invited them to an imperial cable and wireless conference in January 1928, where the merger of the cable and wireless systems was decided. In 1929, the newly formed Imperial and International Communications Ltd. took over the business of Marconi's Wireless Telegraph Company, the Pacific Cable Board, the Imperial Atlantic cables, the Eastern and Associated Companies, and twenty smaller companies. In 1934, the company was renamed as Cable and Wireless, and by the outbreak of the Second World War, it possessed "the most extensive system of world communication ever to exist under control of a single body."[41] In 1932, the ITU also acknowledged this systemic change and renamed itself as the International *Telecommunications* Union.[42]

Today, in the age of Web 3.0 and satellites, ocean cables, in the form of fiber optics, have gained new relevance. Moreover, the wiring of the world is, as the example of the recently laid ocean cable between the Arabian Peninsula and East Africa illustrates, far from over. As in the previous centuries, a global communication system plays a central role for facilitating processes of globalization and allowing not only for economic and political, but also for social and cultural exchange and interdependence. Today, however, people do not play chess via cable, but online and in a virtual global community. In the nineteenth century, this virtual global community was strictly defined according to Euro-American conceptions of race, class, and gender. Weingärtner's vision of a "Torchlight Procession Around the Globe" with Caucasians, Asians, Africans, and Native Americans dancing around the globe with a telegraph cable was highly deceptive. And yet, as this book shows through its actor networks' perspective, there was never just one vision of whom or what a united electric world should entail. Various actors, such as Cyrus W. Field, John Pender, William Siemens, and Henniker Heaton, battled over the ends and means of global communication. Their respective national, professional, and cultural backgrounds played an important role in shaping their interpretation of *Weltcommunication*. In the end, processes of globalization in the nineteenth century, as part of the world, depended just as much on a global coloniality driven by imperial and commercial interests as on its actors. These actors moved within this world and employed their respective cultural and social interpretations to shape it.

APPENDIX

ACTORS OF GLOBALIZATION[1]

WILLIAM ADAMS (1807–1880), United States, Presbyterian Minister

William Adams was a Presbyterian clergyman and close friends with Cyrus W. Field and Samuel F. B. Morse. He was also a member of the Evangelical Alliance and in 1871 visited Russia to plead the cause of the Protestant dissenters in the Baltic provinces.

SIR JAMES ANDERSON (1824–1893), Great Britain, Mariner, Telegraph Manager

The Scotsman James Anderson joined the Merchant Navy at sixteen and in 1851 entered the service of the Cunard Steam Ship Company. In 1865 and 1866, he directed the *Great Eastern* during its Atlantic enterprises. In 1868, he became superintendent of the French Cable Company and thereafter general manager of several companies of the Electra House Group.

JAMES GORDON BENNETT (1841–1918), United States, Newsmaker, Cable Entrepreneur

James Gordon Bennett was born to Irish immigrants in the United States but spent most of his life in Europe. His father, James Gordon Bennett Sr., was not only the owner of the *New York Herald* but also a key player in the development of yellow journalism and the penny press. In his own news reporting, Gordon Bennett made headlines with stories of discovery, such as the search expedition for the missionary David Livingston, and stories of speed, turning car or air balloon races into media events. In the 1880s, together with John W. Mackay, he launched the Commercial Cable Company.

JOHN W. BRETT (1805–1863), Great Britain, Art Collector, Telegraph Engineer

John Brett initially aimed at becoming an artist. In the 1840s, he and his brother Jacob became interested in electric telegraphy. In 1845 they registered a company for uniting Europe and America by telegraph but failed to find support. In 1850 and 1851, they laid a cable across the English Channel. He was one of the key contacts for Cyrus W. Field in Great Britain.

SIR CHARLES T. BRIGHT (1832–1888), Great Britain, Telegraph Engineer

Charles T. Bright was one of the most eminent telegraph engineers of his time. From 1852 on, he worked for the Magnetic Telegraph Company and laid many of Great Britain's terrestrial lines and early submarine lines. In 1856, he founded, together with John Brett and Cyrus W. Field, the Atlantic Telegraph Company and was knighted upon the first cable's success in 1858. In 1860, he formed a partnership with Latimer Clark as consulting engineers. Together, they laid cables in the Mediterranean, the Persian Gulf, and the West Indies. After he contracted malaria, he abandoned his cable work and turned to mining. In 1865–1868, Bright was a member of Parliament for Greenwich. He was a member of the Society of Telegraph Engineers and in 1887–1888 served as its president.

SIR CHARLES BRIGHT (1863–1927), Great Britain, Telegraph Engineer, Writer

Charles Bright was the son of Charles T. Bright and himself an eminent expert on submarine telegraphy. He wrote numerous articles and books and was knighted for his achievements.

SIR JOHN BRIGHT (1811–1889), Great Britain, Politician

John Bright was, together with Richard Cobden, the spearhead of the Anti-Corn Law League of 1840. The success of their campaign marked both their fame and their rise to key figures of the philosophy of Manchester Liberalism. He was close to Cyrus W. Field and supported the cable project.

SIR SAMUEL CANNING (1823–1908), Great Britain, Telegraph & Consulting Engineer

Samuel Canning commenced working as a railway engineer before he turned to the telegraphs. In 1852 he entered the service of the cable manufacturer Glass, Elliot & Co. and laid the landline connection between Newfoundland and the United States in 1855–1856. During the Atlantic cable enterprise, he served as assistant to Charles T. Bright and in 1865 was appointed chief engineer of the Telegraph Construction and Maintenance Company. In 1866, he was knighted and received the Portuguese Order of St. Jago d'Espada. After 1866, he worked as a consulting engineer and laid cables in the Atlantic, the Mediterranean, and the Caribbean. He was a member of the Society of Telegraph Engineers.

GEORGE VON CHAUVIN (1846–192?), Germany/Great Britain, Telegraph Engineer, Manager

George von Chauvin probably came as a Siemens Brothers employee to England in the 1860s. In 1873 he became managing director of the newly established Direct United States Cable Company and between 1879 and 1880 served as engineer-in-chief for the French PQ company. In the 1890s, he became a representative of Western Union in London. His main employer, however, remained Siemens Brothers, for which he worked as managing director until his retirement in 1925. In 1914, he became a naturalized British citizen.

LATIMER CLARK (1822–1898), Great Britain, Telegraph Engineer

Latimer Clark began as a chemist before he changed to railway surveying in the 1840s and to telegraphy in 1850. For some months, he was employed as engineer to the Atlantic Cable Company and in 1860 served on the committee appointed by the government to inquire into the subject of submarine telegraphy. In 1861, he entered into partnership with Charles T. Bright, and after it was dissolved, he formed the contracting firm Clark, Forde, and Taylor. Under Clark's supervision, some 50,000 miles of submarine cable were laid. He was a member of the Society of Telegraph Engineers and in 1874–1875 served as president.

RICHARD COBDEN (1804–1865), Great Britain, Politician

Richard Cobden was part of the Anti-Corn Law League from 1838 on. This established his association with John Bright and his national fame. Cobden is seen as the most important representative of Manchester Liberalism. He was also a leading supporter of the Peace Society.

PETER COOPER (1791–1883), United States, Entrepreneur, Philanthropist

Peter Cooper was the stereotypical American self-made man. Born into poor circumstances and with only one year of schooling, he made a fortune with a glue factory, iron works, and a wire production plant. Cooper was also engaged in various projects of public service, such as the buildup of professional police and fire departments, public education, and clean water supply for the city of New York. In 1859, he launched Cooper Union, which offered scientific and artistic teaching, lectures, and concerts free to the public. In 1876, he ran as presidential candidate for the Greenback Party. Cooper was one of Cyrus W. Field's early supporters. He belonged to the Cable Cabinet, which in 1854 endorsed the project of an Atlantic cable. He was the father-in-law of Abram Hewitt.

BARON EMILE D'ÉRLANGER (1832–1911), Germany/Great Britain, Merchant Banker

Emile d'Érlanger was one of the most successful financiers of the second half of the nineteenth century. In 1859, he established his own banking house in Paris and in the 1860s moved his business to London. In 1864, he received the title of baron as a reward for securing a loan for the Swedish government and its railway construction projects. Together with Paul J. Reuter, he set up the French Atlantic Cable Company in 1868 and from the 1870s on served as director for several telegraph companies of the Electra House group.

GEORGE WILLIAM DES VOEUX (1834–1909), Great Britain, Colonial Governor

William des Voeux was an agent in the British Foreign Service. In 1875 he married John Pender's daughter Marion Denison. In 1877–1878, des Voeux acted as governor of Trinidad and thereafter as governor of Fiji. In 1886, he became governor of Newfoundland, and from 1887 until his retirement from the service in 1891, he was governor of Hong Kong. His correspondence with Emma Pender reveals much about the inner life of the Pender family.

CYRUS W. FIELD (1819–1892), United States, Cable Entrepreneur

Cyrus Field was the son of a Congregationalist minister and made a fortune with paper wholesaling. From 1854 on, he was one of the most ardent promoters of a transatlantic telegraph cable, the success of which made Field an international celebrity. He used his position to engage in transatlantic "cable diplomacy."

In the 1870s, he lobbied unsuccessfully for a Pacific cable. Throughout his life, Field remained engaged in cable matters. He served on the board of directors of several ocean cable companies as well as the Globe Trust. In 1887, Field lost almost his entire fortune in one single day in a complicated financial maneuver against Jay Gould.

DAVID DUDLEY FIELD (1805–1894), United States, Lawyer, Law Reformer

Cyrus W. Field's older brother was a renowned lawyer, law reformer, and codifier. He codified the entire body of law in New York and developed an American civil code. In the 1850s, Field supported Abraham Lincoln's presidency, which probably influenced Lincoln's appointment of his brother Stephen J. Field to the U.S. Supreme Court. In 1873, Field served as first president of the Association of International Law and in 1872 published his *Draft of a Code of International Law*. Dudley Field was a supporter of *l'esprit d'internationalité* and served his brother Cyrus as legal advisor in his cable work.

SIR SANDFORD FLEMING (1827–1915), Canada, Engineer

Sandford Fleming was born in Scotland but moved to Canada in 1845. Throughout his career, he played a central role in the railway development of Canada. From the late 1870s on, he became engaged in the project of an imperial Pacific cable. After almost three decades, he successfully convinced the colonial governments of Canada, New Zealand, and Australia as well as the British imperial government to carry out the scheme. In 1902, he was the first person to send a telegram around the world.

SIR RICHARD A. GLASS (1820–1873), Great Britain, Cable Manufacturer

Richard Glass was a wire-rope manufacturer before he entered the submarine cable business. In 1852, he set up Glass, Elliot & Co., which became one of the most important cable manufacturers of the early period. In 1864, his firm merged with the Gutta Percha and Co. to form the Telegraph Construction and Maintenance Company, probably the most important cable manufacturer of the nineteenth century. In 1866, he was knighted as a reward for his engagement with the Atlantic cable.

SIR DANIEL GOOCH (1816–1889), Great Britain, Railway Engineer, Cable Entrepreneur

Daniel Gooch was a noted railway engineer when he became engaged in the Atlantic cable project in the 1860s. He served as director of the Great Eastern Steamship Company and the Telegraph Construction and Maintenance Company and was one of the key figures in this vertical business network. In 1866, he was awarded with a baronetcy. Despite his new position and his wealth, Gooch continued to work in business. He was one of the major figures behind the submarine telegraph expansion in the 1870s.

JAY GOULD (1832–1896), United States, Entrepreneur

Jay Gould is remembered for being the "robber baron" during America's gilded age. He made a fortune with stock speculation and the America railways before he turned to telegraphy in the 1870s. He was the largest shareholder of Western Union and also controlled several New York newspapers. In 1880, he set up the American Telegraph and Cable Company and had two Atlantic cables laid by Siemens Brothers, which he leased to Western Union.

JAMES GRAVES (1833–1911), Great Britain, Cable Operator

James Graves was from 1858 to 1909 superintendent of the Valentia Telegraph station, the landing place of the Atlantic cable. He was engaged in research, cable testing, and repair. In 1882, when the German Union Telegraph Company laid a telegraph cable from Emden to Valentia, they also appointed Graves as their superintendent. He was a member of the Society of Telegraph Engineers.

ROBERT CHARLES HALPIN (1836–1894), Ireland, Mariner

Robert Halpin was born in Ireland and went to sea in 1847. In 1865/1866, he served as first officer on board the *Great Eastern* during the laying of the Atlantic cable. Thereafter he was promoted to captain of the *Great Eastern* and laid several ocean cables to India, Australia, and Brazil and across the Atlantic. In 1874, he became marine superintendent for the Telegraph Construction and Maintenance Company.

SIR HENNIKER HEATON (1848–1914), Australia, Newspaper Proprietor, Media Reformer

Henniker Heaton was an important media reformer. As a young man, he immigrated to Australia, where he took up the newspaper business. In 1884 he returned to England but maintained close ties to Australia. In 1883, he represented New South Wales as commissioner at the Amsterdam Exhibition of 1883, and in 1885, he represented Tasmania at the Berlin International Telegraphic Conference. He secured an imperial penny press in 1898 and an Anglo-American penny press in 1908. For twenty-five years after 1885, he served as a member of Parliament for the Conservatives. In 1911, he accepted a baronetcy.

ABRAM S. HEWITT (1822–1903), United States, Entrepreneur, Politician

Abram S. Hewitt owned an iron manufacturing business and introduced the first American open-hearth furnace. In the 1870s and 1880s, he was a member of the House of Representatives and Lord Mayor of New York. He was the son-in-law of Peter Cooper, whom he supported in the buildup of the Cooper Union. Through his engagement in the Atlantic enterprise, he formed a lifelong friendship with John Pender.

GEORGE KENNAN (1845–1924), United States, Explorer, Journalist, Author

George Kennan commenced his working career as a telegrapher in the American Civil War. Thereafter, he led Western Union's surveying team in Siberia in the 1860s to prepare the laying of the Collins Overland Line to Europe. This laid the foundation for his career as an expert on Russia. In the 1870s and 1880s, Kennan was assistant manager of the Associated Press and in 1881 oversaw all telegraphic reports on the condition of President Garfield. During the Spanish-American War, he served as a correspondent in Cuba.

JOHN W. MACKAY (1831–1902), United States, Entrepreneur

John W. Mackay was born in Ireland but immigrated to the United States in 1840. In the 1850s, he moved west to try his fortune with mining. In 1873, he struck, together with four other miners, "the big bonanza." This made him a millionaire overnight. Mackay was involved in several development projects in the midwest; he was a director of the Southern Pacific Railway and organized the Bank of Nevada. In 1883, he established, together with James Gordon Bennett, the Commercial Cable Company and remained engaged in ocean cabling. He successfully managed to establish a duopoly on the Atlantic market and had the American Pacific cable laid.

MATTHEW F. MAURY (1806–1873), United States, Naval Officer, Oceanographer

Matthew F. Maury joined the U.S. Navy in the 1820s and became superintendent of the U.S. Naval Observatory and Hydrographic Office in 1842. He acquired renown with his various publications on winds and currents, which effected great savings in sailing time between ports on the Atlantic. In 1854, he discovered the Atlantic Plateau, along which not only the old Atlantic telegraph cables but also the fiber-optic cables of today are laid.

HENRY A. MORIARTY (1815–1906), Great Britain, Mariner

Henry A. Moriarty served in Great Britain's Royal Navy and was, in the 1850s and 1860s, "leased" during several of the Atlantic cable enterprises to the Anglo-American Telegraph Company to serve as staff commander on board the *Great Eastern*.

SAMUEL F. B. MORSE (1791–1872), United States, Artist, Inventor

Samuel F. B. Morse initially aimed at becoming an artist, but then turned to electricity. He was one of the inventors of an electric telegraph in the 1830s and also developed the Morse code. In 1844, he sent the famous message "What hath God wrought" through the first landline telegraph. He was one of Field's early supporters and served as honorary electrician to the Atlantic cable undertaking. In the 1860s, however, he also supported the Collins Overland Line. Otherwise, he only played a minor role in the succeeding history of the telegraph.

EMMA PENDER (1816–1890), Great Britain, Lady

Emma Pender, née Denison, was born in Edgehill, Lancashire. In 1851, she became John Pender's second wife, and together they had two sons, Henry and John, and two daughters, Marion and Anne. Emma Pender was a fervent letter writer and a shrewd observer of her time. Her letter books and diaries give valuable background information on her husband's business dealings and also offer great insights into the life of a lady at the time.

SIR JOHN PENDER (1816–1890), Great Britain, Merchant, Entrepreneur

The Scotsman John Pender was a textile merchant in Glasgow and Manchester, engaged in trades with China, India, and the East, before he turned to ocean telegraphy. In 1856, he became involved with the Great Atlantic Cable project and in subsequent years built up the Eastern and Associated Telegraph Companies network, which by 1900 controlled four-fifths of the world's ocean cables. Pender also was involved in the Metropolitan Electric Supply Company, concerned with the electric lighting of London. He served as member of Parliament for the Liberal Party and was knighted in 1888.

FRANK PERRY, (no dates available), Great Britain, Telegraph Operator

Frank Perry served as telegraph operator at Heart's Content, Newfoundland. In 1884, upon the death of Weedon, he became superintendent. He was close friends with James Graves.

SIR WILLIAM HENRY PREECE (1834–1913), Great Britain, Engineer, Administrator

William Preece was an important telegraph administrator of his time. He commenced working as a telegraph engineer in the 1850s and in 1877 became electrician to the General Post Office. He spent the next two decades directing the expansion and improvement of the British telegraph network; in 1892, he was promoted to engineer-in-chief. He was a member of the Society of Telegraph Engineers and served as its president in 1880 and 1893.

LOUIS RENAULT (1843–1918), France, Law Reformer

Louis Renault was professor of law at Dijon and Paris and an important law reformer. He was a member of the Institute of International Law and wrote many international law codices, such as the early International Telegraph Union contracts or a draft for the International Telegraph Convention for the Protection of Submarine Cables of 1884. He served as legal advisor at the Hague Conventions and was awarded the Nobel Prize for Peace in 1907.

PAUL JULIUS REUTER (1816–1899), Germany/Great Britain, News Agent

Julius Reuter was born in Germany but moved to Great Britain in 1845. In 1851, he set up a small news agency, which developed into the famous Reuters news agency. In 1870, Reuters formed a news cartel with Havas and Wolff, which lasted until the 1930s and divided the globe into three spheres of news influence. In the 1850s and 1860s, Reuter became involved in submarine telegraphy: he laid his own news cable from England to Germany and in 1868 launched the French Cable Company. In 1857, he became a naturalized British citizen and in 1871 was ennobled by Ernst II, Duke of Saxe-Coburg and Gotha, as Baron von Reuter.

WILLIAM SIEMENS (1823–1883), Germany/Great Britain, Engineer

William Siemens was born in Germany but moved to London in the 1840s searching for employment as a civil engineer. In 1847, his brother Werner Siemens (1816–1892) had established the Company of Siemens and Halske of which William was appointed agent in London. They served as subcontractor for the cable manufacturer Newall & Co. before they became independent as Siemens Brothers in 1858. Siemens was one of the most important antagonists to John Pender and the Eastern and Associated Companies. In 1872/1873, he became the first president of the Society of Telegraph Engineers. In 1859, he became a naturalized British subject.

WILLOUGHBY SMITH (1821–1891), Great Britain, Telegraph Engineer, Electrician

Willoughby Smith commenced working as a telegraph engineer in the 1840s and was engaged in the manufacture of the English Channel cable. He was part of the Atlantic cable enterprise and in 1866 appointed chief electrician at the Telegraph Construction and Maintenance Company. In 1881, he additionally became manager of the Gutta Percha works. He was a member of the Society of Telegraph engineers and served as president in 1882–1883.

GEORGE SPRATT, (no dates available), Great Britain, Telegraph Operator

George Spratt was the assistant superintendent at Porthcurno cable station between 1871 and 1909. Throughout his entire career, he kept a diary, and his papers are some of the most important sources with regard to life at a cable station.

WILLIAM THOMSON (1824–1907), Great Britain, Scientist

William Thomson was professor of natural philosophy at Glasgow University. Aside from the important work he did in the mathematical analysis of electricity and in thermodynamics, he was also an eminent telegraph engineer and inventor. He was involved in the Great Atlantic Cable and many subsequent ocean telegraph undertakings. In 1866 he was knighted and in 1892 made Baron Kelvin of Larges. He was a member of the Society of Telegraph Engineers and twice served as its president.

HENRY WEAVER, (no dates available), Great Britain, Telegraph Administrator

Henry Weaver served as general manager of the Anglo-American Telegraph Company. One of his tasks was to supervise the superintendents at the different cable stations on both sides of the Atlantic.

EZRA WEEDON, (no dates available), Great Britain, Telegraph Operator

Ezra Weedon was superintendent at Heart's Content cable station from its opening in 1866 until his death in 1884. He stood in almost constant rivalry with James Graves.

NOTES

INTRODUCTION

1 A. Talexy, "Atlantic Telegraph Polka" (1858), cited in Carter, *Cyrus Field*, p. 169.
2. Samuel Carter, *Cyrus Field: Man of Two Worlds* (New York: Putman, 1968), p. 164; Gillian Cookson, "The Transatlantic Telegraph Cable—Eighth Wonder of the World," *History Today* 50 (2000): p. 44.
3. Carter, *Cyrus Field*, p. 169; see Tiffany's advertisement in the *New York Times*, August 21, 1858.
4. "Completion of the Atlantic Telegraph," *The Caledonian Mercury*, July 30, 1866.
5. Christopher Hanlon, *America's England: Antebellum Literature and Atlantic Sectionalism* (Oxford: Oxford University Press, 2013), pp. 187–88.
6. Hanlon, *America's England*, pp. 187–88.
7. Similarly, David S. Grewal, *Network Power: The Social Dynamics of Globalization* (New Haven: Yale University Press, 2008), and Gary B. Magee and Andrew S. Thompson, *Empire and Globalisation: Networks of People, Goods and Capital in the British World, c. 1850–1914* (Cambridge: Cambridge University Press, 2010).
8. For this book's purposes, "globalization" is used as a perspective. It characterizes nonlinear processes of a global intensification, expansion, and densification of international networks and spaces of transboundary interaction. Sebastian Conrad and Andreas Eckert, "Globalgeschichte, Globalisierung, multiple Modernen: Zur Geschichtsschreibung der modernen Welt," in *Globalgeschichte: Theorien, Ansätze, Themen*, eds. Sebastian Conrad et al. (Frankfurt/Berlin: Campus, 2007), p. 20. On the use and problems of the term "globalization," see Jürgen Osterhammel and Niels P. Petersson, *Globalization: A Short History* (Princeton, NJ: Princeton University Press, 2005), pp. 1–5.
9. "Extension of the Electric Telegraph to France Ireland and America," *Morning Herald*, September 10, 1851, Newspaper Clippings, Jacob Brett Papers, IET Archives.

10. Carter, *Cyrus Field*, p. 168; Cookson, "Transatlantic Telegraph Cable," p. 45.
11. John Tully, "A Victorian Ecological Disaster: Imperialism, the Telegraph, and Gutta-Percha," *Journal of World History* 20, 4 (2009): pp. 559–579, 568; Charles Bright, "Imperial Telegraphic Communication and the 'All-British' Pacific Cable," *The London Chamber of Commerce Pamphlet Series*, No. 40, 1902, NAEST 17/105, IET Archives, p. 39.
12. Cookson, "Transatlantic Telegraph Cable," p. 44.
13. Congratulatory telegrams between President James Buchanan and Queen Victoria in "Telegraph Supplement," *Harper's Weekly. A Journal of Civilization*, September 4, 1858.
14. "Epitome of this Morning's News," *The Pall Mall Gazette*, July 27, 1866.
15. Congratulatory telegrams between President James Buchanan and Queen Victoria in "Telegraph Supplement," *Harper's Weekly. A Journal of Civilization*, September 4, 1858.
16. Charles Maier, "Consigning the Twentieth Century to History: Alternative Narratives for the Modern Era," *American Historical Review* 105 (2000): pp. 813–14; Similarly William H. McNeill, "Globalization: Long Term Process or New Era in Human Affairs?," *New Global Studies* 2 (2008); Christopher A. Bayly, *The Birth of the Modern World: 1780–1914* (Malden, MA: Blackwell, 2009); Emily S. Rosenberg, ed., *A World Connecting: 1870–1945* (Cambridge, MA: Belknap Press of Harvard University Press, 2012); or Jürgen Osterhammel, *Die Verwandlung der Welt: Eine Geschichte des 19. Jahrhunderts* (München: Beck, 2009).
17. Roland Wenzlhuemer, "Editorial: Telecommunication and Globalization in the Nineteenth-Century," *Historical Social Research. Historische Sozialforschung* 35 (2010): p. 10; also see Roland Wenzlhuemer, *Connecting the Nineteenth-Century World: The Telegraph and Globalization* (Cambridge: Cambridge University Press, 2013), chapter 2.
18. Jorma Ahvenainen, "The Role of Telegraphs in the 19th Century Revolution of Communication," in *Kommunikationsrevolutionen: Die neuen Medien des 16. und 19. Jahrhunderts*, ed. Michael North (Köln: Böhlau, 1995), pp. 77–79.
19. The Royal Swedish Academy of Science, "The Nobel Prize in Physics, 2009—Press Release," accessed June 11, 2011, http://nobelprize.org/nobel_prizes/physics/laureates/2009/press.html.
20. Lee Mwiti, "East Africa: Sea Cable Ushers in New Internet Era," *Allafrica.com*, July 23, 2009, accessed August 3, 2009, allafrica.com; Diane McCarthy, "Cable Makes big Promises for African Internet," *CNN.com*, July 23, 2009, accessed June 11, 2011, http://edition.cnn.com/2009/TECH/07/22/seacom.on/index.html.
21. Samuel P. Huntington, *The Clash of Civilizations and the Remaking of World Order* (London: Touchstone Books, 1998).
22. Emma Pender to William des Voeux, September 18, 1877, Emma Pender Papers, Cable and Wireless Archive, Porthcurno Telegraph Museum (hereafter cited CWA).

23. Michael Adas, *Dominance by Design: Technological Imperatives and America's Civilizing Mission* (Cambridge, MA: Belknap Press of Harvard University Press, 2006); Michael Adas, *Machines as the Measure of Men: Science, Technology, and Ideologies of Western Dominance* (Ithaca, NY: Cornell University Press, 1989).
24. Harold A. Innis, *Empire and Communications* (Lanham, MD: Rowman & Littlefield, 2007); Jill Hills, *The Struggle for Control of Global Communication: The Formative Century* (Urbana, IL: University of Illinois Press, 2002); Peter J. Hugill, *Global Communication Since 1844: Geopolitics and Technology* (Baltimore: John Hopkins University Press, 1999); Peter J. Hugill, "The Geopolitical Implications of Communication Under the Seas," in *Communications Under the Seas: The Evolving Cable Network and Its Implications*, eds. Bernard S. Finn and Daqing Yang (Cambridge, MA: MIT Press, 2009).
25. Daniel R. Headrick, *The Tools of Empire: Technology and European Imperialism in the Nineteenth Century* (New York: Oxford University Press, 1981); similarly Daniel R. Headrick, *The Tentacles of Progress: Technology Transfer in the Age of Imperialism, 1850–1940* (New York: Oxford University Press, 1988), or Headrick, *The Invisible Weapon: Telecommunications and International Politics, 1851–1945* (New York: Oxford University Press, 1991); Pascal Griset and Daniel R. Headrick, "Submarine Telegraph Cables: Business and Politics, 1838–1939," *Business History Review* 75 (2001): p. 543; Christopher A. Bayly, *Empire and Information: Intelligence Gathering and Social Communication in India, 1780–1870* (New Delhi: Cambridge University Press, 2007).
26. Robert Boyce, "Submarine Cables as a Factor in Britain's Ascendancy as a World Power, 1850–1914," in *Kommunikationsrevolutionen: Die neuen Medien des 16. und 19. Jahrhunderts*, ed. Michael North (Köln: Böhlau, 1995). The oldest proponent of this imperial thesis is P. M. Kennedy: P. M. Kennedy, "Imperial Cable Communications and Strategy, 1870–1914," *The English Historical Review* 86 (1971): p. 729.
27. See A. Kunert, *Telegraphen-Seekabel* (Köln-Mühlheim: Karl Glitscher, 1962); Kenneth R. Haigh and Edward Wilshaw, *Cableships and Submarine Cables* (London: Coles, 1968); Richard Hennig, "Die historische Entwicklung der deutschen Seekabelunternehmungen," *Beiträge zu Geschichte der Technik und Industrie. Jahrbuch des Vereines Deutscher Ingenieure* 1 (1909).
28. On global coloniality, see Sebastian Conrad and Sorcha O'Hagan, *German Colonialism: A Short History* (Cambridge: Cambridge University Press, 2011).
29. New literature has argued accordingly: Robert M. Pike and Dwayne R. Winseck, "The Politics of Global Media Reform, 1907–1923," *Media, Culture & Society* 26 (2004): p. 4; Dwayne R. Winseck and Robert M. Pike, *Communication and Empire: Media, Markets, and Globalization, 1860–1930* (Durham: Duke University Press, 2007); Harold Jacobson, "International Institutions for Telecommunications: The ITU's Role," in *The International Law of Communications*, ed. Edward McWhinney (Dobbs Ferry, NY: Sijthoff; Oceana, 1971); Jorma Ahvenainen, "The International Telegraph Union: The Cable Companies and the Governments,"

in *Communications Under the Seas: The Evolving Cable Network and Its Implications*, eds. Bernard S. Finn and Daqing Yang (Cambridge, MA: MIT Press, 2009); or Léonard Laborie, *L'Europe Mise en Réseaux: La France et la Coopération Internationale dans les Postes et les Télécommunications (Années 1850-Années 1950)* (Brussels: Peter Lang, 2010).

30. Simone Müller-Pohl, "Working the Nation State: Submarine Cable Actors, Cable Transnationalism and the Governance of the Global Media System, 1858–1914," in *The Nation State and Beyond. Governing Globalization Processes in the Nineteenth and Twentieth Century*, eds. Isabella Löhr and Roland Wenzlhuemer (New York: Springer Press, 2012).
31. On the network concept, see Frederick Cooper, "Networks, Moral Discourse, and History," in *Intervention and Transnationalism in Africa: Global-Local Networks of Power* (Cambridge: Cambridge University Press, 2001). John Law and John Hassard, *Actor Network Theory and After* (Oxford: Blackwell, 2005); Bruno Latour, *Reassembling the Social: An Introduction to Actor-Network-Theory* (Oxford: Oxford University Press, 2007).
32. On global history, see Michael H. Geyer and Charles Bright, "World History in a Global Age," *The American Historical Review* 100 (October, 1995): pp. 1034–60; Osterhammel, *Globalization*; Sebastian Conrad, ed., *Globalgeschichte: Theorien, Ansätze, Themen* (Frankfurt: Campus-Verl., 2007).
33. See Jorma Ahvenainen, *The Far Eastern Telegraphs: The History of Telegraphic Communications Between the Far East, Europe, and America Before the First World War.* (Helsinki: Suomalainen Tiedeakatemia, 1981); Jorma Ahvenainen, *The European Cable Companies in South America: Before the First World War* (Helsinki: Finnish Academy of Science and Letters, 2004); Michaela Hampf and Simone Müller-Pohl, eds., *Global Communication Electric: Business, News and Politics in the World of Telegraphy* (Frankfurt/ Berlin: Campus, 2013). On technology-in-use, see David Edgerton, "From Innovation to Use: Ten Eclectic Theses on the Historiography of Technology," *History and Technology* 16 (1999).
34. Euro-American does not denote Americans of European descent, but the community of European imperial powers and the United States, which dominated the world at the time.
35. Bruno Latour, "Die Logistik der Immutable Mobiles," in *Mediengeographie: Theorie—Analyse—Diskussion*, eds. Jörg Döring and Tristan Thielmann (Bielefeld: Transcript-Verlag, 2009).
36. Osterhammel, *Verwandlung der Welt*, pp. 1064–81; see also Jürgen Kocka and Allan Mitchell, eds., *Bourgeois Society in Nineteenth-Century Europe* (Oxford: Berg, 1993); Jürgen Kocka, "The Middle Classes in Europe," in *The European Way: European Societies During the 19th and 20th Centuries*, ed. Hartmut Kaelble (Oxford: Berghahn, 2004); for an overview on a global social history see Christof Dejung, "Towards a Global Social History? New Studies on the Global History of the Middle Class," *Neue Politische Literatur* 59 (2014): pp. 229–53.

37. Richard R. John, "Field, Cyrus West," accessed September 18, 2008, http://www.anb.org/articles/10/10-00542.html.
38. On gatekeepers, see Peter J. Bowler and Iwan R. Morus, *Making Modern Science: A Historical Survey* (Chicago: University of Chicago Press, 2005), p. 319.
39. James Anderson cited in "Telegraph Companies and Charges," *Daily News*, February 17, 1873. Already in 1877, Ernst Kapp coined in his book *Philosophy of Technology* the term *Weltcommunication* in reference to the submarine telegraphs. Although the term remained far from an elaborate theory of global communication, it suggested that communication via ocean telegraphs integrates the entire world. Ernst Kapp, *Grundlinien einer Philosophie der Technik: Zur Entstehungsgeschichte der Cultur aus neuen Gesichtspunkten* (Braunschweig, 1877).
40. See Ferguson's definition of the "West" versus "the rest" constellation as a political-economic system: Niall Ferguson, *Civilization: The West and the Rest* (London: Lane, 2011).
41. Sebastian Conrad and Jürgen Osterhammel, *Das Kaiserreich Transnational: Deutschland in der Welt 1871–1914* (Göttingen: Vandenhoeck & Ruprecht, 2006); Sebastian Conrad, *Globalisation and the Nation in Imperial Germany* (Cambridge: Cambridge University Press, 2010).
42. Shmuel N. Eisenstadt, *Multiple Modernities* (New Brunswick, NJ: Transaction Publishers, 2002).
43. Osterhammel, *Verwandlung der Welt*, pp. 143–44.
44. Simone Müller-Pohl, "'By Atlantic Telegraph': A Study on Weltcommunication in the 19th Century," *Medien & Zeit* 4 (2010): p. 42.
45. Tom Standage, *The Victorian Internet: The Remarkable Story of the Telegraph and the Nineteenth Century's Online Pioneers* (London: Phoenix, 2003).
46. "A Baronet of the World," *New York Times*, February 1, 1912. J. H. Heaton, "Wanted, Cheap Imperial Telegraphs," *Times*, July 13, 1899. In 2011, an American-based nongovernmental organization made international headlines with its endeavor to purchase a satellite through its crowd-funded initiative "Buy This Satellite," move it over Central Africa, and thereby "build up a free communication network available anywhere in the world." Ahumanright.org (last accessed June 11, 2011).
47. See the concept of mental maps in Christoph Conrad, "Vorbemerkung," *Geschichte und Gesellschaft* 28 (2002): pp. 339–42.
48. Eric J. Hobsbawm, *The Age of Capital: 1848–1875* (New York: Vintage Books, 1996), p. 60; also see Frank Bösch's chapter on colonial scandals in *Öffentliche Geheimnisse: Skandale, Politik und Medien in Deutschland und Großbritannien 1880–1914* (München: Oldenbourg, 2009).
49. David Edgerton, "The Contradictions of Techno-Nationalism and Techno-Globalism: A Historical Perspective," *New Global Studies* 1 (2007).
50. "Ships and Seafaring," in *Encyclopedia of the Ancient Greek World*, eds. David Sacks, Oswyn Murray, and Lisa R. Brody, Rev. ed. (New York: Facts On File, 2005), p. 310.

51. Charles Bright, *Submarine Telegraphs: Their History, Construction and Working* (London: C. Lockwood, 1898), p. xiii.
52. On the American system, see Richard R. John, *Network Nation: Inventing American Telecommunications* (Cambridge, MA: Belknap Press of Harvard University Press, 2010), and David Hochfelder, *The Telegraph in America: 1832–1920* (Baltimore: Johns Hopkins University Press, 2012). On the German system, see Kunert, *Telegraphen-Seekabel*; on the British, see Haigh and Wilshaw, *Submarine Cables*; and on the French, see Pascal Griset, *Entreprise, Technologie et Souveraineté: Les Télécommunications Transatlantiques de la France, XIXe-XXe Siècles* (Paris: Editions Rive droite, 1996).
53. On transnationalism, see Ian R. Tyrrell, "What Is Transnational History?", accessed August 28, 2011, http://iantyrrell.wordpress.com/what-is-transnational-history/. This study draws from Patricia Clavin's concept of transnationalism as a honeycomb, a structure that also contains hollowed-out spaces where organizations, individuals, and ideas can wither away to be replaced by new groups, people, or innovations; see Patricia Clavin, "Defining Transnationalism," *Contemporary European History* 14 (2005): pp. 421–39.
54. Steven Vertovec, *Transnationalism and Identity* (Torpoint: Taylor & Francis, 2001); Bruce Robbins, "Introduction Part I: Actually Existing Cosmopolitanism," in *Cosmopolitics: Thinking and Feeling Beyond the Nation*, ed. Pheng Cheah (Minneapolis, MN: University of Minnesota Press, 2000).
55. Hartmut Böhme, "Einleitung," in *Netzwerke: Eine Kulturtechnik der Moderne*, eds. Jürgen Barkhoff, Hartmut Böhme, and Jeanne Riou (Köln: Böhlau, 2004).
56. See Frank Bösch on "global" media events, Frank Bösch, *Medialisierte Ereignisse: Performanz, Inszenierung und Medien seit dem 18. Jahrhundert* (Frankfurt/Berlin: Campus, 2010).
57. See Griset for a periodization of communications technologies that marks 1866 to 1911 as a period of British monopoly in submarine telegraphy: Griset, *Entreprise, Technologie et Souveraineté.*
58. Haigh and Wilshaw, *Submarine Cables*, p. 152; on wireless, see Jonathan R. Winkler, *Nexus: Strategic Communications and American Security in World War I* (Cambridge, MA: Harvard University Press, 2008), or Heidi Tworek, "'The Path to Freedom'? Transocean and German Wireless Telegraphy, 1914–1922," *Historical Social Research. Historische Sozialforschung* 35 (2010): pp. 209–236.
59. Sally Ledger, *The New Woman: Fiction and Feminism at the Fin de Siècle* (Manchester: Manchester University Press, 1997), and Angelique Richardson and Chris Willis, *The New Woman in Fiction and Fact: Fin-de-Siècle Feminisms* (New York: Palgrave, 2002).
60. Paul Virilio, *The Aesthetics of Disappearance* (Los Angeles: Semiotext(e), 2009); David Harvey, "Between Space and Time: Reflections on the Geographical Imagination," *Annals of the Association of American Geographers* 80 (1990): pp. 418–34. Anthony Giddens speaks of a "time-space-distanciation"; see Lars B. Kaspersen, *Anthony Giddens: An Introduction to a Social Theorist* (Oxford: Blackwell, 2000), p. 54.

1. NETWORKING THE ATLANTIC

1. Henry M. Field, *History of the Atlantic Telegraph* (New York: C. Scribner & Co, 1866), p. 26.
2. Charles Bright, *Submarine Telegraphs: Their History, Construction and Working* (London: C. Lockwood, 1898), p. 26.
3. Field, *Atlantic Telegraph*, p. 30.
4. John S. Gordon, *Thread Across the Ocean: The Heroic Story of the Transatlantic Cable* (New York: Walker, 2002), blurb.
5. William H. Russell to James Anderson, June 7, 1868, cited in Isabella F. Judson, *Cyrus W. Field: His Life and Work* (New York: Harper & Brothers Publishers, 1896), p. 245.
6. Jürgen Osterhammel, *Die Verwandlung der Welt: Eine Geschichte des 19. Jahrhunderts* (München: Beck, 2009), p. 1011.
7. Field, *Atlantic Telegraph*, p. 36; E. B. Grant, *The Western Union Telegraph Company: Its Past, Present and Future* (New York: Hotchkiss, Burnham & Co, 1883), p. 38.
8. Eric J. Hobsbawm, "The Example of the English Middle Class," in *Bourgeois Society in Nineteenth-Century Europe*, ed. Jürgen Kocka and Allan Mitchell (Oxford, Providence: Berg, 1993), pp. 127–50.
9. *Eliza Cook's Journal*, November 29, 1851, Newspaper Clippings, Jacob Brett Papers, IET Archives, p. 126.
10. *Spectatory*, "Transmarine Telegraph," August 31, 1850; *Journal du Calais*, 1851, Newspaper Clippings, Jacob Brett Papers, IET Archives, p. 157.
11. Eric J. Hobsbawm, *The Age of Capital: 1848–1875* (New York: Vintage Books, 1996), pp. 31–33.
12. T. P. Shaffner, "Communication with America, via the Faeroes, Iceland, and Greenland," *Proceedings of the Royal Geographical Society of London* 4, no. 3 (1859–1860): p. 101.
13. Charles Bright, *The Story of the Atlantic Cable* (New York: D. Appleton and Company, 1903), p. 159.
14. Bright, *Atlantic Cable*, p. 165. For the South Atlantic Telegraph Company, see *The Atlantic and South Atlantic Telegraphs, by a Member of the Institution of Civil Engineers*, London 1859, The Atlantic Telegraph, 1858–1859, CWA; *South Atlantic Telegraph Company Ltd., 1858–1884*, Post Office Papers, National Archives, Kew.
15. Philip Jenkins, *A History of the United States* (Basingstoke: Palgrave Macmillan, 2007), p. 117.
16. Ray A. Billington and Martin Ridge, *Westward Expansion: A History of the American Frontier* (Albuquerque: University of New Mexico Press, 2001).
17. Ernest N. Paolino, *The Foundation of the American Empire: William Henry Seward and U.S. Foreign Policy* (Ithaca: Cornell University Press, 1973), p. 53; Samuel Carter, *Cyrus Field: Man of Two Worlds* (New York: Putman, 1968), p. 192; On the

Western Union project, see John B. Dwyer, *To Wire the World: Perry M. Collins and the North Pacific Telegraph Expedition* (Westport: Praeger, 2001).

18. Paolino, *American Empire*, p. 59.
19. E. B. Grant, *Report on the Western Union Telegraph Company* (New York: The Sun Job Print, Printing House Square, 1869), p. 21. In 1871, the Russian government revived the project. Bright, *Submarine Telegraphs*, p. 114.
20. Chester G. Hearn, *Circuits in the Sea: The Men, the Ships, and the Atlantic Cable* (Westport: Praeger, 2004), p. xi.
21. Field, *Atlantic Telegraph*, p. 28.
22. Matthew F. Maury, *The Physical Geography of the Sea* (London: Harper & Brothers, 1855), p. 37.
23. Matthew F. Maury to Cyrus W. Field, February 22, 1854, cited in Field, *Atlantic Telegraph*, p. 30; on the myth of the Atlantic plateau, see Christian Holtorf, *Das erste transatlantische Telegraphenkabel von 1858 und seine Auswirkungen auf die Vorstellungen von Zeit und Raum* (Berlin: Humboldt-Universität Dissertation, 2009).
24. Samuel F. B. Morse to John C. Spenser, Secretary of the Treasury, August 10, 1843, cited in Field, *Atlantic Telegraph*, p. 34.
25. The cable cabinet consisted of Peter Cooper, Chandler White, Marshall O. Roberts, and Moses Taylor. Robert Lowber replaced White after his death. Later on, Wilson G. Hunt joined the company; Carter, *Cyrus Field*, p. 118.
26. Field, *Atlantic Telegraph*, p. 70.
27. Bright, *Submarine Telegraphs*, p. 28.
28. The Telegraph Submarine Cable Question, p. 3.
29. Carter, *Cyrus Field*, p. 110.
30. Carter, *Cyrus Field*, pp. 95, 124–26; Dwayne R. Winseck and Robert M. Pike, *Communication and Empire: Media, Markets, and Globalization, 1860–1930* (Durham: Duke University Press, 2007), p. 21.
31. Field, *Atlantic Telegraph*, p. 98; Roger Bridgman, "Brett, John Watkins (1805–1863)," in *Oxford National Biography Online*, ed. H. C. G. Matthew et al. (Oxford: Oxford University Press, 2004), accessed September 15, 2008, http://www.oxforddnb.com/view/article/3345.
32. Telegraph Construction and Maintenance Company, *The Telcon Story 1850–1950* (London: Telegraph Construction and Maintenance Company, 1950), p. 46.
33. James Graves, *Thirty-Six Years in the Telegraphic Service 1852–1888 Being a Brief Autobiography of James Graves MSTE*, transcribed by Dominic de Cogan, ed. Donard de Cogan (1909/2008), p. 43; Franklin Parker, *George Peabody: A Biography* (Nashville: Vanderbilt University Press, 1995), pp. 85, 146.
34. Dorceta E. Taylor, *The Environment and the People in American Cities, 1600–1900s: Disorder, Inequality, and Social Change* (Durham: Duke University Press, 2009), p. 232; Carole Klein, *Gramercy Park: An American Bloomsbury* (Baltimore: Johns Hopkins University Press, 1999).

35. Richard R. John, "Field, Cyrus West," in *American National Biography*, accessed September 18, 2008, http://www.anb .org/articles/10/10-00542.html.
36. Field, *Atlantic Telegraph*, p. 34.
37. Pierre Bourdieu, "The Forms of Capital," in *Handbook of Theory and Research for the Sociology of Education*, ed. J. Richardson (New York: Greenwood Press, 1986), pp. 241–58; Gary B. Magee and Andrew S. Thompson, *Empire and Globalisation: Networks of People, Goods and Capital in the British World, c. 1850–1914* (Cambridge: Cambridge University Press, 2010), Chapter 3.
38. Field, *Atlantic Telegraph*, p. 50.
39. Field, *Atlantic Telegraph*, p. 51.
40. A. C. Lynch, "Bright, Sir Charles Tilston (1832–1888)," in *Oxford National Biography Online*, ed. H. C. G. Matthew et al. (Oxford: Oxford University Press, 2004), accessed September 15, 2008, http://www.oxforddnb.com/view/article/3415.
41. Gillian Cookson, "The Transatlantic Telegraph Cable: Eighth Wonder of the World," *History Today* 50 (2000): p. 44.
42. *Twenty-Seventh Anniversary of the First Atlantic Cable: Mr. Cyrus W. Field's Banquet at the Star and Garter Hotel, Richmond, on August 5th 1885* (London: Printed for Private Circulation, 1886), p. 11, Western Union Collection SI.
43. Carter, *Cyrus Field*, p. 164; *Epitome of Proceedings at a Telegraphic Soirée Given by Samuel Gurney, Esq., M.P., at 25, Prince's Gate, Hyde Park, March 26, 1862* (London: Thomas Piper, at the Electrician Office, 1862), p. 4.
44. Telegraph Construction and Maintenance Company, *Telcon Story*, p. 51.
45. Bright, *Submarine Telegraphs*, p. 80; Carter, *Cyrus Field*, p. 195.
46. Telegraph Construction and Maintenance Company, *Telcon Story*, p. 55; Daniel Gooch, *Diaries of Sir Daniel Gooch Baronet: With an Introductory Notice by Sir Theodore Martin K.C.B.* (London: Kegan Paul, Trench Trübner & Co., 1892), p. xiv.
47. Markus Krajewski, *Restlosigkeit: Weltprojekte um 1900* (Frankfurt am Main: Fischer-Taschenbuch-Verl., 2006); Henry Petroski, *Remaking the World: Adventures in Engineering* (New York: Vintage Books, 1999).
48. On the Irish as "other" in the construction of British nationalism, see Mary J. Hickman and Bronwen Walter, "Deconstructing Whiteness: Irish Women in Britain," *Feminist Review* 50, The Irish Issue: The British Question (1995).
49. Hobsbawm, "English Middle Class," p. 95.
50. John F. C. Harrison, *Late Victorian Britain, 1875–1901* (London: Routledge, 2000), p. 29; François Bédarida and A. S. Forster, *A Social History of England 1851–1990* (London: Routledge, 1998), p. 48.
51. William H. Russell and Robert Dudley, *The Atlantic Telegraph by W.H. Russell L.L.D. Illustrated by Robert Dudley: Dedicated by Special Permission to His Royal Highness Albert Edward Prince of Wales* (London: Day & Son Limited, 1865), p. 29; also see Bright, *Atlantic Cable*, p. 174.
52. Bright, *Atlantic Cable*, p. 175.
53. Field, *Atlantic Telegraph*, pp. 299–300.

54. Jules Verne, *A Floating City* (London: Sampson Low, Marston & Company, 1871).
55. Gordon, *Thread Across the Ocean*, p. 177; Field, *Atlantic Telegraph*, p. 300.
56. Ibid.; Robert A. Buchanan and Isambard K. Brunel, *Brunel: The Life and Times of Isambard Kingdom Brunel* (London: Hambledon Continuum, 2008).
57. James Dugan, *The Great Iron Ship* (London: Hamilton, 1953), p. 194; "The Great Eastern: Serious Accident," *Manchester Guardian*, September 12, 1859.
58. Buchanan and Brunel, *Brunel*, p. 114.
59. The Great Ship Company, *The Great Eastern Steam Ship. The Past—The Future*, London 1858, CWA, pp. 4–5.
60. Daniel Gooch was a friend of Isambard Kingdom Brunel. Both worked together on the Great Western Railway Company. Introductory Notice by Sir Theodore Martin in Gooch, *Diaries*, p. viii. On Thomas Brassey (1805–1870), see David Brooke, "Brassey, Thomas (1805–1870)," in *Oxford National Biography Online*, ed. H. C. G. Matthew et al. (Oxford: Oxford University Press, 2004), accessed September 15, 2008, http://www.oxforddnb.com/view/article/3289.
61. Cookson, "Transatlantic Telegraph Cable," p. 50; Gooch, *Diaries*, p. 83. Telcon is a fusion of two London cable manufacturing companies. In 1864, Glass, Elliot & Co. amalgamated with the Gutta Percha Company. John Pender initiated the merger. Daniel Gooch was his co-director. Anita McConnell, "Pender, Sir John (1816–1896)," in *Oxford National Biography Online*, ed. H. C. G. Matthew et al. (Oxford: Oxford University Press, 2004), accessed September 15, 2008, http://www.oxforddnb.com/view/article/21831; Bright, *Submarine Telegraphs*, p. 156.
62. The Great Ship Company, *Great Eastern*, pp. 7–12.
63. Field, *Atlantic Telegraph*, p. 86.
64. Gooch, *Diaries*, p. 124; See the notice on deserted sailors in "The Atlantic Telegraph Expedition: From our Special Correspondent," *Daily News*, July 19, 1866.
65. Field, *Atlantic Telegraph*, p. 81.
66. Geoffrey Channon, "Gooch, Sir Daniel, (1816–1889)," in *Oxford Dictionary of National Biography*, ed. H. C. G. Matthew et al. (Oxford: Oxford University Press, 2004), accessed September 15, 2008, http://www.oxforddnb.com/view/article/10939; A. M. Spear, "Canning, Sir Samuel (1823–1908)," in *Oxford Dictionary of National Biography*, ed. H. C. G. Matthew et al. (Oxford: Oxford University Press, 2004), accessed September 15, 2008, http://www.oxforddnb.com/view/article/32279.
67. Anita McConnell, "Glass, Sir Richard Atwood (1820–1873)," in *Oxford Dictionary of National Biography*, ed. H. C. G. Matthew et al. (Oxford: Oxford University Press, 2004), accessed September 24, 2008, http://www.oxforddnb.com/view/article/10801; Bruce J. Hunt, "Varley, Cromwell Fleetwood (1828–1883)," in *Oxford Dictionary of National Biography*, ed. H. C. G. Matthew et al. (Oxford: Oxford University Press, 2004), accessed September 24, 2008, http://www.oxforddnb.com/view/article/28114; Anita McConnell, "Smith Willoughby (1828–1891)," in *Oxford Dictionary of National Biography*, ed. H. C. G. Matthew et al.

(Oxford: Oxford University Press, 2004), accessed September 15, 2008, http://www.oxforddnbcom/view/article/25942.

68. T. P. Shaffner, *The Telegraph Manual: A Complete History and Description of the Semaphoric, Electric and Magnetic Telegraphs of Europe, Asia, Africa, and America, Ancient and Modern with Six-Hundred-and-Twenty-Five Illustrations* (New York: Pudney & Russell, Publishers, 1859), pp. 198–202; Kenneth G. Beauchamp, *History of Telegraphy* (London: Institution of Electrical Engineers, 2001), pp. 73–78.
69. Kenneth R. Haigh and Edward Wilshaw, *Cableships and Submarine Cables* (London: Coles, 1968), pp. 28, 34, 41.
70. Hunt, "Varley"; A. F. Pollard, "Clark, (Josiah) Latimer (1822–1898)," in *Oxford Dictionary of National Biography*, ed. H. C. G. Matthew et al. (Oxford: Oxford University Press, 2004), accessed September 15, 2008, http://www oxforddnb com/view/article/5469; Lynch, "Bright."
71. Cromwell F. Varley (1828–1883), founding member of the Society of Telegraph Engineers, member of the joint committee of 1861, and from 1865 on chief electrician of the Anglo-American Telegraph Company, also worked at the Electric and International Telegraph Company. In 1865 he formed a patent partnership with William Thomson and Fleeming Jenkin. Hunt, "Varley."
72. Graves, *Autobiography*, p. 40.
73. Lynch, "Bright"; A. F. Pollard, "Clark"; Bruce J. Hunt, "Preece, Sir William Henry (1834–1913)," in *Oxford Dictionary of National Biography*, ed. H. C. G. Matthew et al. (Oxford: Oxford University Press, 2004), accessed May 7, 2010, http://www.oxforddnb.com/view/article/35605.
74. Michael Fry, *The Scottish Empire* (East Linton: Tuckwell Press, 2001).
75. Murray Pittock, *The Invention of Scotland: The Stuart Myth and the Scottish Identity, 1638 to the Present* (London: Routledge, 1991), p. 100; Arthur Herman, *The Scottish Enlightenment: The Scots' Invention of the Modern World* (London: Fourth Estate, 2001).
76. Olive Checkland and Sydney Checkland, *Industry and Ethos: Scotland, 1832–1914* (Edinburgh: Edinburgh University Press, 1997), Chapter 1.
77. "John Pender, the Cable King, Dies," *Chicago Daily Tribune*, July 8, 1896.
78. Telegraph Construction and Maintenance Company, *Telcon Story*, p. 46. On Manchester, see Hartmut Berghoff, *Englische Unternehmer 1870–1914: Eine Kollektivbiographie führender Wirtschaftsbürger in Birmingham, Bristol und Manchester* (Göttingen: Vandenhoeck & Ruprecht, 1991).
79. Contrary to P. J. Cain and A. G. Hopkins's theory of "gentlemanly capitalism," which inserts London as the motor of British overseas expansion; P. J. Cain and A. G. Hopkins, "The Political Economy of British Expansion Overseas, 1750–1914," *The Economic History Review* 33, no. 4 (1980): pp. 463–90.
80. Field, *Atlantic Telegraph*, p. 290. Samuel Cunard and Edward Cunard were directors of the Atlantic Telegraph Company. Atlantic Telegraph Company, *Reports of Proceedings at General Meetings, 1858–1868*, CWA.

81. Field, *Atlantic Telegraph*, p. 300.
82. Jane Chapman, *Comparative Media History: An Introduction; 1789 to the Present* (Cambridge: Polity Press, 2008), p. 50.
83. Field, *Atlantic Telegraph*, p. 304; Russell and Dudley, *Atlantic Telegraph*.
84. Field, *Atlantic Telegraph*; Bright, *Atlantic Cable*; Gooch, *Diaries*; George Saward, *The Trans-Atlantic Submarine Telegraph: A Brief Narrative of the Principal Incidents in the History of the Atlantic Telegraph Company* (London, 1878); Russell and Dudley, *Atlantic Telegraph*.
85. Gooch, *Diaries*; Verne, *Floating City*, p. 6.
86. Gooch, *Diaries*, p. 114.
87. Russell and Dudley, *Atlantic Telegraph*, p. 75.
88. Ibid., p. 95.
89. Jim Rees, *The Life of Capt. Robert Halpin* (Arklow, Co. Wicklow, Ireland: Dee-Jay Publications, 1992), p. 60; Henry O' Neil, *The Atlantic Telegraph, 1865*, CWA.
90. "Willoughby Smith," *The Electrician*, January 31, 1885, p. 239.
91. For a recording of service messages, see *S.S. Great Eastern, Signal Log whilst Laying Atlantic Cable of 1866*, Atlantic Telegraph 1866/67, CWA.
92. Robert Dudley, *The Great Eastern Telegraph 1866 and Test-Room Chronicle. Containing all the Latest Home, Colonial, and Continental News and Being an Epitome of Political, Social, Ecclesiastical, Military, Naval, Legal, Medical, Artistic, Literary, Dramatic, Commercial, Electrical, Sporting, Geographical, Geological, Zoological, Mythological and General Intelligence, 1866*, CWA.
93. "The Atlantic Telegraph Expedition," *Daily News*.
94. Gooch, *Diaries*, p. 100.
95. Gooch, *Diaries*, p. 102.
96. O'Neil, *Atlantic Telegraph*.
97. Ibid.
98. Ibid.
99. Bright, *Submarine Telegraphs*, p. 90; Russell and Dudley, *Atlantic Telegraph*, pp. 92–93.
100. Field, *Atlantic Telegraph*, pp. 330–33.
101. Graves, *Telegraphic Service*, p. 43.
102. Bright, *Submarine Telegraphs*, p. 91.
103. J. C. Deane, G. V. Poore, and illustrated by Robert Dudley, *Great Atlantic Haul: Contentina—or the ROPE!! The GRAPNELL !!!! and the YANKEE DOODLE !!!!!!!! Being a Great Eastern Mystery Typifying the Instructive Story of Cyrus in Search of His Love*, September 17, 1866, Atlantic Telegraph 1866/67, CWA; N. A. Woods and J. C. Parkinson, *Being a CABLEISTIC and Great Eastern EXTRAVAGANZA*, July 1866, CWA.
104. Deane and Poore, *Contentina*, p. 1.
105. "Completion of the Atlantic Telegraph," *The Caledonian Mercury*, July 30, 1866.
106. Carter, *Cyrus Field*, p. 169.

107. Woods and Parkinson, *Being a CABLEISTIC*, p. 2, CWA.
108. Henry A. Moriarty to the Editor of the Standard, March 8, 1870, *Private Letter Book of Captain Henry August Moriarty, Aug. 1865–Dec. 1884*, CWA (hereafter cited as Moriarty Letter Book CWA).
109. Ibid.
110. Field, *Atlantic Telegraph*, p. 36.
111. Telcon, for instance, at its incorporation led off with a subscription of £100,000. Ten gentlemen, all cable people, each put down £10,000. Among them were Richard Glass, Cyrus W. Field, Daniel Gooch, Thomas Brassey, and John Pender. Ibid., p. 339.
112. David P. Nickles, "Telegraph Diplomats: The United States' Relations with France in 1848 and 1870," *Technology and Culture* 40.1 (1999): p. 9.
113. "Wednesday, August 1, 1866: The Atlantic Telegraph," *Trewman's Exeter Flying Post*, August 1, 1866.
114. Gooch, *Diaries*, p. 144.
115. Abram Hewitt to J. C. Neilson, Esq., Baltimore, August 12, 1858, Hewitt Papers, CUA.
116. "Atlantic Telegraph," *Times*, July 27, 1866.
117. G. C. Boase and rev. Anita McConnell, "Lampson, Sir Curtis Miranda, First Baronet (1806–1885)," in *Oxford Dictionary of National Biography*, ed. H. C. G. Matthew et al. (Oxford: Oxford University Press, 2004), accessed August 26, 2011, http://www.oxforddnb.com /view/article/15957.
118. Field, *Atlantic Telegraph*, p. 243; Judson, *Field*, p. 235.
119. Spear, "Canning"; Lynch, "Bright."
120. "The French Atlantic Cable: Speech by Sir James Anderson," *Glasgow Herald*, August 13, 1869.
121. Buchanan, "Brunel." Until 1874, the *Great Eastern* served primarily as a cable-laying ship. Then she remained docked for twelve years. Some wished to use her for meat storage, whereas others suggested she be used as a smallpox hospital, a poor house, or even a sewage disposal ship for the city of London. Eventually she served as a floating billboard with slogans painted on her massive hull. In 1889 she was broken up. It took two hundred men working around-the-clock shifts two years to finish the job. Rees, *Robert Halpin*, pp. 124–25.
122. Hunt, "Varley."
123. Bright, *Submarine Telegraphs*, p. 155.
124. Charles Bright, *Sir Charles Tilston Bright Civil Engineer: With Which Is Incorporated the Story of the Atlantic Cable, and the First Telegraph to India and the Colonies* (London: Archibald Constable & Co, 1908), pp. 202, 308.
125. Channon, "Gooch."
126. Bright, *Charles Tilston Bright*, p. 265.
127. Nabyraouy Bey to Cyrus W. Field, September 25, 1869, Cyrus Field Papers, New York Public Library (hereafter cited Field Papers, NYPL); Gordon, *Thread Across the Ocean*, p. 175.

128. Emma Pender to William des Voeux, November 23, 1875, Emma Pender Papers, CWA.
129. The Globe Telegraph Company, *Proceedings*, p. 18; similarly Bright, *Charles Tilston Bright*, p. 234. On the culture of friendship see Frank Schimmelfennig, *Internationale Politik* (Paderborn: Schöningh, 2010), p. 180.
130. Abram Hewitt to Sir John Pender, November 27, 1872, Abram Hewitt Papers, CUA.
131. Allan Nevins, *Abram S. Hewitt. With Some Account of Peter Cooper* (New York/London, 1935), p. 450.
132. Gooch, *Diaries*, p. 125.
133. John Pender to Lord Wilton, October 10, 1866, CWA.
134. Telegraph Construction and Maintenance Company, *Telcon Story*, p. 79.
135. Bright, *Submarine Telegraphs*, pp. 156, 380.
136. Gillian Cookson, *The Cable: The Wire That Changed the World* (Strout: Tempus Publishing, 2003); Tom Standage, *The Victorian Internet: The Remarkable Story of the Telegraph and the Nineteenth Century's Online Pioneers* (London: Phoenix, 2003).
137. Gordon, *Thread Across the Ocean*; Vary T. Coates and Bernard S. Finn, *A Retrospective Technology Assessment: Submarine Telegraphy—The Transatlantic Cable of 1866* (San Francisco: San Francisco Press, 1979).
138. Emma Pender to William des Voeux, April 15, 1877, Emma Pender Papers, CWA.
139. "Occasional Notes," *Pall Mall Gazette*, October 27, 1866.
140. Graves, *Autobiography*, p. 66. See also William Thomson, "Atlantic Telegraph Cable. Address of Professor William Thomson Delivered before the Royal Society of Edinburgh, December 18th, 1865, with Other Documents, 1866," Atlantic Telegraph Company, March 1862–September 1870, CWA.
141. "Occasional Notes," *Pall Mall Gazette*.
142. 41st Congress, 2nd Session, Bill H.R. 1778, April 14, 1870; 41st Congress, 2nd Session. Bill S. 958. May 31, 1870.
143. Frank Bösch, *Medialisierte Ereignisse: Performanz Inszenierung und Medien seit dem 18. Jahrhundert* (Frankfurt: Campus-Verl., 2010).
144. Bright, *Submarine Telegraphs*, p. 89; Russell and Dudley, *Atlantic Telegraph*, p. 98.
145. John Mullaly, *The Laying of the Cable, or the Ocean Telegraph* (New York: D. Appleton and Company, 1858); Bright, *Submarine Telegraphs*, p. 89.
146. Saward, *Trans-Atlantic Submarine Telegraph*; Willoughby Smith, *The Rise and Extension of Submarine Telegraphy* (New York: Arno Press, 1974); Field, *Atlantic Telegraph*; Bright, *Charles Tilston Bright*.
147. Smith, *Submarine Telegraphy*, p. vi.
148. "A Telegraphic Evening Party," *Illustrated London News*, July 2, 1870.
149. The Globe Telegraph Company, *Proceedings*, p. 11.
150. Ibid., p. 2.

151. "A Submarine Telegraph Memorial," *New York Times*, October 7, 1896.
152. Winseck and Pike, *Communication and Empire*, p. 17.
153. Carl Siemens to Werner Siemens, December 17, 1872, Collection of the Brothers' Letters 1842–1892, Siemens Corporate Archive (hereafter cited as Letters SCA).
154. James A. Scrymser, *Personal Reminiscence of James A. Scrymser: In Times of Peace and War* (New York, 1915), p. 68.

2. THE BATTLE FOR CABLE SUPREMACY

1. Charles Bright, *Submarine Telegraphs: Their History, Construction and Working* (London: C. Lockwood, 1898), p. 32. Puck describes his journey to fetch the magical flower in the terms that he is able "to put a girdle round about the earth in forty minutes"; William Shakespeare, *A Midsummer Night's Dream* (Stuttgart: Reclam, 2011), act 2, scene 1.
2. Daniel R. Headrick, *The Tentacles of Progress: Technology Transfer in the Age of Imperialism, 1850–1940* (New York: Oxford University Press, 1988), p. 100; Great Britain and Joint Committee to Inquire into the Construction of Submarine Telegraph Cables, *Report of the Joint Committee appointed by the Lords of the Committee of Privy Council for Trade and the Atlantic Telegraph Company to Inquire into the Construction of Submarine Telegraph Cables* (London: G.E. Eyre and W. Spottis-Woode, 1861).
3. Jorma Ahvenainen, "The Role of Telegraphs in the 19th Century Revolution of Communication," in *Kommunikationsrevolutionen: Die neuen Medien des 16. und 19. Jahrhunderts*, ed. Michael North (Köln: Böhlau, 1995), p. 74; Robert Boyce, "Submarine Cables as a Factor in Britain's Ascendancy as a World Power, 1850–1914," in *Kommunikationsrevolutionen: Die neuen Medien des 16. und 19. Jahrhunderts*, ed. Michael North (Köln: Böhlau, 1995), p. 82.
4. John Tully, "A Victorian Ecological Disaster: Imperialism, the Telegraph, and Gutta-Percha," *Journal of World History* 20 (2009): pp. 559–79.
5. William Montague-Hay took over the cable business from his brother John Hay, who had served as director of the Telegraph Construction and Maintenance Company. Henry M. Field, *History of the Atlantic Telegraph* (New York: C. Scribner, 1866), p. 367.
6. Pascal Griset, *Entreprise, Technologie et Souveraineté: Les télécommunications transatlantiques de la France, XIXe-XXe siècles* (Paris: Editions Rive Droite, 1996), p. 6.
7. "Our London Correspondence," *Glasgow Herald*, September 2, 1875.
8. Anita McConnell, "Pender, Sir John," in *Oxford Dictionary of National Biography*, ed. H. C. G. Matthew et al. (Oxford: Oxford University Press, 2004), accessed September 15, 2008, http://www.oxforddnb.com/view /article/21831.
9. Headrick, *Tentacles of Progress*, p. 97.
10. "A Submarine Telegraph to India," *Times*, May 3, 1853; H. B. Rawlinson and William Boutcher, "The Telegraph to India," *Times*, May 31, 1858.

11. Headrick, *Tentacles of Progress*, p. 99; Daniel R. Headrick, *The Invisible Weapon: Telecommunications and International Politics, 1851–1945* (New York: Oxford University Press, 1991), p. 19.
12. G. R. M. Garratt, *One Hundred Years of Submarine Cables* (London: H. M. Stationary Office, 1950), p. 28–29; K. C. Baglehole, *A Century of Service: A Brief History of Cable and Wireless, 1868–1968* (Essex: Anchor Brendon Limited, 1969), p. 3.
13. Robert A. Buchanan, *The Engineers: A History of the Engineering Profession in Britain, 1750–1914* (London: Kingsley, 1989), p. 108.
14. Charles Bright, "The Extension of Submarine Telegraphy in a Quarter-Century: Reprinted from *Engineering Magazine* (December 1898)," in *Development of Submarine Cable Communications*, ed. B. S. Finn, pp. 417–28 (New York: Arno Press, 1980), p. 420.
15. J. Brown, *The Cable and Wireless Communications of the World: A Survey of Present Day Means of International Communication by Cable and Wireless* (London: Sir Isaac Pitman & Sons, 1927), p. 11; Headrick, *Tentacles of Progress*, p. 105.
16. Bright, *Submarine Telegraphs*, p. 167.
17. Headrick, *Tentacles of Progress*, p. 105; Robert Boyce, "Imperial Dreams and National Realities: Britain, Canada and the Struggle for a Pacific Telegraph Cable, 1879–1902," *The English Historical Review* 115, no. 460 (2000): p. 43.
18. Gillian Cookson, *The Cable: The Wire That Changed the World* (Strout: Tempus Publishing, 2003), p. 100.
19. "Memoriam Booklet for Sir James Anderson," 1893, James Anderson Papers, CWA.
20. See James Anderson, "Anglo-Mediterranean Telegraphs: To the Editor of the Times," *Times*, September 24, 1868; "The Duplex System in Telegraphing: To the Editor of the Times," *Times*, November 15, 1878; "Submarine Cables: To the Editor of the Times," *Times*, November 21, 1881.
21. "The Brazilian Submarine Telegraph Company," *Daily News*, October 15, 1874.
22. Eric J. Hobsbawm, *The Age of Capital: 1848–1875* (New York: Vintage Books, 1996), p. 1.
23. Bright, "Extension of Submarine Telegraphy," p. 417.
24. See Emma Pender to Marion and William des Voeux dated April 19, 1878; July 4 & 16, 1878; March 20, 1879, Emma Pender Papers, CWA.
25. Tully, "Ecological Disaster," p. 563.
26. Helmuth Pfitzner, *Seekabel und Funktelegraphie: Im überseeischen Schnellnachrichtenwesen* (Leipzig: Curt Böttger Verlag, 1931), pp. 47–48; Jill Hills, *The Struggle for Control of Global Communication: The Formative Century* (Urbana: University of Illinois Press, 2002), p. 16; Headrick, *Invisible Weapon*, p. 28.
27. Henniker Heaton, "The Cable Telegraph System of the World," *The ARENA* 38, no. 214, (September 1907): pp. 226–29; "Telegraphy and Submarine Cables," Sandford Fleming Papers, National Archives Canada (hereafter cited as Fleming Papers, NAC).

28. Simultaneously to the Eastern and Associated Companies' telegraph system, other submarine telegraph companies were formed. While the Eastern system stretched eastward to India and Southeast Asia, the Great Northern Telegraph Company, an 1869 amalgamation of Danish, Norwegian, Russian, and English interests, established telegraphic communication via the Baltic Sea and the Russian landlines with Shanghai, where it met the telegraph lines of the Eastern. The Great Northern Company became the fiercest competitor of the Eastern and Associated Companies with regards to the Chinese and Japanese telegraph markets. In southern Europe, the Direct Spanish Telegraph Company (1872) established telegraphic communication between Spain and England with a cable between Falmouth and Bilbao. In South America, two telegraph companies, namely the Brazilian Submarine Telegraph Company (1873) and the Western and Brazilian Telegraph Company (1873), connected Brazil with other South American countries. Both companies had an agreement to "work in unison," and their cables linked with the system of the West India and Panama Telegraph Company and North America. G. R. M. Garratt, *One Hundred Years of Submarine Cables* (London: H. M. Stationery Office, 1950), pp. 29–30; Bright, *Submarine Telegraphs*, pp. 125–26; On the Great Northern, see Daqing Yang, "Submarine Cables and the Two Japanese Empires," in *Communications Under the Seas: The Evolving Cable Network and Its Implications*, ed. Bernard S. Finn and Daqing Yang (Cambridge, MA: MIT Press, 2009), p. 230.
29. Bright, *Submarine Telegraphs*, p. 128.
30. "Memoriam Booklet for Sir James Anderson," 1893, James Anderson Papers, CWA.
31. Bright, *Submarine Telegraphs*, p. 46.
32. "Private Correspondence," *Birmingham Daily Post*, October 29, 1866; Atlantic Telegraph Company, "Report of the Proceedings at the Ordinary General Meeting of the Shareholders, Held at the London Tavern, Bishopsgate Street, within the City of London, on Thursday, the 3rd Day of February, 1870," Atlantic Telegraph Company, CWA, p. 3.
33. "The French Atlantic Telegraph," *Daily News*, February 1, 1869.
34. New Atlantic Telegraph Company Ltd., 1872, Public Record Office, National Archives Kew; Direct Atlantic Telegraph Company Ltd., 1872, Public Record Office, National Archives Kew; For the U.S. context, see the numerous cable bills brought before the 41st Congress.
35. Ibid.
36. Gillian Cookson, "Ruinous Competition: The French Atlantic Telegraph of 1869," *Entreprises et Histoire* 23 (1999): p. 99.
37. "The French Atlantic Telegraph," *Daily News*, February 1, 1869.
38. Bright, *Submarine Telegraphs*, p. 107.
39. James Graves, *Thirty Six Years in the Telegraphic Service 1852–1888 Being a Brief Autobiography of James Graves MSTE*, ed. Donard de Cogan (1909/2008), p. 43, accessed July 16, 2015, https://dandadec.files.wordpress.com/2013/07/technical-autobiography.pdf; Dwayne R. Winseck and Robert M. Pike, *Communication*

and Empire: Media, Markets, and Globalization. 1860–1930 (Durham: Duke University Press, 2007), p. 50.

40. George Saward, *The Trans-Atlantic Submarine Telegraph: A Brief Narrative of the Principal Incidents in the History of the Atlantic Telegraph Company* (London: 1878), p. 78.
41. Cookson, "Ruinous Competition," p. 103; Bright, *Submarine Telegraphs*, p. 130.
42. Winseck and Pike, *Communication and Empire*, p. 52.
43. Pascal Griset and Daniel R. Headrick, "Submarine Telegraph Cables: Business and Politics, 1838–1939," *Business History Review* 75 (2001): pp. 543–78.
44. Wilfried Feldenkirchen, ed., *Siemens: Von der Werkstatt zum Weltunternehmen*, (München: Piper, 2003); Georg Siemens, *History of the House of Siemens* (Freiburg: Alber, 1957).
45. E. F. Bamber, ed., *The Scientific Works of William Siemens, Civil Engineer: A Collection of Addresses, Lectures, Etc.* III (London: John Murray, Albemarle Street, 1889), pp. 49–51; James Anderson, "The Telegraph to India: To the Editor of the Times," *Times*, December 4, 1868; John M. Champain, "To the Editor of the Times," *Times*, December 4, 1868; James Anderson, "The Telegraph to India: To the Editor of the Times," *Times*, December 7, 1868.
46. Carl Siemens to Werner Siemens, December 17, 1872, Letters SCA.
47. Bamber, *Scientific Works of William Siemens*, pp. 19, 38.
48. E. W. Cook to William Siemens, cited in *A Collection of Letters to Charles William Siemens 1823–1883: With a foreword by Sir George H. Nelson*, ed. George H. Nelson, May 30, 1878 (London: The English Electric Company Limited, 1953), p. 108.
49. J. D. Scott, *Siemens Brothers 1858–1958: An Essay in the History of Industry* (London: Weidenfeld & Nicholson, 1958), pp. 32–35.
50. Sigfrid von Weiher, *Die englischen Siemens-Werke und das Siemens-Überseegeschäft in der zweiten Hälfte des 19. Jahrhunderts* (Berlin: Duncker u. Humblot, 1990), p. 62; When Halske withdrew, the Siemens brothers formed a holding company of their three businesses in Berlin, London, and St. Petersburg. Each company retained its own administration and accountancy, but gains and losses were amalgamated. Werner von Siemens, *Lebenserinnerungen von Werner von Siemens* (Berlin: Julius Springer, 1908), p. 262.
51. Scott, *Siemens Brothers*, p. 88.
52. Ibid.
53. "Death of Sir John Pender," *Times*, July 8, 1898.
54. Werner Siemens to William and Carl Siemens, March 7, 1871, cited in von Weiher, *Siemens-Werke*, p. 98.
55. "A Cheap Ocean Telegraph," *Daily News*, May 2, 1874; "The Cheap Ocean Telegraph," *The Leeds Mercury*, May 5, 1874.
56. Carl Siemens to William Siemens, January 31, 1873, cited in von Weiher, *Siemens-Werke*, p. 98.

57. Mira Wilkins, *The History of Foreign Investment in the United States to 1914* (Cambridge, MA: Harvard University Press, 1989), p. 494.
58. The Globe Telegraph Company, *Proceedings*, pp. 6–7.
59. Wilkins, *U.S. Foreign Investment*, p. 495.
60. Scott, *Siemens Brothers*, p. 56.
61. "The Amalgamation of Cable Companies," *Glasgow Herald*, May 5, 1873; Field in: The Globe Telegraph Company, *Proceedings*, pp. 6–7.
62. Ibid., p. 8.
63. von Weiher, *Siemens-Werke*, p. 87.
64. "Glasgow," *Glasgow Herald*, March 13, 1873.
65. von Siemens, *Lebenserinnerungen*, p. 263. Translation mine.
66. Youssef Cassis, *Capitals of Capital: A History of International Financial Centres, 1780–2005* (Cambridge: Cambridge University Press, 2006).
67. William Kennedy, "Notes on Economic Efficiency in Historical Perspective: The Case of Great Britain, 1870–1914," *Research in Economic History* 9 (1984): pp. 109–41.
68. Direct United States Cable Company Ltd., "Annual List of Members and Summary of Capital and Shares of the Direct United States Cable Company Limited, made up to the 23rd day of July 1873," 1873, Public Record Office, National Archives Kew (hereafter cited as DUSC, Shareholders year).
69. Jehanne Wake, "Kleinwort, Alexander Frederick Henry (1815–1886)," in *Oxford Dictionary of National Biography*, ed. H. C. G. Matthew et al. (Oxford: Oxford University Press, 2004), accessed November 26, 2010, http://www.oxforddnb.com/view/article/48912; Albrecht Blank, "Ausgewählte Familien und Personen," accessed June 15, 2009, http://www.albrecht-blank.de/ahnenblan/pafg172.htm; Ferdinand Jezler, "Johann Conrad Im Thurn," in *Schaffhauser Beiträge zur Geschichte*, ed. Historischer Verein des Kantons Schaffhausen, I (1956): pp. 311–16.
70. Gary B. Magee and Andrew S. Thompson. *Empire and Globalisation: Networks of People, Goods and Capital in the British World, c. 1850–1914* (Cambridge: Cambridge University Press, 2010), p. 171.
71. Hobsbawm, *Age of Capital*, p. 40.
72. William Siemens to Werner Siemens, October 18, 1875, cited in von Weiher, *Siemens-Werke*, p. 101.
73. von Siemens, *Lebenserinnerungen*, pp. 242–44; Winseck and Pike, *Communication and Empire*, p. 53; "Atlantic Cable Companies," *Liverpool Mercury*, February 15, 1877.
74. von Weiher, *Siemens-Werke*, pp. 100–102.
75. Carl Siemens to Werner Siemens, December 13, 1875, cited in von Weiher, *Siemens-Werke*, p. 102.
76. Carl Siemens to Werner Siemens, December 20, 1875, cited in von Weiher, *Siemens-Werke*, p. 102. Translation mine.

77. Ibid.; William Siemens to Werner Siemens, December 13, 1875, cited in von Weiher, *Siemens-Werke*, p. 102. Translation mine.
78. "General News," *Bristol Mercury*, March 4, 1876. George von Chauvin to William Thomson, January 1876, SAA/30/Lm 286, SCA.
79. "Summary of this Morning's News," *Pall Mall Gazette*, March 29, 1877.
80. Emma Pender to William des Voeux, March 30, 1877, Emma Pender Papers, CWA.
81. "Summary of This Morning's News," *Pall Mall Gazette*, March 6, 1877.
82. "Direct United States Cable Company," *Daily News*, February 3, 1877; "Direct United States Cable Company," *Daily News*, February 6, 1877.
83. Emma Pender to William des Voeux, March 3, 1877, Emma Pender Papers, CWA.
84. "Summary of This Morning's News," *Pall Mall Gazette*, April 7, 1877; "Dangerous Legislation—If Allowed," *Daily Citizen*, April 12, 1880.
85. William Siemens to Werner Siemens, April 7, 1877, cited in von Weiher, *Siemens-Werke*, p. 104.
86. Carl Siemens to Werner Siemens, April 7, 1877, cited in von Weiher, *Siemens-Werke*, p. 104. Translation mine.
87. Emma Pender to William des Voeux, April 15, 1877, Emma Pender Papers, CWA.
88. "Special Resolution of the Direct United States Cable Company Limited, Passed 26th June, 1877," Public Record Office, National Archives Kew.
89. Winseck and Pike, *Communication and Empire*, p. 55.
90. *The Leeds Mercury*, "The Proposed Amalgamation of Telegraph Companies," February 3, 1876.
91. Werner Siemens to William Siemens, April 28, 1873, cited in von Weiher, *Siemens-Werke*, p. 104.
92. William Siemens, "Address as President of the Society of Telegraph Engineers: Delivered on January 23, 1878," in *The Scientific Works of William Siemens, Civil Engineer: A Collection of Addresses, Lectures, Etc.*, ed. E. F. Bamber (London: John Murray, Albemarle Street, 1899), pp. 177–78.
93. Scott, *Siemens Brothers*, p. 57.
94. See William and Werner Siemens' vast oeuvre on telegraphy: William Siemens, "On Determining the Depth of the Sea Without the Use of Sounding-Line," *Philosophical Transactions of the Royal Society of London* 166 (1876); William Siemens, "On Certain Means of Measuring and Regulating Electric Currents," *Proceedings of the Royal Society of London* 20 (1878–1879); William Siemens, "Deep-Sea Telegraphs" and "Testing Electric Cables," in *The Scientific Works of William Siemens, Civil Engineer*; or generally Werner von Siemens, ed., *Wissenschaftliche und Technische Arbeiten* (Berlin: Julius Springer, 1889).
95. Bernard S. Finn, "Submarine Telegraphy: A Study in Technical Stagnation," in *Under the Seas: The Evolving Cable Network and its Implications*, eds. Bernard S. Finn and Daqing Yang (Cambridge, MA: MIT Press, 2009), pp. 9, 19.

96. Rollo Appleyard, *The History of the Institution of Electrical Engineers (1871–1931)* (London: The Institution of Electrical Engineers, 1939), pp. 105–6.
97. D. H. Aldcroft, "The Entrepreneur and the British Economy, 1870–1914," *The Economic History Review* 17 (1964): pp. 113–15; David Edgerton, *Science, Technology and the British Industrial "Decline": 1870–1970* (Cambridge: Cambridge University Press, 1996).
98. David Edgerton and S. M. Horrocks, "British Industrial Research and Development Before 1945," *The Economic History Review* 47 (1994): pp. 213–38; Richard Noakes, "Industrial Research at the Eastern Telegraph Company, 1872–1929," *The British Journal for the History of Science* 47 (2014): 119–46.
99. Edgerton, *British Industrial "Decline,"* p. 1.
100. Bright, *Submarine Telegraphs*, p. 156.
101. William Siemens to Werner Siemens, October 18, 1875, Letters SCA.
102. Bright, *Submarine Telegraphs*, p. 154.
103. Scott, *Siemens Brothers*, p. 95.
104. George Robb, "Ladies of the Ticker: Women, Investment, and Fraud in England and America 1850–1930," in *Victorian Investments: New Perspectives on Finance and Culture*, eds. Nancy Henry and Cannon Schmitt (Bloomington: Indiana University Press, 2009), pp. 120–42, 120; David C. Itkowitz, "Fair Enterprise or Extravagant Speculation: Investment, Speculation, and Gambling in Victorian England," *Victorian Studies* 45 (2002): pp. 121–147, 121; Hobsbawm, *Age of Capital*, pp. 214–15; Niall Ferguson, *The Ascent of Money* (New York: Penguin Press, 2008), pp. 120–22.
105. David S. Landes, *The Unbound Prometheus: Technological Change and Industrial Development in Western Europe from 1750 to the Present* (Cambridge: Cambridge University Press, 2008), p. 206.
106. Bright, *Submarine Telegraphs*, p. 154.
107. "Banquet to Sir John Pender," *Times*.
108. Scott, *Siemens Brothers*, p. 60; see also Pfitzner, *Seekabel und Funktelegraphie*, p. 57.
109. Bright, *Submarine Telegraphs*, p. 165; Itkowitz, "Fair Enterprise or Extravagant Speculation," p. 132.
110. Bright, *Submarine Telegraphs*, p. 154.
111. A. Kunert, *Telegraphen-Seekabel* (Köln-Mühlheim: Karl Glitscher, 1975), p. 68.
112. "General News," *Birmingham Daily Post*, September 29, 1875; Emma Pender to William des Voeux, December 25, 1875, Emma Pender Papers, CWA.
113. "This Evening's News," *Pall Mall Gazette*, November 3, 1875.
114. "London, Monday December 13," *Daily News*, December 13, 1875.
115. Anglo-American Telegraph Company Ltd., "Information for Shareholders, January 14, 1868," Atlantic Telegraph Co., CWA.
116. Bright, *Submarine Telegraphs*, p. 153.
117. Pfitzner, *Seekabel und Funktelegraphie*, pp. 56–57.

118. The analysis is based on July 1873; April 1877, the last census before the refounding of the company; November 1877, the first census of the new company; 1887, the year of the last cable war between the Atlantic Pool and the Commercial Cable Company; and 1909, marking the time period after the completion of a Pacific cable and a girdle around the globe as well as the year when Guglielmo Marconi received the Nobel Prize in Physics for his accomplishments in wireless telegraphy. DUSC, "Shareholders 1873"; DUSC, "Shareholders April 1877"; DUSC, "Shareholders 1887"; DUSC, "Shareholders 1909," Public Record Office, National Archives Kew.
119. George Robb, *White-Collar Crime in Modern England. Financial Fraud and Business Morality, 1845–1929* (Cambridge: Cambridge University Press, 1992), pp. 3, 91.
120. Alex Preda, "The Rise of the Popular Investor: Financial Knowledge and Investing in England and France, 1840–1880," *The Sociological Quarterly* 42 (2001): pp. 207, 227–28; M. C. Reed, *Investment in Railways in Britain, 1820–1844: A Study in the Development of Capital Market* (Oxford: Oxford University Press, 1975).
121. Preda, "Popular Investor," p. 208; see also Alex Preda, "Socio-Technical Agency in Financial Markets: The Case of the Stock Ticker," *Social Studies of Science* 36 (2006): pp. 753–82.
122. Jim Rees, *The Life of Capt. Robert Halpin* (Wicklow, Ireland: Dee-Jay Publications, 1992) pp. 106–7.
123. Gooch cited in "The Amalgamation of Cable Companies," *Glasgow Herald*, May 5, 1873.
124. Details of the joint-purse agreement of 1877 are obscure, but they were presumably similar to the joint-purse agreement of the Anglo-American, French Atlantic, and Western Union Telegraph Companies of 1870; according to this agreement, the French Atlantic Company would, with one Atlantic cable, receive 36 ⅔ percent of the income, and the Anglo-American Company, which owned two Atlantic cables, would receive 63 ⅓ percent of the income. Once it had a second cable, the French Atlantic Company would receive 48 percent. Newfoundland &. T. C. New York, "Notice for Subscription of Shares," November 30, 1872, Field Papers, NYPL.
125. Bright, *Submarine Telegraphs*, p. 123; Bill Glover, "Direct United States Cable Company," accessed July 23, 2011, http://www.atlantic-cable.com/CableCos/DirectUS/index.htm.
126. *Sun*, 1893 cited in Glover, "Direct United States Cable Company."
127. Julia Ott, *When Wall Street Met Main Street* (Cambridge, MA: Harvard University Press, 2011).
128. In November 1877, the Globe held 7,945 shares, and in both 1887 and 1909, it held 9,945 shares.

129. In 1888, Frank Perry, superintendent at Heart's Content, Newfoundland, reported on "some half dozen shareholders in Heart's Content." F. Perry to T. H. Wells "Anglo-American Dividends," February 28, 1888, Vol. 8/Reel 10, *Letter Books Heart's Content Letter Book from Anglo-American Telegraph Co. Station at Heart's Content, Newfoundland. 1870–1876: Originals Preserved at Heart's Content Cable Museum, Under the Historic Resource Division, Department of Provincial Affairs, Government of Newfoundland and Labrador, St. John's Newfoundland,* ed. Anglo-American Telegraph Company, 11 vols. (reproduced 1973 by B. Finn, Smithsonian Institutions), Smithsonian Institutions Archives (hereafter cited as Letter Book SI).
130. E. Weedon to H. Weaver, July 14, 1874, Vol. 1/Reel 2, Letter Book SI.
131. Field, *Atlantic Telegraph*, p. 55.
132. Philip B. McDonald, *A Saga of the Seas: The Story of Cyrus W. Field and the Laying of the First Atlantic Cable* (New York: Wilson-Erickson, 1937), p. 144.
133. Donald R. Tarrant, *Atlantic Sentinel: Newfoundland's Role in Transatlantic Cable Communications* (St. John's, Newfoundland: Flanker Press, 1999), p. 92. On landlines, see Thomas C. Jepsen, *My Sisters Telegraphic: Women in the Telegraph Office, 1846–1950* (Athens: Ohio University Press, 2000).
134. Peter J. Bowler and Iwan R. Morus, *Making Modern Science: A Historical Survey* (Chicago: University of Chicago Press, 2005), p. 487.
135. Robert Beachy, ed., *Women, Business, and Finance in Nineteenth-Century Europe: Rethinking Separate Spheres* (Oxford: Berg, 2006), or Carmen D. Deere and Cheryl R. Doss, eds., *A Special Issue on Women and Wealth*, in *Feminist Economics* 12 (2006).
136. Josephine Maltby and Janette Rutterford, "'She Possessed Her Own Fortune': Women Investors from the Late Nineteenth to the Early Twentieth Century," *Business History* 48 (2006): pp. 220–53; Lance E. Davis and Robert A. Huttenback, *Mammon and the Pursuit of Empire: The Political Economy of British Imperialism, 1860–1912* (Cambridge: Cambridge University Press, 2010); Maltby Josephine and Janette Rutterford, "'The Widow, the Clergyman and the Reckless': Women Investors in England, 1830–1914," in *A Special Issue on Women and Wealth*, eds. Carmen D. Deere and Cheryl R. Doss, *Feminist Economics* 12 (2006): pp. 111–38, 113.
137. Emma Pender Papers, CWA.
138. "Visitors by Old Electric," *The Telegraphist. A Monthly Journal for Postal, Telephone and Railway Telegraph Clerks*, January (1886): p. 14.
139. Of these 444 women, we can distinguish 227 spinsters, 114 widows, 100 married women, 1 lady, and 1 princess.
140. Robb, "Ladies of the Ticker," p. 140; see also Mary B. Combs, "CUI BONO? The 1870 British Married Women's Property Act, Bargaining Power, and the Distribution of Resources within Marriage," in *A Special Issue on Women and*

Wealth, eds. Carmen D. Deere and Cheryl R. Doss, *Feminist Economics* 12 (2006): pp. 51–83.

141. In comparison, in April 1877, the number of female shareholders only increased to 28 women (12 spinsters, 14 widows, and 2 married women). From a total of 922 investors, they represent 3 percent of all shareholders.
142. Of those 784 shareholders, we can distinguish the following: 395 spinsters = 50.4 percent, 225 married women = 28.7 percent, 158 widows = 20.2 percent, and 4 ladies, 2 misses = 0.7 percent.
143. Sally Ledger, *The New Woman: Fiction and Feminism at the Fin de Siècle* (Manchester: Manchester University Press, 1997); Angelique Richardson and Chris Willis, *The New Woman in Fiction and Fact: Fin-de-Siècle Feminisms* (New York: Palgrave, 2002).
144. Robb, "Ladies of the Ticker," p. 120.
145. Ibid., p. 122.
146. Robb, "Ladies of the Ticker," p. 121; Preda, "Popular Investor," p. 216.
147. "Summary of This Morning's News," *Pall Mall Gazette*, January 27, 1876.

3. THE IMAGINED GLOBE

1. David Dudley Field to Cyrus W. Field, telegram, August 9, 1858, cited in *Harper's Weekly. A Journal of Civilization*, September 4, 1858, p. 16.
2. On the response in Germany and France, see Michael Geistbeck, *Der Weltverkehr: Telegraphie und Post, Eisenbahnen und Schiffahrt in ihrer Entwicklung dargestellt* (Freiburg: Herbersche Verlagshandlung, 1895), p. 486; Christian Holtorf, *Das erste transatlantische Telegraphenkabel von 1858 und seine Auswirkungen auf die Vorstellungen von Zeit und Raum* (Berlin: Humboldt-Universität Dissertation, 2009); Regine Buschauer, *Mobile Räume: Medien- und diskursgeschichtliche Studien zur Tele-Kommunikation*, (Bielefeld: Transcript-Verl., 2010), pp. 75–85; Pascal Griset, *Entreprise, Technologie et Souveraineté: Les télécommunications transatlantiques de la France, XIXe-XXe siècles* (Paris: Editions Rive Droite, 1996).
3. C. van Rensselaer, ed., *Miscellaneous Sermons, Essays and Addresses: by the Rev. Cortland van Resselaer* (Philadelphia: J.B. Lippincott, 1861), p. 207; Benedict Anderson, *Imagined Communities: Reflections on the Origin and Spread of Nationalism* (London: Verso, 2006).
4. Martin H. Geyer and Johannes Paulmann, eds., *The Mechanics of Internationalism: Culture, Society, and Politics from the 1840s to the First World War* (Oxford: Oxford University Press, 2008).
5. Queen Victoria to President James Buchanan, telegram, August 18, 1858, cited in Telegraph Construction and Maintenance Company, *The Telcon Story 1850–1950* (London: Telegraph Construction and Maintenance Company, 1950), p. 49.
6. King James Bible.

7. "Europa and America. Report of the Proceedings at an Inauguration Banquet, given by Mr. Cyrus W. Field, of New York, at the Palace Hotel, Buckingham Gate, London 1864," April 15, 1864, Isambard Kingdom Brunel, Papers, Bristol University Archive, p. 12.
8. Henry M. Field, *History of the Atlantic Telegraph* (New York: C. Scribner & Co, 1866), p. 217.
9. Eric J. Hobsbawm, *The Age of Capital: 1848–1875* (New York: Vintage Books, 1996), p. 65.
10. Charles Bright, *The Story of the Atlantic Cable* (New York: D. Appleton and Company, 1903), pp. 21–22.
11. Ronald Wenzlhuemer, *Connecting the Nineteenth-Century World: The Telegraph and Globalization* (Cambridge: Cambridge University Press, 2013), p. 37.
12. "Monday Morning, July 30," *Glasgow Herald*, July 30, 1866; "The Atlantic Telegraph," *Liverpool Mercury*, July 30, 1866.
13. "Wednesday, August 1, 1866: The Atlantic Telegraph," *Trewman's Exeter Flying Post*, August 1, 1866; "Monday Morning, July 30," *Glasgow Herald*.
14. "The Atlantic Telegraph," *Liverpool Mercury*.
15. Waldemar Kaempffert, "Communication and World Peace," *Proceedings of the American Philosophical Society* 70 (1931): p. 281; Hamid Mowlana, *Global Information and World Communication: New Frontiers in International Relations* (London: Sage, 1998), p. 157.
16. Richard R. John, *Network Nation: Inventing American Telecommunications* (Cambridge MA: Belknap Press of Harvard University Press, 2010), p. 16.
17. Ibid., p. 106.
18. Napoleon III, cited in David P. Nickles, *Under the Wire: How the Telegraph Changed Diplomacy* (Cambridge, MA: Harvard University Press, 2003), p. 6.
19. "Peace and Good-will between England and France," *Punch*, September 14, 1851, Newspaper Clippings, Jacob Brett Papers, IET Archives.
20. Samuel F. B. Morse to Norvin Green, July 1855, cited in Samuel F. B. Morse and Edward L. Morse, *Samuel F. B. Morse: His Letters and Journals* (New York: Da Capo Press, 1973), p. 345.
21. "Peace and Good-will between England and France," *Punch*, September 14, 1851, Newspaper Clippings, Jacob Brett Papers, IET Archives.
22. *Harper's Weekly. A Journal of Civilization*, September 4, 1858, p. 16.
23. David P. Nickles, "Submarine Cables and Diplomatic Culture," in *Communications under the Seas: The Evolving Cable Network and Its Implications*, eds. Bernard S. Finn and Daqing Yang (Cambridge MA: MIT Press, 2009), p. 210.
24. Anthony S. Travis, "Engineering and Politics: The Channel Tunnel in the 1880s," *Technology and Culture* 32 (1991): p. 461.
25. "Neighbours Getting Over Their Distance to One Another," *Punch* 41 (1861).
26. Cortlandt van Rensselaer, *Signals from the Atlantic Cable: An Address Delivered at the Telegraphic Celebration, September 1st, 1858, in the City Hall, Burlington, New Jersey* (Philadelphia: J. M. Wilson, 1858), p. 210.

27. Travis, "Engineering and Politics," p. 462.
28. Ferdinand de Lesseps to Maxime Hélène, cited in Daniel R. Headrick, *The Tentacles of Progress: Technology Transfer in the Age of Imperialism, 1850–1940* (Oxford: Oxford University Press, 1988), p. 113.
29. Iestyn Adams, *Brothers Across the Ocean: British Foreign Policy and the Origins of the Anglo-American "Special Relationship" 1900–1905* (London: Tauris Academic Studies, 2005), p. 10.
30. Queen Victoria to President Andrews Johnson, telegram, cited in Daniel Gooch, *Diaries of Sir Daniel Gooch Baronet: With an Introductory Notice by Sir Theodore Martin K.C.B.* (London: Kegan Paul, Trench Trübner & Co., 1892), p. 118.
31. *Harper's Weekly. A Journal of Civilization*, September 4, 1858, p. 15 [italics mine].
32. Reginald Horsman, *Race and Manifest Destiny: The Origins of American Racial Anglo-Saxonism* (Cambridge, MA: Harvard University Press, 1994), p. 62.
33. Philip E. Myers, *Caution and Cooperation: The American Civil War in British-American Relations* (Kent, OH: Kent State University Press, 2008), pp. 1–2.
34. John, *Network Nation*, p. 104.
35. Myers, *Caution and Cooperation*, pp. 1–2.
36. Adams, *Brothers Across the Ocean*, p. 10.
37. Geyer and Paulmann, "Introduction," in idem, *The Mechanics of Internationalism: Culture, Society, and Politics from the 1840s to the First World War* (Oxford: Oxford University Press, 2008, p 1-26, p. 13.
38. "English Enterprise and Irish Abuse," *The Nation*, June 1852.
39. "L'Arrivée à Paris, par le tunnel de la manche," 1884, Sidney M. Edelstein Library, Hebrew University of Jerusalem.
40. "Atlantic Telegrams," *Pall Mall Gazette*, August 4, 1866.
41. Ibid.
42. Istvan Kende, "History of Peace: Concept and Organization from the Late Middle Ages to the 1870s," *Journal of Peace Research* 26 (1989), p. 236.
43. Ibid., p. 237.
44. Adam Smith, *The Wealth of Nations: Adam Smith; ed. with Notes and Marginal Summary by Edwin Cannan* (New York: Bantam Dell, 2003); Paul Hirst, "Politics: Territorial or Non-Territorial?" in *Habitus: A Sense of Place*, ed. Jean Hillier (Aldershot: Ashgate, 2008), p. 78.
45. Jürgen Osterhammel, *Europe, the "West" and the Civilizing Mission* (London: German Historical Institute, 2006), p. 22.
46. Gregory Claeys, *Imperial Sceptics: British Critics of Empire, 1850–1920* (Cambridge: Cambridge University Press, 2010), p. 28.
47. Michael E. Howard, *War and the Liberal Conscience* (New York: Columbia University Press, 2008), p. 33.
48. David Nicholls, "The Manchester Peace Conference of 1853," *Manchester Region History Review* 5 (1991): p. 11; Martin Caedel, "Cobden and Peace," in *Rethinking Nineteenth-Century Liberalism: Richard Cobden Bicentenary Essays*, eds. Anthony

Howe, Simon Morgan, and Richard Cobden (Aldershot: Ashgate, 2006), p. 194.

49. *Punch* cited in Howard, *Liberal Conscience*, p. 33.
50. Bright, *Atlantic Cable*, p. 174.
51. John Bright to Cyrus W. Field, September 10, 1865, Field Papers, NYPL; John Bright, February 1863, in George Trevelyan, *The Life of John Bright* (Boston: Houghton Mifflin, 1913), p. 321; Frank Moore, ed., *Speeches of John Bright, M.P. on the American Question: With an Introduction by Frank Moore* (Boston: Little, Brown and Company, 1865); Isabella F. Judson, *Cyrus W. Field: His Life and Work* (New York: Harper & Brothers Publishers, 1896), pp. 283–92.
52. Judson, *Field*, p. 243.
53. Hartmut Berghoff, *Englische Unternehmer 1870–1914: Eine Kollektivbiographie führender Wirtschaftsbürger in Birmingham, Bristol und Manchester* (Göttingen: Vandenhoeck & Ruprecht, 1991), p. 53.
54. Nicholls, "Manchester Peace Conference," p. 14. David Nicholls, "Richard Cobden and the International Peace Congress Movement, 1848–1853," *Journal of British Studies* 30 (1991): pp. 351–76.
55. John Watts, *The Facts of the Cotton Famine* (London: Simpkin, Marshall & Co, 1866), p. iii.
56. James Anderson, "Telegraph Reform, being a Paper read by Sir James Anderson before a Congress of the Chambers of Commerce of the British Empire, at the Indian and Colonial Exhibition, South Kensington, on July 6th (and the Discussion upon the same on July 6th and 7th) 1886," July 6, 1886, in *Cables in Time of War Etc.*, ed. Anderson James, pp. 3–26 (E. C.: George Tucker, 1886), James Anderson Papers, CWA, p. 8.
57. Herbert Hovenkamp, *Enterprise and American Law 1836–1937* (Cambridge, MA: Harvard University Press, 1991), chapter 15.
58. William Spitznasski, "Festival Song at the Celebration of the Laying of the Atlantic Telegraph. Dedicated to the Atlantic Telegraph Company—Translated from the German," September 1, 1858, America Singing Collection, Library of Congress.
59. Rensselaer, "Signals from the Atlantic Cable," in *Miscellaneous Sermons*, p. 207.
60. William Adams to Cyrus W. Field, December 4, 1868, Field Papers, NYPL.
61. Peter J. Bowler and Iwan R. Morus, *Making Modern Science: A Historical Survey* (Chicago: University of Chicago Press, 2005), p. 341.
62. Samuel Carter, *Cyrus Field: Man of Two Worlds* (New York: Putman, 1968), p. 166.
63. DUSC, "Shareholders 1887," and DUSC, "Shareholders 1909," Public Record Office, The National Archives, Kew.
64. Samuel F. B. Morse to his wife, July 28, 1857, in Morse, *Samuel F. B. Morse*, p. 376.
65. Peter Cooper, "Pamphlets, No. 7: A Letter from Peter Cooper to the Delegates of the Evangelical Alliance," September 24, 1873, Cooper Union Archives (hereafter cited CUA).

66. Peter Cooper, *A Sketch of the Early Days and Business Life of Peter Cooper. An Autobiography* (New York: Cooper Union, 1877); Gano Dunn, *Peter Cooper (1791–1883) "A Mechanic of New York"* (New York: Newcomen Society of New York, 1949), p. 14.
67. Dunn, *Peter Cooper*, p. 11.
68. Philip E. A. Schaff to Cyrus W. Field, December 24, 1870, Field Papers, NYPL; Akira Iriye, *Global Community: The Role of International Organizations in the Making of the Contemporary World* (Berkeley, Calif.: University of California Press, 2004), p. 17.
69. Phillip E. A. Schaff to Cyrus W. Field, December 24, 1870, Field Papers, NYPL.
70. Carter, *Cyrus Field*, p. 58.
71. U.S. Sanitary Commission, "Scrap Book," U.S. Sanitary Commission Records/ English Branch Archives Box 940, NYPL.
72. Dunn, *Peter Cooper*, pp. 16–17.
73. U.S. Sanitary Commission, "Minutes Printed," U.S. Sanitary Commission Records/ English Branch Archives Box 940, NYPL.
74. The Eastern Telegraph Company, "Extracts from London 'Daily Telegraph' December 25–26, 1872. Special Submarine Telegrams from Our Own Correspondents: Christmas round the World," December 25, 1872, James Anderson Papers, CWA, p. 3.
75. Common Council of the City of New York, "Address of the Common Council of the City of New York to the New York, Newfoundland and London and Atlantic Telegraph Companies 1858," CUA.
76. The Eastern Telegraph Company, "Extracts from London 'Daily Telegraph' December 25–26, 1872. Special Submarine Telegrams from Our Own Correspondents: Christmas round the World," December 25, 1872, James Anderson Papers, CWA, p. 3.
77. "English Enterprise and Irish Abuse," *The Nation*, June 1852.
78. Iwan R. Morus, "The Nervous System of Great Britain: Space, Time and the Electric Telegraph in the Victorian Age," *The British Journal for the History of Science* 33 (2000): p. 474.
79. George Squier, "The Influence of Submarine Cables upon Military and Naval Supremacy," *Proceedings of the U.S. Naval Institute* XXVI (1901): p. 600.
80. Franz Scholz, *Krieg und Seekabel: Eine völkerrechtliche Studie* (Berlin: Verlag von Franz Vahlen, 1904), p. 1.
81. Madeleine Herren, *Hintertüren zur Macht: Internationalismus und modernisierungsorientierte Außenpolitik in Belgien, der Schweiz und den USA 1865–1914* (München: Oldenbourg, 2000), pp. 13–82.
82. Martti Koskenniemi, *The Gentle Civilizer of Nations: The Rise and Fall of International Law 1870–1960* (Cambridge: Cambridge University Press, 2010), p. 13.
83. "The Atlantic Telegraph," *Pall Mall Gazette*, July 3, 1868.
84. "Epitome of Opinion in the Morning Journals: The Atlantic Telegraph," *Pall Mall Gazette*, October 3, 1866.
85. "English Enterprise and Irish Abuse," *The Nation*, June 1852.

86. Pastor Rev. J. Bray, "On the Atlantic Cable," *New York Morning Express*, August 17, 1858.
87. Nickles, *Under the Wire*, p. 79. Similarly, Jonathan R. Winkler, *Nexus: Strategic Communications and American Security in World War I* (Cambridge, MA: Harvard University Press, 2008), and John Britton, "The Confusion Provoked by Instantaneous Discussion: The New International Communications Network and the Chilean Crisis of 1891–1892 in the United States," *Technology and Culture* 48 (October 2007): p. 732.
88. "Epitome of Opinion in the Morning Journals," *Pall Mall Gazette*.
89. "The French Atlantic Cable," *Glasgow Herald* [italics mine].
90. Gooch, *Diaries*, p. 208.
91. Léonard Laborie, *L'Europe mise en réseaux: La France et la coopération internationale dans les postes et les télécommunications (années 1850-années 1950)* (Brussels: Peter Lang, 2010).
92. Scholz, *Krieg und Seekabel*, p. 6; Samuel F. B. Morse to Cyrus W. Field, December 4, 1871, cited in Judson, *Field*, p. 280.
93. L. Oppenheim and Ronald Roxburgh, *International Law: A Treatise* (Clark, NJ: Lawbook Exchange, 2005), p. 447; A. Kunert, *Telegraphen-Seekabel* (Köln-Mühlheim: Karl Glitscher, 1962), p. 95.
94. Cyrus W. Field to Samuel F. B. Morse, December 28, 1871, cited in Judson, *Field*, p. 280.
95. Jorma Ahvenainen, "The International Telegraph Union: The Cable Companies and the Governments," in *Communications under the Seas: The Evolving Cable Network and Its Implications*, eds. Bernard S. Finn and Daqing Yang (Cambridge, MA: MIT Press, 2009), p. 69.
96. Judson, *Field*, p. 280.
97. On international cooperation concerning the government-run landlines, see Günther Krause, *Der Internationale Fernmeldeverein* (Frankfurt/Berlin: Alfred Metzner Verlag, 1960), p. 8.
98. Judson, *Field*, p. 281.
99. Jan Kropholler, *Internationales Einheitsrecht: Allg. Lehren* (Tübingen: Mohr, 1975), pp. 86–88.
100. Scholz, *Krieg und Seekabel*, p. 14.
101. Louis Renault, *Etudes sur les Rapports Internationaux: La Poste et le Télégraphes* (Paris: 1877).
102. Henry M. Field, *The Life of David Dudley Field* (New York: Charles Scribner's Sons, 1898), pp. 226–27.
103. Renault, *Etudes sur les Rapports Internationaux*, pp. 5–6 [translation mine].
104. Koskenniemi, *Gentle Civilizer*, p. 277.
105. Field, *David Dudley Field*, p. 246; on his ambition for an international law code, see ibid., pp. 219–42, or David D. Field, *An International Code: Address on this Subject, before the Social Science Association, at Manchester, October 5, 1866*, ed. YA Pamphlet Collection (New York: W. J. Read, 1867); David D. Field, *Draft Outlines*

of an International Code (New York: Diossy, 1872); David D. Field and Albéric Rolin, *Projet d'un code international* (Paris: Gand, 1881).

106. Renault, *Etudes sur les Rapports Internationaux*, pp. 6–7; Field, *David Dudley Field*, p. 245.
107. Kunert, *Telegraphen-Seekabel*, p. 98.
108. "Current Foreign Topics," *New York Times*, October 22, 1883.
109. Kunert, *Telegraphen-Seekabel*, pp. 99–104.
110. Charles Bright, "An All-British or Anglo-American Pacific Cable," in *Imperial Telegraphic Communication*, ed. Charles Bright (London: P.S. King & Son, 1911), p. 7; Canberra Department of Foreign Affairs and Trade, *Convention for the Protection of Submarine Telegraph Cables, 1884* (1901), Article II.
111. Oppenheim and Roxburgh, *International Law*, pp. 447–48.
112. "The Submarine Cable Conference," *Pall Mall Gazette*, October 23, 1883.
113. David D. Field, "Address at the de Lesseps Banquet: Given at Delmonicos, March 1, 1880," in *Speeches, Arguments and Miscellaneous Papers of David Dudley Field*, eds. David D. Field and A. P. Sprague (New York: D. Appleton and Company, 1884), p. 359.
114. Madeleine Herren, "Governmental Internationalism and the Beginning of a New World Order in the Late Nineteenth Century," in *The Mechanics of Internationalism: Culture, Society, and Politics from the 1840s to the First World War*, eds. Martin H. Geyer and Johannes Paulmann (Oxford: Oxford University Press, 2008), p. 128.
115. James Anderson cited in J. H. Heaton, "Imperial Telegraph Cables: To the Editor of the Times," *Times*, March 3, 1902.
116. Dutch delegate, cited in Kunert, *Telegraphen-Seekabel*, p. 96.
117. George E. Walsh, "Cable Cutting in War," *The North American Review* 167 (1898): p. 500.
118. "Submarine Cables in War," *Pall Mall Gazette*, May 14, 1886.
119. Squier, "Submarine Cables," p. 600.
120. Kunert, *Telegraphen-Seekabel*, p. 114.
121. Squier, "Submarine Cables," p. 621.
122. Ibid., pp. 111–12.
123. Squier, "Submarine Cables," p. 616; see Edgar Russel, Adolphus W. Greely, and Samuel Reber, *Handbook of Submarine Cables* (Washington, DC: Government Printing Office, 1905).
124. War Department, Office of the Chief of Staff, *Rules of Land Warfare* (Washington, DC: Government Printing Office, 1914), p. 118.
125. Oppenheim and Roxburgh, *International Law*, p. 448; War Department, Office of the Chief of Staff, *Rules of Land Warfare*, Article 344.
126. Dwayne R. Winseck and Robert M. Pike, *Communication and Empire: Media, Markets, and Globalization. 1860–1930* (Durham: Duke University Press, 2007), p. 14.

127. Kunert, *Telegraphen-Seekabel*, p. 115.
128. Winkler, *Nexus*, p. 5; Pascal Griset and Daniel R. Headrick, "Submarine Telegraph Cables: Business and Politics, 1838–1939," *Business History Review* 75 (2001): pp. 568–69.
129. James Bryce cited in Kaempffert, "Communication and World Peace," p. 283.
130. Geyer and Paulmann argue that proponents of internationalism were all missionaries of civilization; see Geyer and Paulmann, "Introduction," p. 9.
131. *Harper's Weekly*, September 4, 1858, p. 15.
132. Squier, "Submarine Cables," p. 600.
133. Rose Pender, *No Telegraph: Or, a Trip to Our Unconnected Colonies. 1878* (London: Gilbert & Rivington, 1879), p. 1.
134. William H. Preece, "The Functions of the Engineer. An Address Delivered to the Glasgow Association of the Students of the Institution of Civil Engineers," February 8, 1900, Jacob Brett Papers, IET Archives.
135. Boris Barth and Jürgen Osterhammel, eds., *Zivilisierungsmissionen: Imperiale Weltverbesserung seit dem 18. Jahrhundert* (Konstanz: UVK-Verl.-Ges., 2005).
136. Jürgen Osterhammel, *Die Verwandlung der Welt: Eine Geschichte des 19. Jahrhunderts* (München: Beck, 2009), p. 1173.
137. Harald Fischer-Tiné and Michael Mann, eds., *Colonialism as Civilizing Mission: Cultural Ideology in British India* (London: Anthem Press, 2004), p. 4.
138. "Epitome of Opinion in the Morning Journals," *Pall Mall Gazette*, July 27, 1866.
139. John W. Wayland, *The Pathfinder of the Seas: The Life of Matthew Fontaine Maury* (Richmond: Garrett & Massie, 1930).
140. Matthew F. Maury and L. Warrington, *Wind and Current Chart of the North Atlantic* (Washington, DC: U.S. Hydrographical Office, 1847/1852).
141. E. A. Marland, "British and American Contributions to Electrical Communication," *The British Journal for the History of Science* 1 (1962): pp. 44–45.
142. J. A. Stevens, "Report of the Proceedings of a Meeting Called to Further the Enterprise of the Atlantic Telegraph. Held at the Hall of the Chamber of Commerce, New York, March 4, 1863," Dibner Library, SI, pp. 6–7.
143. Michael Adas, *Machines as the Measure of Men: Science, Technology, and Ideologies of Western Dominance* (Ithaca: Cornell University Press, 1989), p. 141; also see the debate on the "great divergence": Kenneth Pomeranz, *The Great Divergence: China, Europe, and the Making of the Modern World Economy* (Princeton, NJ: Princeton University Press, 2000); Niall Ferguson, *Civilization: The West and the Rest* (London: Lane, 2011).
144. Morus, "Nervous System of Great Britain," p. 474.
145. See the visit of the Prince of Wales or visits to the Siemens' works or at Valentia Bay: "The Atlantic Telegraph Expedition: From our Special Correspondent," *Daily News*, July 19, 1866; "The Cheap Ocean Telegraph," *Leeds Mercury*, May 5, 1874.
146. Morus, "Nervous System of Great Britain," p. 474.

147. George Kennan, *Tent Life in Siberia: And Adventures among the Koraks and Other Tribes in Kamtchatka and Northern Asia* (New York: G.P. Putnam's Sons, 1877), pp. iii–iv.
148. John, *Network Nation*, p. 100.
149. George Kennan to his mother, December 26, 1866, George Kennan Papers, NYPL.
150. Ibid. Similarly, Yakup Bektas, "Displaying the American Genius: The Electromagnetic Telegraph in the Wider World," *The British Journal for the History of Science* 34 (2001): pp. 199–232.
151. "Epitome of Opinion in the Morning Journals," *Pall Mall Gazette*.
152. Adas, *Machines*, p. 144.
153. Perry Collins to Henry Seward, October 13, 1862, cited in Ernest N. Paolino, *The Foundation of the American Empire: William Henry Seward and U.S. Foreign Policy* (Ithaca: Cornell University Press, 1973), p. 48.
154. Ibid.
155. J. C. Parkinson, *The Ocean Telegraph to India* (London: William Blackwood and Sons, 1870), pp. 47–61.
156. Paul Gilmore, "The Telegraph in Black and White," *ELH* 69 (2002): pp. 805–33.
157. Jim Rees, *The Life of Capt. Robert Halpin* (Wicklow, Ireland: Dee-Jay Publications, 1992), p. 101.
158. E. Weedon, "Extract from Memo to General Manager Dated April 22, '76 'Land Staff,'" August 1885, Vol. 7/Reel 9, Letter Book SI.
159. Ronald Takaki, *Iron Cages: Race and Culture in 19th-Century America* (New York: Oxford University Press, 1979), p. 148.
160. Jeanette Choisi, *Wurzeln und Strukturen des Zypernkonfliktes 1878 bis 1990: Ideologischer Nationalismus und Machtbehauptung im Kalkül konkurrierender Eliten* (Stuttgart: Steiner, 1993), pp. 82–83.
161. Emma Pender to Marion des Voeux, July 4, 1878, Emma Pender Papers, CWA.
162. W. P. Granville, "Diary. Reminiscences of the Cyprus Alexandria Expedition 1878," CWA.
163. Don C. Seitz, *The James Gordon Bennetts—Father and Son, Proprietors of the New York Herald* (Ann Arbor: University of Michigan Library, 1974), p. 302; Henry M. Stanley, *The Autobiography of Sir Henry Morton Stanley* (Boston/New York: Houghton Mifflin Company, 1890), p. 318; Chauvin Municipal Office, "The Village of Chauvin: Our History," accessed August 9, 2011, http://www.villageofchauvin.ca/history.htm; Judson, *Field*, p. 311.
164. Charles Bright, *Submarine Telegraphs: Their History, Construction and Working* (London: C. Lockwood, 1898), p. 368.
165. David Harvey, "Between Space and Time: Reflections on the Geographical Imagination," *Annals of the Association of American Geographers* 80 (1990): p. 419.
166. Preece, "The Functions of the Engineer."
167. Lynch, *Charles Tilston Bright*, p. 223.

168. Similarly, Daniel R. Headrick, "A Double-Edged Sword: Communications and Imperial Control in British India," *Historical Social Research/Historische Sozialforschung* 35 (2010): pp. 51–65.
169. "From FUN," *Manchester Times*, August 18, 1866.
170. "The Belfast Newsletter," *Belfast Newsletter*, January 4, 1867; "London, Thurs. Jan. 3," *Daily News*, January 3, 1867.
171. Osterhammel, *Verwandlung der Welt*, pp. 143–44.
172. "Our Contemporaries: The Atlantic Cable," *Lloyds Weekly Newspaper*, September 9, 1866.
173. Harry Cranbrook Allen, *Great Britain and the United States: A History of Anglo-American Relations (1783–1952)* (Hamden, CT: Archon Books, 1969), p. 161.
174. John T. Flanagan, "Review: Domestic Manners of the Americans, by Frances Trollope," *American Literature*, March 1951, p. 159.
175. "Our Contemporaries," *Lloyds Weekly Newspaper*, September 9, 1866.
176. Michael Adas, *Dominance by Design: Technological Imperatives and America's Civilizing Mission* (Cambridge, MA: Belknap Press of Harvard University Press, 2006), p. 74.
177. Ralph Waldo Emerson, "The Young American: A Lecture Read before the Mercantile Library Association," Boston, MA, February 7, 1844.
178. Adas, *Dominance by Design*, pp. 73–74.
179. Ibid., p. 68.
180. Ibid., p. 84.
181. Francis Parkman and David Levin, *The Oregon Trail* (Harmondsworth: Penguin Books, 1983), p. 337; Adas, *Dominance by Design*, p. 84.
182. John, *Network Nation*, p. 45.
183. U.S. Senate, "Telegraph. Catalogue of Fine Art," accessed September 18, 2011, http://www.senate.gov/artandhistory/art/artifact/Painting_33_00019.htm#bio, p. 363.
184. Ibid.
185. "Europa and America," p. 22.
186. Adas, *Machines*, pp. 141–47.

4. WELTCOMMUNICATION

1. Stefan Zweig, *Sternstunden der Menschheit: Zwölf historische Miniaturen* (Frankfurt: Fischer, 1982).
2. Roland Wenzlhuemer, "The Dematerialization of Telecommunication: Communication Centres and Peripheries in Europe and the World, 1850–1920," *Journal of Global History* 2 (2007): pp. 345–72.
3. Roland Wenzlhuemer, "Editorial: Telecommunication and Globalization in the Nineteenth Century," *Historical Social Research. Historische Sozialforschung* 35, no. 1 (2010): pp. 7–18, p. 9; Jorma Ahvenainen, "The Role of Telegraphs in the

19th Century Revolution of Communication," in *Kommunikationsrevolutionen: Die neuen Medien des 16. und 19. Jahrhunderts*, ed. Michael North (Köln: Böhlau, 1995), pp. 74–79.

4. Frank Hartmann, *Globale Medienkultur: Technik, Geschichte, Theorien* (Wien: WUV, 2006), pp. 11, 79; Ernst Kapp, *Grundlinien einer Philosophie der Technik: Zur Entstehungsgeschichte der Cultur aus neuen Gesichtspunkten* (Braunschweig, 1877).
5. Tom Standage, *The Victorian Internet: The Remarkable Story of the Telegraph and the Nineteenth Century's Online Pioneers* (London: Phoenix, 2003); Charles H. Cooley, *Social Organization: A Study of the Larger Mind* (New York: C. Scribner's, 1909), chapter 8; Rudolf Stichweh, "The Genesis and Development of a Global Public Sphere," revised version of a paper first published in *Development* 46 (2003): pp. 26–29, http://www.fiw.uni-bonn.de/demokratieforschung/personen/stichweh/pdfs/51_2stwglobalpublic.pdf; Marshall McLuhan, *The Global Village: Transformations in World Life and Media in the 21st Century* (New York: Oxford University Press, 1992).
6. Marshall McLuhan, *Understanding Media: The Extensions of Man* (London: Routledge, 2001).
7. Eric J. Hobsbawm, *The Age of Capital: 1848–1875* (New York: Vintage Books, 1996), p. 60.
8. "The Atlantic Telegraph Expedition," *Daily News*, July 21, 1866.
9. "Atlantic Telegraph," *Birmingham Daily Post*, October 8, 1866.
10. *Belfast News-Letter*, December 13, 1866.
11. Jacob Brett Papers, Newspaper Clippings, IET Archives, p. 146.
12. Charles Bright, *Submarine Telegraphs: Their History, Construction and Working* (London: C. Lockwood, 1898), p. 143–44.
13. Anglo-American Telegraph Company Limited, "Tariff Book," p. 9.
14. H. Thurn, *Die Seekabel unter besonderer Berücksichtigung der deutschen Seekabeltelegraphie: In technischer, handelswirtschaftlicher, verkehrspolitischer und strategischer Beziehung dargestellt* (Leipzig: S. Hirzel, 1909), p. 193.
15. Anglo-American Telegraph Company limited, "Tariff Book," pp. 1, 6.
16. Terhi Rantanen, *When News Was New* (Malden, MA: Wiley-Blackwell, 2009), chapter 1.
17. "A New Atlantic Cable," *Western Mail*, May 22, 1874.
18. "Our London Correspondence," *Glasgow Herald*, September 2, 1875.
19. See also PQ's advertisement, La Compagnie Française du Télégraphe de Paris à New York, "Advertisement," *Freeman's Journal and Daily Commercial Advertiser*, March 2, 1880.
20. "Commercial News," *Glasgow Herald*, June 2, 1880. The difference in tariffs between the Anglo-American and its working partner DUSC of 1.5 shillings is probably due to the faster working speed of the latter's cables.
21. La Compagnie Française du Télégraphe de Paris à New York, "Advertisement," *Liverpool Mercury*, September 20, 1880; "The Financial Position," *Pall Mall Gazette*, April 20, 1886.

22. "Latest News," *Freeman's Journal and Daily Commercial Advertiser*, September 24, 1880; "The Anglo-American Telegraph Company," *Leeds Mercury*, January 15, 1881.
23. "Saturday Morning, September 25," *Glasgow Herald*, September 25, 1880.
24. John Pender, "Letter to Jay Gould, December 11th: The New American Telegraph Company," *Glasgow Herald*, December 30, 1880.
25. Pascal Griset and Daniel R. Headrick, "Submarine Telegraph Cables: Business and Politics, 1838–1939," *Business History Review* 75 (2001): p. 556.
26. Bright, *Submarine Telegraphs*, pp. 143–44; Dwayne R. Winseck and Robert M. Pike, *Communication and Empire: Media, Markets, and Globalization. 1860–1930* (Durham: Duke University Press, 2007), p. 146; According to Bernhard Finn, there was another tariff reduction in 1904, for which, however, I can find no verification. Vary T. Coates and Bernard S. Finn, *A Retrospective Technology Assessment: Submarine Telegraphy—The Transatlantic Cable of 1866* (San Francisco: San Francisco Press, 1979), p. 87.
27. S. A. Goddard, "The Atlantic Telegraph: To the Editor of the Daily Post," *Birmingham Daily Post*, August 2, 1866; Richard R. John, *Network Nation: Inventing American Telecommunications* (Cambridge, MA: Belknap Press of Harvard University Press, 2010), p. 185.
28. John F. C. Harrison, *Late Victorian Britain, 1875–1901* (London: Routledge, 2000), p. 30.
29. J. H. Heaton, "The Cable Telegraph System of the World," *Financial Review of Reviews* (1908), p. 228.
30. Ibid.
31. Jürgen Wilke, "The Telegraph and Transatlantic Communication Relations," in *Atlantic Communications: The Media in American and German History from the Seventeenth to the Twentieth Century*, eds. Norbert Finzsch and Ursula Lehmkuhl (Oxford: Berg, 2004), p. 119.
32. J. W. A. Blundell, *The Manual of Submarine Telegraph Companies 1872* (London: published by the author, 1872), p. 18.
33. Wilke, "Transatlantic Communication Relations," p. 119.
34. General Post Office, International Telegraph Convention, London, 1879. Proposition of German Administration, Receiver and Accountant General's Observation, November 27, 1877, POST 30/361, Part I, BT Archives, pp. 13–14.
35. "International Telegrams," *Daily News*, July 26, 1879.
36. William Maver, "Ocean Telegraphy: Extract from the Electrical World [American]," *The Telegraphist. A Monthly Journal for Postal, Telephone and Railway Telegraph Clerks* 2 (1884): pp. 87–88.
37. Ibid.
38. International Telegraph Union, *Internationale Telegraphen-Conferenz von St. Petersbourg: Vertrag, Dienstreglement und Tarifftabellen* (Bern: International Telegraph Union, 1875), p. 16.
39. Ibid.

40. Maver, "Ocean Telegraphy," p. 87.
41. For example, see Emma Pender to Marion des Voeux, May 7, 1877, Emma Pender Papers, CWA.
42. Emma Pender to Marion des Voeux, February 18, 1877, Emma Pender Papers, CWA.
43. William des Voeux to Emma Pender, January 28, 1878, Emma Pender Papers, CWA.
44. Emma Pender to Marion des Voeux, April 14, 1878, Emma Pender Papers, CWA.
45. Anglo-American Telegraph Company Limited, "Tariff Book," p. 1.
46. Albert B. Chandler, *A New Code, or Cipher Specially Designed for Important Private Correspondence by Telegraph. Applicable as well to Correspondence by Mail* (Washington, DC, 1869); Robert Slater, *Telegraphic Code to Ensure Secrecy in the Transmission of Telegrams, Second Edition* (London, 1897).
47. Brandon Dupont, Alka Gandhi, and Thomas J. Weiss, "The American Invasion of Europe: The Long-Term Rise in Overseas Travel, 1820–2000," *NBER Working Paper Series* (May 2008): p. 19.
48. Dwight Golder, *Official Cable Code and General Information for European Tourists Including French and German Phrases with English Pronunciation* (New York, 1887); Anglo-American Telegraphic Code and Cypher Co., *The Anglo-American Telegraphic Code to Cheapen Telegraphy and to Furnish a Complete Cypher* (New York, 1891); J. E. Palmer, *Palmer's European Travelers and Telegraph Code* (1880).
49. Golder, *Official Cable Code and General Information for European Tourists,* p. 111.
50. The Globe Telegraph Company, *The Globe Telegraph Company: Report of the Proceedings at an Anniversary Banquet given by Mr. Cyrus Field of New York at The Buckingham Palace Hotel, London, on Monday, the 10th March, 1873. In Commemoration of the Signature of the Agreement on the 10th of March, 1854 for the Establishment of a Telegraph Across the Atlantic,* March 10, 1873, James Anderson Papers, Porthcurno Cable and Wireless Archive, p. 10.
51. Anglo-American Telegraph Company Limited, "Tariff Book," p. 1.
52. James Anderson, "Telegraph Reform, Being a Paper Read by Sir James Anderson Before a Congress of the Chambers of Commerce of the British Empire, at the Indian and Colonial Exhibition, South Kensington, on July 6th (and the Discussion upon the Same on July 6th and 7th) 1886, July 6, 1886," in *Cables in Time of War Etc.*, ed. Anderson James (E.C.: George Tucker, 1886), pp. 3–26; James Anderson Papers, CWA, p. 7.
53. John Pender, "Speech to the Delegates" cited in "International Telegraph Conference," *Times,* June 18, 1879. Jay Gould offered to sell Western Union for $80 million. John, *Network Nation,* p. 193.
54. Daniel R. Headrick, *The Invisible Weapon: Telecommunications and International Politics, 1851–1945* (New York: Oxford University Press, 1991), p. 13.
55. "The Belfast News-Letter," *Belfast News-Letter,* July 28, 1866.

56. Wilke, "Transatlantic Communication Relations," pp. 119, 130; John, *Network Nation*, p. 129.
57. John, *Network Nation*, p. 79.
58. Michael Geistbeck, *Der Weltverkehr: Telegraphie und Post, Eisenbahnen und Schiffahrt in ihrer Entwicklung dargestellt* (Freiburg: Herbersche Verlagshandlung, 1895), p. 487.
59. Oliver Boyd-Barrett and Terhi Rantanen, "The Globalization of News," in *The Globalization of News* (New York: Sage, 2005), p. 5.
60. Wolff in the *National-Zeitung*, November 27, 1849, cited in Rantanen, *When News Was New*, p. 32.
61. Graham Storey, *Reuters' Century, 1851–1951* (London: Parrish, 1951), p. 53; further see Winseck and Pike, *Communication and Empire*, p. 5; Mark D. Alleyne, *News Revolution: Political and Economic Decisions about Global Information* (New York: St. Martin's Press, 1997), p. 6.
62. John, *Network Nation*, p. 81. Michael Palmer, Oliver Boyd-Barrett, and Terhi Rantanen, "Global Financial News," in *The Globalization of News* (New York: Sage, 2005), p. 61.
63. Kenneth G. Beauchamp, *History of Telegraphy* (London: Institution of Electrical Engineers, 2001), p. 80; K. M. Shrivastava, *News Agencies from Pigeon to Internet* (Elgin, IL: New Dawn Press, 2007), p. 102.
64. Jill Hills, *The Struggle for Control of Global Communication: The Formative Century* (Urbana, IL: University of Illinois Press, 2002), p. 34.
65. T. C. Gawford to Cyrus W. Field, April 13, 1884, Field Papers, NYPL. The *Washington Post* suggests that Field's purchase was a response to difficulties between the Atlantic Cable Company and the Associated Press (AP). With the *New York Evening Mail*, he controlled a newspaper interested in the news of a rival press association to hold AP in check. "A Newspaper's Tribulation," *Washington Post*, May 20, 1878.
66. J. S. Ogilvie, *Life and Death of Jay Gould, and How He Made His Millions* (New York: Arno Press, 1981), pp. 106–7.
67. "A Napoleon of Finance," *Pall Mall Gazette*, May 12, 1881.
68. John, *Network Nation*, pp. 158–64.
69. John, *Network Nation*, pp. 157, 171. For details on Gould's Western Union takeover, see ibid., pp. 156–70.
70. Western Union entered the field of submarine telegraphy in 1873 with their purchase of a majority shareholding in the International Ocean Telegraph Company. Kenneth R. Haigh and Edward Wilshaw, *Cableships and Submarine Cables* (London: Coles, 1968), p. 251; on the financial scheme, see "The New American Telegraph Company," *Glasgow Herald*, December 30, 1880.
71. Julius Grodinsky, *Jay Gould: His Business Career* (Philadelphia: University of Philadelphia Press, 1957), p. 279.

72. Griset and Headrick, "Submarine Telegraph Cables," p. 556; Bright, *Submarine Telegraphs*, p. 132.
73. On Gordon Bennett, see Albert S. Crockett, *When James Gordon Bennett Was Caliph of Bagdad* (New York: Funk and Wagnalls, 1926); Don C. Seitz, *The James Gordon Bennetts—Father and Son, Proprietors of the New York Herald* (Ann Arbor: University of Michigan Library, 1974); Richard O'Connor, *The Scandalous Mr. Bennett* (Garden City, NY: Doubleday, 1962).
74. David R. Spencer, *The Yellow Journalism: The Press and America's Emergence as a World Power* (Evanston, IL: Northwestern University Press, 2007), p. 29; see also Michael Schudson, *Discovering the News: A Social History of American Newspapers* (New York: Basic Books, 1998).
75. Michael A. Longinow, "News Gathering," in *American Journalism: History, Principles, Practices*, eds. William D. Sloan and Lisa M. Parcell (Jefferson, NC: McFarland, 2002), p. 147.
76. "James Gordon Bennett: The Cheerless Treatment Accorded to Him by His Sweetheart's Brother," *Chicago Daily Tribune*, January 5, 1877; "Mr. Bennett's Ocean Journey: Picked Up from the 'Herald' Yacht and a Pleasant Trip Across from New York World," *Boston Daily Globe*, February 15, 1877.
77. "James Gordon Bennett Becomes a Benedict," *Boston Daily Globe*, September 11, 1914.
78. "J.G. Bennett Dies in Beaulieu Villa," *New York Times*, May 15, 1918.
79. "John William Mackay," *Dictionary of American Biography: Gale Biography in Context*, accessed October 5, 2009, http://ic.galegroup.com/ic/bic1/ReferenceDetailsPage/ReferenceDetailsWindow?displayGroupName =K12-Reference&prodId=BIC1&action=e&windowstate=normal&catId=&documentId=GALE%7CBT23 10008540&mode=view&userGroupName=nypl&jsid=a8d625d8ea4009a39925cec4967e1d6f.
80. Ethel Manter, *Rocket of the Comstock: The Story of John William Mackay* (Caldwell, ID: Caxton Printers, 1950), p. 200.
81. John, *Network Nation*, p. 178.
82. Bright, *Submarine Telegraphs*, pp. 135–36; Direct United States Cable Company Report, "Reports & Accounts," December 31, 1884, CWA. Ludwig Löffler to Carl and Werner Siemens, February 5, 1884. SAA/30 Lm 257 SCA.
83. Ludwig Löffler, Letter to Carl Siemens, St. Petersburg, April 11, 1886, SAA/30 Lm 257, 2 SCA.
84. Jay Gould, cited in John, *Network Nation*, p. 179.
85. "The War of Transatlantic Cable Rates," *Aberdeen Weekly Journal*, April 30, 1886.
86. Griset and Headrick, "Submarine Telegraph Cables," p. 557.
87. "Friday, August 3," *Liverpool Mercury*, August 3, 1888.
88. *Dictionary of American Biography*, "John William Mackay."
89. Griset and Headrick, "Submarine Telegraph Cables," p. 557.

90. "An Independent Cable," *New York Times*, July 3, 1882.
91. Ogilvie, *Jay Gould*, p. 125.
92. Richard A. Schwarzlose, *The Nation's Newsbrokers: The Rush to Institution, from 1865 to 1920* (Evanston, IL: Northwestern University Press, 1990), p. 14; Ogilvie, *Jay Gould*, p. 125; Menahem Blondheim, *News over the Wires: The Telegraph and the Flow of Public Information in America, 1844–1897* (Cambridge, MA: Harvard University Press, 1994), p. 165.
93. For a detailed account on Gould's news scheme in the early 1880s, see Blondheim, *News over the Wires*, pp. 157–68.
94. Maury Klein, *The Life and Legend of Jay Gould* (Baltimore: Johns Hopkins University Press, 1986), p. 394.
95. "His Majesty Jay Gould," *New York Times*, February 23, 1881. Similarly "It Can't Be Done: The News of the New York Associated Press Cannot Be Manipulated by Field and Gould," *Chicago Daily Tribune*, December 19, 1881.
96. *New York Herald* quoted in "The 'Herald' Deplores the Extraordinary Ambition of the Great Stock-Gambler," *Chicago Daily Tribune*, September 23, 1882.
97. See Klein, *Jay Gould*, p. 395.
98. "Gould's Censorship of Cable Messages," *New York Times*, November 24, 1882.
99. Seitz, *James Gordon Bennetts*, p. 340.
100. Ogilvie, *Jay Gould*, p. 128.
101. Official statement by Bennett and Mackay. "Mr. Mackay on His New Cable," *New York Times*, February 5, 1884.
102. Klein, *Jay Gould*, p. 312; the full story from Mr. Gould's perspective is told in "Mr. Gould and Mr. Bennett, *New York Tribune*, March 27, 1888.
103. From the *Herald* 1882, cited in ibid., p. 313.
104. See New York *Herald* reprinted in "Jay Gould's Methods: His Audacity in Declaring a Stock Fraudulent," *Chicago Daily Tribune*, November 25, 1886; "Jay Gould's Latest Coup: By Commercial Cable to the Herald," *Paris Herald*, October 4, 1887. "London by Wire," *Paris Herald*, October 10, 1887; "A New Atlantic Cable: The Faraday Lays the Commercial Cable Company's Third Line Under the Ocean," *Paris Herald*, July 16, 1894.
105. Upon James Gordon Bennett's birth in 1841, the *Sun* suggested that the elder Bennett was not the father. Charles L. Robertson, *The International Herald Tribune: The First Hundred Years* (New York: Columbia University Press, 1987), p. 32.
106. Klein, *Jay Gould*, pp. 408–10; "To Sue Gould and Sage for Millions," *New York Tribune*, October 19, 1887.
107. Seitz, *James Gordon Bennetts*, p. 364; "Mr. Gould to Mr. Bennett," *New York Tribune*, April 1, 1888.
108. "Mr. Gould to Mr. Bennett: A Caustic Open Letter," *Washington Post*, April 2, 1888.
109. Klein, *Jay Gould*, p. 410.

110. Richard John suggests that an AP takeover had never been part of Gould's intention. John, *Network Nation*, p. 184.
111. Haigh and Wilshaw, *Submarine Cables*, p. 246; Hills, *Struggle for Control*, p. 140.
112. Klein, *Jay Gould*, p. 408.
113. David R. Roediger, *The Wages of Whiteness: Race and the Making of the American Working Class* (London: Verso, 1991); Noel Ignatiev, *How the Irish Became White* (New York: Routledge, 1995).
114. Seitz, *James Gordon Bennetts*, p. 218.
115. Klein, *Jay Gould*, p. 12.
116. D. de Cogan, "The Commercial Cable Co. and Their Waterville Station," Paper presented at IEE History of Technology weekend, Trinity College Dublin, July 1987, p. 2.
117. Al Laney, *Paris Herald: The Incredible Newspaper* (New York: Greenwood Press Publishers, 1947), p. 17; "Didn't See Mr. Bennett: Experiences of a Sunday Editor Who Went to Paris," *Washington Post*, July 8, 1900. Simon J. Potter, *News and the British World: The Emergence of an Imperial Press System, 1876–1922* (Oxford: Clarendon Press, 2003), p. 16.
118. "James Gordon Bennett," *New York Times*.
119. See, for instance, *Daisy Miller* (1879) or *Portrait of a Lady* (1881); also William W. Stowe, *Going Abroad: European Travel in Nineteenth-Century American Culture* (Princeton, NJ: Princeton University Press, 1994).
120. Robertson, *International Herald Tribune*, p. 12.
121. "Bennett a Figure in Many Anecdotes," *New York Times*, May 15, 1918.
122. Spencer, *Yellow Journalism*, p. 38.
123. Ibid.
124. "The Jeannette's Long Cruise," *New York Times*, December 21, 1881. Further, see Victor Slocum, *Castaway Boats* (Dobbs Ferry, NY: Sheridan House, 2001) and A. A. Hoehling, *The Jeannette Expedition: An Ill-Fated Journey to the Arctic* (London: Abelard-Schuman, 1969).
125. Seitz, *James Gordon Bennetts*, p. 350.
126. See Henry S. Villard, *Blue Ribbon of the Air: The Gordon Bennett Races* (Washington, DC: Smithsonian Institution Press, 1987), or Don Berliner, *Airplane Racing: A History; 1909-2008* (Jefferson, NC: McFarland, 2009).
127. See "Belgian Elections: By the Herald's Special Telephone," *Paris Herald*, October 17, 1887.
128. Seitz, *James Gordon Bennetts*, p. 372.
129. "Completion of Bennett-Mackay Cable," *Birmingham Daily Post*, October 21, 1884.
130. "The New Atlantic Cable," *Daily News*, December 24, 1884.
131. "Completion of Bennett-Mackay Cable," *Birmingham Daily Post*.
132. "Telegraphing Extraordinary," *Pall Mall Gazette*, January 23, 1888; "Chess by Cable," *Penny Illustrated Paper and Illustrated Times*, March 16, 1895; "The Cable Chess

Match—Great Britain v. United States. The Trophy Won Back," *Glasgow Herald*, February 15, 1897; "Our London Correspondence," *Glasgow Herald*, April 6, 1897.

133. "Gleanings," *Birmingham Daily Post*, October 15, 1886.
134. "London Correspondence," *Freeman's Journal and Daily Commercial Advertiser*, September 7, 1895.
135. "America Cup," *Glasgow Herald*, September 11, 1895.
136. Robertson, *International Herald Tribune*, p. 26.
137. Hobsbawm, *Age of Capital*, p. 60.
138. "James Gordon Bennett," *New York Times*, May 15, 1918.
139. For the term *global media reform*, see Sandford Fleming to Henniker Heaton, March 11, 1886, Fleming Papers NCA.
140. "Occasional Notes," *Pall Mall Gazette*.
141. "Topics of the Week," *The Era*, July 18, 1869.
142. H. L., "Telegraph Companies and Charges," *Daily News*, February 14, 1873.
143. James Anderson, "Manifesto on Telegraph Charges, 1873," *Daily News*, February 14, 1873.
144. "Telegraph Companies and Charges," *Daily News*, February 17, 1873.
145. John, *Network Nation*, p. 182.
146. "To the Editor of the Daily News," *Daily News*, February 19, 1873.
147. H. Thurn, *Die Seekabel*, p. 123.
148. H. L., "Telegraph Companies and Charges," *Daily News*, February 18, 1873.
149. Henry A. Moriarty to the Editor of the *Standard*, March 8, 1870, Moriarty letter book CWA.
150. "Telegraph Companies and Charges," *Daily News*, February 18, 1873.
151. Jürgen Osterhammel, *Die Verwandlung der Welt: Eine Geschichte des 19. Jahrhunderts* (München: Beck, 2009), pp. 235–37; Eric J. Hobsbawm, *The Age of Empire: 1848–1875* (New York: Vintage Books, 1996), pp. 36–37.
152. Griset and Headrick, "Submarine Telegraph Cables," p. 557; Auswärtiges Amt, Vorschriften Kabel von Emden über die Azoren nach Nord-Amerika, March 15, 1901, BArch R 901/80740, Bundesarchiv.
153. "The Anglo-American Telegraph Company," *Leeds Mercury*.
154. James Anderson, "Manifesto on Telegraph Charges, 1873," cited in H. L., "Telegraph Companies and Charges."
155. Direct United States Cable Company Report, "Reports & Accounts," July 1886.
156. "Cable Rate Abuses: J. Henniker Heaton's Call for a Universal 2-Cent Charge Stirs the Whole World," *New York Times*, November 29, 1908.
157. "Cheaper Ocean Postage Wanted: The Mission of Mr. Heaton to This Country from Great Britain," *New York Times*, October 4, 1890.
158. Timothy L. Alborn, *Conceiving Companies: Joint-Stock Politics in Victorian England* (London: Routledge, 2003), p. 250; Jean-Guy Rens, *The Invisible Empire: A History of the Telecommunications Industry in Canada 1846–1956* (Montréal: McGill-Queen's University Press, 2001), p. 92.

159. See Lewis L. Gould, *America in the Progressive Era, 1890–1914* (Harlow: Longman, 2001).
160. Alvin F. Harlow, cited in Schwarzlose, *Newsbrokers*, p. 12.
161. According to Heaton, in 1895, "only one in ten" Anglo-Americans cabled one word a year. J. H. Heaton, "A Cable Post: The Possibilities of Atlantic Submarine Communication," *The North American Review* 160 (1895): p. 664.
162. Adrian Porter, *The Life and Letters of Sir John Henniker Heaton Bt.: By His Daughter Mrs. Adrian Porter with Numerous Illustrations* (New York: John Lane Company, 1916), p. 10.
163. See his speech before the Royal Colonial Institute in 1887. J. H. Heaton, "The Postal and Telegraphic Communication of the Empire," *Proceedings of the Royal Colonial Institute* 19 (1887–1888).
164. Anderson and Heaton had a feud over a cable from Natal to Australia touching upon Mauritius. Heaton had proposed the cable in 1884 but received a rebuff from Anderson. When fifteen years later the cable was laid, Heaton used that anecdote of his late triumph to support his crusade for a penny system. The relationship between Heaton, Anderson, and John Pender was further strained by Heaton's close friendship with the newspaper publisher Henry Labouchère, who ever since the DUSC affair bore a grudge against the two. Porter, *John Henniker Heaton*, pp. 27, 49, 81.
165. Heaton, "A Cable Post"; Heaton, "Cable Telegraph System"; J. H. Heaton, "The World's Cables and the Cable's Rings," *Financial Review of Reviews* (May 1908). His 1908 essay on the cable ring sparked a series of letters to the editor in the London *Times*: Charles Bright, "Universal Penny-a-Word Telegraphy: Letter to the Editor," *Times*, November 18, 1908; J. H. Heaton, "Universal Penny-a-Word Telegrams: Letters to the Editor," *Times*, November 23, 1908.
166. "The State Ownership of Cables: Mansion-House Meeting," *Times*, December 12, 1908; "Cable Reform," *Times*, December 12, 1908.
167. J. H. Heaton, "Penny-a-Word Telegrams Throughout the Empire," *Proceedings of the Royal Colonial Institute* 40 (1908–1909): pp. 3–37, 15.
168. Ibid., p. 16.
169. Porter, *John Henniker Heaton*, p. 223.
170. Heaton, "Penny-a-Word Telegrams," p. 12.
171. See Barry Wellman and Wenhong Chen, "The Global Digital Divide—Within and Between Countries," *IT & Society* 1 (2004): pp. 39–45.
172. Adrian Porter, "The Life and Letters of Sir John Henniker Heaton Bt.," *Times*, September 9, 1914.
173. Heaton, "A Cable Post," p. 660; Pierre-Joseph Proudhon, *Qu'est-ce que la propriété? Ou recherches sur le principe du droit et du gouvernement* (Paris: A la Librarie de Prévot, 1841).
174. Heaton, "A Cable Post," p. 660.
175. "Cable Rate Abuses," *New York Times*.

176. "The People and the Peers: To the Editors of Reynold's Newspaper," *Reynold's Newspaper*, September 9, 1866.
177. Friedrich Engels, "Herrn Eugen Dührings Umwälzung der Wissenschaft," in *Karl Marx, Friedrich Engels Ausgewählte Werke*, ed. Mathias Bertram (Berlin: Directmedia, 2004), p. 507.
178. Heaton, "A Cable Post," p. 661.
179. Ibid., p. 665; "Cable Rate Abuses," *New York Times*. Heaton still distinguished between "civilized" and "uncivilized" masses.
180. Heaton, "A Cable Post," pp. 664–66.
181. James Anderson, cited in "Telegraph Companies and Charges," *Daily News*, February 18, 1873. On the doubts concerning the technical feasibility, see the response of G. R. Neilson of the East India Cable Company, which followed Heaton's presentation before the Royal Colonial Institute in 1908. Heaton, "Penny-a-Word Telegrams," pp. 25–30.
182. Anderson, as quoted by Heaton in "A Cable Post," p. 661.
183. John, *Network Nation*; Claude S. Fischer, *America Calling: A Social History of the Telephone to 1940* (Berkeley: University of California Press, 1992).
184. See "Cable Rates Soar if No Competition," *New York Times*, December 11, 1908.
185. Heaton, "Penny-a-Word Telegrams," p. 6.
186. Ibid.
187. Ibid., p. 13; also Heaton, "The Postal and Telegraphic Communication of the Empire," p. 172.
188. Heaton, "Penny-a-Word Telegrams," p. 13.
189. James Anderson, cited in Heaton, "The Postal and Telegraphic Communication of the Empire," p. 211.
190. Hartmann, *Globale Medienkultur*, p. 63.

5. THE PROFESSIONALIZATION OF THE TELEGRAPH ENGINEER

1. "Foreign Telegraph Companies," *The Telegraphist*, April (1884): p. 70.
2. Wendy Gagen, "The Manly Telegrapher: The Fashioning of a Gendered Company Culture in the Eastern and Associated Telegraph Companies," in *Global Communication Electric: Business, News, and Politics in the World of Telegraphy*, eds. Michaela Hampf and Simone Müller-Pohl (Frankfurt/Berlin: Campus, 2013).
3. Peter J. Bowler and Iwan R. Morus, *Making Modern Science: A Historical Survey* (Chicago: University of Chicago Press, 2005), p. 319.
4. Bryan H. Bunch and Alexander Hellemans, *The History of Science and Technology: A Browser's Guide to the Great Discoveries, Inventions, and the People Who Made Them, from the Dawn of Time to Today* (Boston: Houghton Mifflin, 2004), p. 309.

5. Eric J. Hobsbawm, *Industry and Empire: The Birth of the Industrial Revolution* (New York: The New Press, 1999), p. 135.
6. Ibid., p. 42.
7. Margaret Gowing, "Science, Technology and Education: England in 1870," *Oxford Review of Education* 4, no. 1 (1978): p. 3.
8. Kees Gispen, "Der gefesselte Prometheus: Die Ingenieure in Großbritannien und den Vereinigten Staaten 1750–1945," in *Geschichte des Ingenieurs: Ein Beruf in sechs Jahrtausenden*, eds. Walter Kaiser and Wolfgang König (München: Hanser, 2006), pp. 127–78, 140.
9. William J. Reader, *A History of the Institution of Electrical Engineers, 1871–1971* (London: P. Peregrinus on behalf of the Institution of Electrical Engineers, 1987), p. 13.
10. "Minutes of Council Meetings of the Society of Telegraph Engineers, 1871," cited in Rollo Appleyard, *The History of the Institution of Electrical Engineers (1871–1931)* (London: The Institution of Electrical Engineers, 1939), p. 29.
11. William Siemens served as the society's president in 1871 and 1878; Samuel Canning, Latimer Clark, Willoughby Smith, and Cromwell Varley represented four of the twelve initial council members. Future presidents also came for the Class of 1866: William Thomson (1874; 1889; 1907), Latimer Clark (1875), Willoughby Smith (1883), and Charles T. Bright (1887). "List of Presidents," IET Archives; Charles Bright, *Submarine Telegraphs: Their History, Construction and Working* (London: C. Lockwood, 1898), pp. 180–81.
12. Society of Telegraph Engineers, "Rules and Regulations," in *Journal of the Society of Telegraph Engineers*, eds. Frank Bolton and George E. Preece (London, 1872/1873), p. 10.
13. William Siemens, "The Society of Telegraph Engineers: Inaugural Address," in *Journal of the Society of Telegraph Engineers: Including Original Communications on Telegraphy and Electrical Science*, eds. Frank Bolton and George E. Preece (London, 1872/1873), p. 21.
14. Latimer Clark to President and Council, March 22, 1879, cited in Reader, *Electrical Engineers*, p. 20.
15. Reader, *Electrical Engineers*, p. 27.
16. William Thomson, "The President's Inaugural Address," *Journal of the Society of Telegraph Engineers* 3, no. 7 (1874): p. 1.
17. The Institution of Engineering and Technology, "Hertha Ayrton," accessed May 2, 2010, http://www.theiet.org/about/libarc/archives/biographies/ayrtonh.cfm, p. 10.
18. Siemens, "Inaugural Address," p. 22.
19. Immanuel Kant and Michael Friedman, *Metaphysical Foundations of Natural Science* (Cambridge: Cambridge University Press, 2004); Michael Friedman and Alfred Nordmann, eds., *The Kantian Legacy in Nineteenth-Century Science* (Cambridge, MA: MIT Press, 2006).

20. See Pheng Cheah, "The Cosmopolitical—Today," in *Cosmopolitics: Thinking and Feeling Beyond the Nation*, ed. Pheng Cheah (Minneapolis: University of Minnesota Press, 2000), pp. 22–30.
21. Linda Colley, *Britons: Forging the Nation; 1707–1837* (New Haven: Yale University Press, 1992).
22. Siemens, "Inaugural Address," p. 21.
23. Bright, *Submarine Telegraphs*, pp. x–xi.
24. Bright, *Submarine Telegraphs*, pp. 3–4.
25. Werner Siemens, for example, discussed his *Legungstheorie* (theory of laying submarine cables) with Lewis Gordon: Werner Siemens to Lewis Gordon, September 26, 1857, and September 28, 1857, cited in A. Kunert, *Telegraphen-Seekabel* (Köln-Mühlheim: Karl Glitscher, 1962), p. 5.
26. Robert Sabine, *The Electric Telegraph* (London: Virtue Brothers, 1867), pp. 21–45.
27. Dionysius Lardner, *The Electric Telegraph Popularised: With One Hundred Illustrations* (London: Walton and Maberly, 1855) [emphasis mine].
28. Similarly, Edward Highton, *The Electric Telegraph: Its History and Progress* (London: John Weale, 1852); the author lists British, German, and French authors such as L. Bregust, H. Schellen, Clemens Gerke, or Werner Siemens (in French translation), ibid., pp. v–vii.
29. Siemens, "Inaugural Address," p. 22.
30. Ibid., p. 23.
31. Ibid.
32. Ibid., pp. 23–26.
33. Reader, *Electrical Engineers*, p. 27.
34. Siemens, "Inaugural Address," p. 27.
35. Ibid.
36. F. H. Webb, ed., *Journal of the Institution of Electrical Engineers: Later the Society of Telegraph Engineers and Electricians XVIII* (London: E. and F. N. Spon, 1889).
37. Society of Telegraph Engineers, "Annual Report for 1873," *Journal of the Society of Telegraph Engineers* 2, no. 6 (1873): pp. 414–15.
38. Society of Telegraph Engineers, "Annual Report for 1874," p. 447.
39. Ibid.
40. Society of Telegraph Engineers, "Annual Report for 1874," pp. 446–47.
41. Siemens, "Inaugural Address," p. 22.
42. Marie Anchordoguy, "Nippon Telegraph and Telephone Company (NTT) and the Building of a Telecommunications Industry in Japan," *The Business History Review* 75 (2001): pp. 508–12; also see Marius B. Jansen, *The Emergence of Meiji Japan* (Cambridge: Cambridge University Press, 1997).
43. Anchordoguy, "Nippon Telegraph and Telephone Company," p. 508; "IET Membership Lists," IET Archives; Reader, *Electrical Engineers*, p. 4.
44. Charles T. Bright to Cyrus W. Field, March 4, 1858, Letter Book Charles T. Bright (June 1857–May 14, 1858), CWA, p. 289.

45. Bright, *Submarine Telegraphs*, p. 83.
46. Society of Telegraph Engineers, "Rules and Regulations," p. 10.
47. George Spratt, November 25, 1874, Spratt Diaries, CWA.
48. See Letter Book SI; James Graves, *The Autobiography of James Graves*, ed. Donard de Cogan (Electronic version of two early autobiographies, 1880s/2005).
49. Bruce J. Hunt, "The Ohm Is Where the Art Is: British Telegraph Engineers and the Development of Electrical Standards," *Osiris* 2 (1994): p. 51.
50. Bright, *Submarine Telegraphs*, p. 20; Werner von Siemens, *Lebenserinnerungen von Werner von Siemens: Mit dem Bildnis des Verfassers in Kupferätzung* (Berlin: Julius Springer, 1908), pp. 126–28; F. C. Webb, "On the Practical Operations Connected with the Paying-Out and Repairing of Submarine Telegraph Cables," *Minutes of Proceedings of the Institution of Civil Engineers* XVII (1858).
51. Bright, *Submarine Telegraphs*, pp. 93–96.
52. Anglo-American Telegraph Company, "General Orders, Rules and Regulations to Be Observed by the Officers, Clerks and Servants of the Company," August 1880, SI, p. 35.
53. Bernard S. Finn, "Introduction," in *Development of Submarine Cable Communications*, ed. B. S. Finn (New York: Arno Press, 1980).
54. Ibid., p. 6.
55. B. S. Finn, "Submarine Telegraphy: A Study in Technical Stagnation," in *Communications Under the Seas: The Evolving Cable Network and Its Implications*, eds. Bernard S. Finn and Daqing Yang (Cambridge, MA: MIT Press, 2009).
56. Richard Noakes, "Industrial Research at the Eastern Telegraph Company, 1872–1929," *The British Journal for the History of Science* 47 (2014): pp. 119–46.
57. James Graves, *Thirty Six Years in the Telegraphic Service 1852–1888 Being a Brief Autobiography of James Graves MSTE.*, ed. Donard de Cogan (1909/2008), p. 80; E. Weedon to H. Weaver, February 19, 1879, Vol. 2/Reel 1, Letter Book SI; Bright, *Submarine Telegraphs*, pp. 583, 639.
58. F. Perry to Alexander Muirhead, January 19, 1891, Vol. 9/Reel 11, Letter Book SI; F. Perry to Alexander Muirhead, June 29, 1894, Vol. 11/Reel 13, Letter Book SI; F. Perry to Alexander Muirhead, Fall 1890, Vol. 9/Reel 11, Letter Book SI.
59. Graves, *Telegraphic Service*, p. 89.
60. Graves, *Telegraphic Service*, pp. 95–96; George Spratt, March 5, 1871, Spratt Diaries, CWA.
61. George Spratt, May 4, 1871, and May 25–27, 1872, Spratt Diaries, CWA.
62. On the contemporary debate, see Edward O. W. Whitehouse, *Reply to the Statement of the Directors of the Atlantic Telegraph Company: Published in the "Daily News" of September 20, and "Times" September 22, 1858* (London: Bradbury & Evans, 1858); Edward O. W. Whitehouse, *Recent Correspondence Between Mr. Wildman Whitehouse and the Atlantic Telegraph Company* (London: Bradbury & Evans, 1858); Edward O. W. Whitehouse, *Atlantic Telegraph: Letter from a Shareholder to Mr. Whitehouse, and His Reply* (London: Bradbury & Evans, 1858).

On today's debate, see Gillian Cookson, "The Transatlantic Telegraph Cable: Eighth Wonder of the World," *History Today* 50 (2000): p. 45, or Alan Green, "Further Evidence in the Defence of Wildman Whitehouse? Porthcurno Research Blog," accessed May 7, 2013, http://researchatporthcurno.blogspot.com/2012/01/further-evidence-in-defence-of-wildman.html.

63. Graves, *Telegraphic Service*, p. 47.
64. Commercial Cable Company, "How Submarine Cables Are Made, Laid, Operated and Repaired," accessed March 8, 2011, http://atlantic-cable.com/Article/1915CCC/index.htm.
65. Eastern and Associated Companies' staff member W. P. Granville reports about testing in his diary; W. P. Granville, Private Diary kept by W. P. Granville on board SS *Caroline*, CWA. Similarly Graves and Richard Collett, an electrician of the company then at Heart's Content, discussed via cable and the exchange of letters which condenser to use; Graves, *Telegraphic Service*, p. 51. Spratt also reports of duplex tests; Spratt Diaries, 1873 CWA.
66. Atlantic Telegraph Company, Newfoundland Station Service Messages, 1858, CWA.
67. Graves, *Telegraphic Service*, pp. 51–54; Benjamin A. Gould, *Transatlantic Longitude as Determined by the Coast Survey Expedition of 1866: A Report to the Superintendent of the U.S. Coast Survey* (Washington, DC: Smithsonian Institution, 1869).
68. George Spratt, May 24, 1874, and November 14–16, 1874, Spratt Diaries, CWA.
69. George Spratt, March 15, 1875.
70. Graves, *Autobiography*, p. 65; Graves, *Telegraphic Service*, p. 18. Similarly, also at Heart's Content: E. Weedon to H. W. Howgate, Washington, DC, November 9, 1871, Vol. 1/Reel 1, Letter Book SI.
71. Christopher A. Bayly, *The Birth of the Modern World: 1780–1914* (Malden, MA: Blackwell, 2009), p. 314.
72. Siemens, "Inaugural Address," p. 22.
73. Thomson, "Inaugural Address," p. 7.
74. Ibid.
75. Ibid., p. 10.
76. H. U. B. Bonnerjea, "The Aurora Borealis in Folk-Lore," *Notes and Queries* 175 (1938): p. 113.
77. James Graves, "Earth Currents, and the Aurora Borealis of 4th February 1872," *Journal of the Society of Telegraph Engineers* 1 (1872): p. 102.
78. Svante Lindquist, "The Spectacle of Science: An Experiment in 1744 Concerning the Aurora Borealis," *Configurations* 1 (1993): pp. 57–94.
79. Carol Quinn, "Dickinson, Telegraphy, and the Aurora Borealis," *The Emily Dickinson Journal* 13 (2004): p. 66.
80. Graves, "Earth Currents," p. 102.
81. George Drapner, "Notes of Earth Currents on February 4th, 1872," *Journal of the Society of Telegraph Engineers* 1 (1872): p. 110.

82. Graves, "Earth Currents," pp. 103–4.
83. Drapner, "Notes of Earth Currents on February 4th, 1872," p. 111.
84. Thomson, "Inaugural Address," p. 11.
85. Ibid., pp. 13–14.
86. Anonymous, Note Book from a Telegraph Construction & Maintenance Company Employee, March 2, 1870–October 23, 1873, CWA.
87. Graves, *Autobiography*, p. 60. At Heart's Content, a Mr. Dickinson was particularly inventive, see E. Weedon to H. Weaver, February 23, 1878, Vol. 2/Reel 1, Letter Book SI; F. Perry to General Managers on "Siphon Recorder," June 24, 1878, Vol. 2/Reel 1, Letter Book SI.
88. Graves, *Telegraphic Service*, p. 93.
89. Anderson reports on a "running battle" between the superintendents at Carcavelos and Gibraltar over duplexing cables. L. T. I. Tyson, "Army or Navy?" January 19, 1980, James Anderson Papers, CWA, p. 2. Similarly, Graves claims that a Mr. George had filed a patent with the company for a key switch, which was very much like his "own child." The result of the subsequent dispute was that the company ordered that neither key was to be proceeded with. Graves, *Telegraphic Service*, p. 68.
90. E. Weedon to H. Weaver, May 18, 1875, Vol. 1/Reel 2, Letter Book SI; E. Weedon to H. Weaver, February 19, 1879, Vol. 2/Reel 1, Letter Book SI; Graves, *Telegraphic Service*, p. 88.
91. Graves, *Telegraphic Service*, p. 53.
92. Graves, *Autobiography*, p. 66; George Spratt, November 20, 1874, Spratt Diaries, CWA.
93. Graves, *Telegraphic Service*, p. 100.
94. James Graves, "On Curbed Signals on Long Cables," *Journal of the Society of Telegraph Engineers* 8 (1879).
95. *Journal of the Society of Telegraph Engineers*, February 12, 1879, p. 92; Graves, *Telegraphic Service*, pp. 95–96.
96. E. Weedon to H. Weaver, May 11, 1879, Vol. 2/Reel 1, Letter Book SI. A similar encounter is documented with regard to the "Allan-Brown relay," which tested on the Eastern's cables. At the same time, Porthcurno's superintendent, Bull, worked on a similar instrument. George Spratt, March 2, 1875, Spratt Diaries, CWA.
97. George Spratt, February 15, 1876, Spratt Diaries, CWA.
98. Charles T. Bright to Cyrus W. Field, March 6, 1858, Letter Book Charles T. Bright (June 1857–May 14, 1858), CWA, pp. 293–96.
99. Jürgen Osterhammel, *Die Verwandlung der Welt: Eine Geschichte des 19. Jahrhunderts* (München: Beck, 2009), p. 1147.
100. John Tully, "A Victorian Ecological Disaster: Imperialism, the Telegraph, and Gutta-Percha," *Journal of World History* 20 (2009): pp. 566–71.
101. "The Atlantic Telegraph Cable," *Birmingham Daily Post*, June 14, 1866.

102. Tully, "A Victorian Ecological Disaster," pp. 572–573; J. N. Dean, "Gutta Percha and Balata, with Particular Reference to Their Use in Submarine Cable Manufacture," *Journal of the Royal Society of Arts* 92 (June 1944): pp. 367–386, 470.
103. J. H. Heaton, "The Postal and Telegraphic Communication of the Empire: A Paper Read Before the Royal Colonial Institute on Tuesday, March 13, 1888," London, 1888, p. 15.
104. See, for example, Appleyard, *Institution of Electrical Engineers*; Reader, *Electrical Engineers*.
105. Siemens, "Inaugural Address," p. 19.
106. Cromwell Varley cited in Siemens, "Inaugural Address," p. 34.
107. Ibid., p. 19.
108. Ibid., p. 20.
109. Ibid.
110. W. H. G. Armytage, *A Social History of Engineering* (London: Faber and Faber, 1961), p. 4.
111. "Atlantic Telegraph," *Times*, July 27, 1866; "Epitome of Opinion in the Morning Journals," *Pall Mall Gazette*, July 27, 1866; E. B. Grant, *The Western Union Telegraph Company: Its Past, Present and Future* (New York: Hotchkiss, Burnham & Co., 1883), p. 23.
112. Reader, *Electrical Engineers*, p. 23.
113. Ibid., p. 45.
114. Ibid., pp. 45–46.
115. David S. Landes, *The Unbound Prometheus: Technological Change and Industrial Development in Western Europe from 1750 to the Present* (Cambridge: Cambridge University Press, 2008), pp. 193–230. For Germany, see Hans-Werner Hahn, *Die Industrielle Revolution in Deutschland*, (München: Oldenbourg Wissenschaftsverlag, 2011), pp. 24–50. For America, see Thomas P. Hughes, *American Genesis: A Century of Invention and Technological Enthusiasm, 1870–1970* (Chicago: University of Chicago Press, 2004).
116. Appleyard, *Institution of Electrical Engineers*, p. 65.
117. Latimer Clark, "On a Standard Voltaic Battery," *Journal of the Society of Telegraph Engineers* 7 (1878).
118. Emphasis mine.
119. Latimer Clark to Edward Graves, February 20, 1880, William Preece Papers, IET Archives.
120. William H. Preece, "Minutes of the Ordinary General Meeting and the Annual Meeting Held on December 22, 1880," *Journal of the Society of Telegraph Engineers* 10, no. 35 (1881): p. 2; Appleyard, *Institution of Electrical Engineers*, p. 65.
121. Appleyard, *Institution of Electrical Engineers*, pp. 105–6.
122. R. E. Crompton, November 24, 1887, cited in ibid., p. 106.
123. Edward Graves, "Inaugural Address of the New President, Mr. Edward Graves," *Journal of the Society of Telegraph Engineers and Electricians* 17, no. 70 (1888): p. 30.

124. Appleyard, *Institution of Electrical Engineers*, p. 109.
125. George Spratt, February 9, 1888, Spratt Diaries, CWA.
126. R. A. Buchanan, "Institutional Proliferation in the British Engineering Profession, 1847–1914," *The Economic History Review* 38 (1985): pp. 44, 53.
127. Bright, *Submarine Telegraphs*, p. 155.
128. Telegraph Construction and Maintenance Company, *Telcon Story, 1850–1950* (London: Telegraph Construction and Maintenance Company, 1950), p. 79.
129. Siemens, "Inaugural Address," p. 28.
130. Similarly, William H. Preece, "The Progress of Telegraphy: 15. February 1883," in *The Practical Applications of Electricity. A Series of Lectures Delivered at the Institution of Civil Engineers, Session 1882/3*, ed. Institution of Civil Engineers (London: Institution of Civil Engineers, 1884), p. 1.
131. Seteney Shami, "Prehistories of Globalization: Circassian Identity in Motion," in *Globalization*, ed. Arjun Appadurai (Durham, NC: Duke University Press, 2001), p. 220.
132. Willoughby Smith, "Inaugural Address of President for 1883, Willoughby Smith," *Journal of the Society of Telegraph Engineers* 12, no. 46 (1883): p. 5.
133. William H. Preece, "The Progress of Telegraphy: 15. February 1883," in *The Practical Applications of Electricity. A Series of Lectures Delivered at the Institution of Civil Engineers, Session 1882/3* (London: Institution of Civil Engineers, 1884), p. 1.
134. Lars Bluma, "Progress by Technology? The Utopian Linkage of Telegraphy and the World Fairs, 1851–1880," in *Global Communication Electric: Business, News, and Politics in the World of Telegraphy*, eds. Michaela Hampf and Simone Müller-Pohl (Frankfurt/Berlin: Campus, 2013).
135. Elisabeth Crawford, *Nationalism and Internationalism in Science, 1880–1939: Four Studies of the Nobel Population* (Cambridge: Cambridge University Press, 1992), p. 32.
136. Hobsbawm, cited in Benedict Anderson, *Imagined Communities: Reflections on the Origin and Spread of Nationalism* (London: Verso, 2006), p. 71. On the nationalization of science, see Ludmilla Jordanova, "Science and Nationhood: Cultures of Imagined Communities," in *Imagining Nations*, ed. Geoffrey Cubitt (Manchester: Manchester University Press, 1998).
137. *Journal of the Society of Telegraph Engineers*, 1873, pp. 6, 385–421; Alexander G. Bell, *The Telephone: A Lecture entitled Researches in Electric Telephony: Delivered before the Society of Telegraph Engineers, October 31, 1877* (London: E. and F. N. Spon, 1878).
138. William H. Preece, Report of Mr. Preece's Visit to the United States of America, 1893, TCB 306/38, BT Archives.
139. William H. Preece, "U.S.A. Report by Mr. Preece on Visit to Study Telephones and Telegraphs 1884," POST 30/474 B, BT Archives.
140. Siemens, "Inaugural Address," p. 22.
141. Bright, *Submarine Telegraphs*, p. 177.

142. Hunt, "The Ohm Is Where the Art Is," p. 54.
143. Bright, *Submarine Telegraphs*, p. 61.
144. Thomas A. Loya and John Boli, "Standardization in the World Polity: Technical Rationality over Power," in *Constructing World Culture: International Nongovernmental Organizations SINCE 1875*, eds. John Boli and George M. Thomas (Stanford, CA: Stanford University Press, 1999), p. 172.
145. Hans Görges, *50 Jahre Elektrotechnischer Verein: Festschrift zum fünfzigjährigen Bestehen des Elektrotechnischen Vereins. 1879–1929* (Berlin: Elektrotechnischer Verein, 1929), pp. 4–54.
146. American Institute of Electrical Engineers, *Historical Sketch of Its Organization and Work: Founded 1884* (New York: American Institute of Electrical Engineers, 1892), p. 4.
147. Görges, *Elektrotechnischer Verein*, p. 6.
148. Reader, *Electrical Engineers*, p. 27.
149. Appleyard, *Institution of Electrical Engineers*, p. 185.
150. Crawford, *Nationalism and Internationalism in Science*.
151. IEE and AIEE Combined Meeting in Paris, Program 1900, IET Archives; Visit to Germany Program with List of Attendees 1901, IET Archives; Visit to Italy Program with List of Attendees, 1903, IET Archives.

6. CABLE DIPLOMACY AND IMPERIAL CONTROL

1. "Cable Cutting in War Time. From the *Pall Mall Gazette*," *New York Times*, May 11, 1898.
2. Ibid.
3. This was different for terrestrial telegraph lines. Most European systems were national, whereas the American system was private. Jill Hills, *The Struggle for Control of Global Communication: The Formative Century* (Urbana, IL: University of Illinois Press, 2002), pp. 22, 26, 42.
4. 34th Congress, 3rd session, House of Representatives, "Submarine Telegraph—Again," in *The Congressional Globe*, ed. U.S. Congress, pp. 586–89 (1857), p. 587.
5. A. Kunert, *Telegraphen-Seekabel* (Köln-Mühlheim: Karl Glitscher, 1962), pp. 200–201.
6. Chandler Hale, "The Projected Cable-Line to the Philippines," *The North American Review* 171 (1900): p. 84.
7. Charles Bright, *Submarine Telegraphs: Their History, Construction and Working* (London: C. Lockwood, 1898), p. 154. The English Channel cable was the only one in joint government possession.
8. On ocean sounding, see E. A. Marland, "British and American Contributions to Electrical Communication," *The British Journal for the History of Science* 1 (June 1962): pp. 44–45; on staff, see the negotiations between Charles T. Bright and the British Royal Navy, Charles T. Bright to Cyrus W. Field, April 1, 1858, Charles T. Bright

Letter Book (June 1857–May 14, 1858) CWA, p. 346; Charles T. Bright to the Directors of the Atlantic Telegraph Company, March 8, 1858, Charles T. Bright Letter Book (June 1857–May 14, 1858) CWA, p. 305; Charles T. Bright to Cyrus W. Field, January 1858, Charles T. Bright Letter Book (June 1857–May 14, 1858) CWA, p. 248.

9. Telegraph Construction and Maintenance Company, *Telcon Story 1850–1950* (London: Telegraph Construction and Maintenance Company, 1950), p. 48; Bruce J. Hunt, "The Ohm Is Where the Art Is: British Telegraph Engineers and the Development of Electrical Standards," *Osiris* 2 (1994): pp. 53–54; Great Britain and Joint Committee to Inquire into the Construction of Submarine Telegraph Cables, *Report of the Joint Committee Appointed by the Lords of the Committee of Privy Council for Trade and the Atlantic Telegraph Company to Inquire into the Construction of Submarine Telegraph Cables* (London: G.E. Eyre and W. Spottis-Woode, 1861).
10. Bright, *Submarine Telegraphs*, p. 167.
11. Samuel Carter, *Cyrus Field: Man of Two Worlds* (New York: Putman, 1968), p. 22.
12. Kenneth R. Haigh and Edward Wilshaw, *Cableships and Submarine Cables* (London: Coles, 1968), pp. 317–19; Daniel R. Headrick, "Review: Entreprise, technologie, et souveraineté," *Technology and Culture* 40 (1999): p. 168.
13. Translation mine. Reichspostamt, Vorschriften Kabel von Emden über die Azoren nach Nord-Amerika. Brief vom Reichspostamt an das Auswärtige Amt, December 25, 1908, BArch R 901/80740, Bundesarchiv.
14. Pascal Griset and Daniel R. Headrick, "Submarine Telegraph Cables: Business and Politics, 1838–1939," *Business History Review* 75 (2001): p. 543.
15. Bright, *Submarine Telegraphs*, p. 176; "West Indies," *New York Times*, November 4, 1872.
16. Newfoundland & T. C. New York, "Company's description, July 1872," Field Papers, NYPL.
17. Haigh and Wilshaw, *Submarine Cables*, p. 258.
18. Donald R. Tarrant, *Atlantic Sentinel: Newfoundland's Role in Transatlantic Cable Communications* (St. John's, Newfoundland: Flanker Press, 1999), p. 117; D. de Cogan, "The Commercial Cable Co. and Their Waterville Station," Paper presented at IEE History of Technology Weekend, Trinity College Dublin, July 1987, p. 3.
19. U.S. Congress, 40th Congress, 3rd Session, Bill 197, January 14, 1869; Hills, *Struggle for Control*, pp. 52, 138; Bright, *Submarine Telegraphs*, p. 178.
20. James A. Scrymser, *Personal Reminiscence of James A. Scrymser: In Times of Peace and War* (New York: HardPress, 1915), p. 106; Similarly, James Anderson in "The French Atlantic Cable: Speech by Sir James Anderson," *Glasgow Herald*, August 13, 1869.
21. Scrymser, *Personal Reminiscence*, p. 106.
22. Scrymser, *Personal Reminiscence*, p. 102.
23. Ulysses Grant, "Monday, December 6, 1869," *Journal of the Senate of the United States of America, 1789–1873* (1869); 41st Congress, House of Representatives,

"Friday, December 10, 1869," *Journal of the House of Representatives of the United States* (1869–1870).

24. Also, the issue of whether landing rights could be granted by state or federal legislature was unresolved. "Occasional Notes," *Pall Mall Gazette*, June 26, 1869; "The French Atlantic Cable," *Daily News*, June 11, 1869.
25. Scrymser, *Personal Reminiscence*, pp. 107–8.
26. Auswärtiges Amt, Letter by James Scrymser to Robert Bacon, Assistant Secretary of State, January 12, 1907, BArch R/901/80740, Bundesarchiv; Auswärtiges Amt, Embassy of the United States of America to the German Foreign Office, February 13, 1907, BArch R 901/80740, Bundesarchiv; Auswärtiges Amt, James Scrymser to Robert Bacon, December 13, 1906, BArch R/901/80740, Bundesarchiv.
27. Bright, *Submarine Telegraphs*, p. 177.
28. Ibid.; Jorma Ahvenainen, "The International Telegraph Union: The Cable Companies and the Governments," in *Communications Under the Seas: The Evolving Cable Network and Its Implications*, eds. Bernard S. Finn and Daqing Yang (Cambridge, MA: MIT Press, 2009), pp. 64–65; Bureau International de l'Union Télégraphique, *L'Union Télégraphique Internationale (1865–1915)* (Bern: Bureau International de l'Union Télégraphique, 1915), p. 9.
29. International Telecommunications Union, "From Morse to Multimedia: A History of the ITU," in *International Telecommunications Union: Celebrating 130 Years, 1865–1995*, ed. International Telecommunications Union, pp. 38–69 (London: International Systems and Communication, 1995), p. 39.
30. Dwayne R. Winseck and Robert M. Pike, *Communication and Empire: Media, Markets, and Globalization. 1860–1930* (Durham: Duke University Press, 2007), p. 17; Bright, *Submarine Telegraphs*, p. 178.
31. International Telecommunications Union, "A History of the ITU," p. 39.
32. "The International Telegraph Conference," *Times*, July 20, 1875.
33. Letter by Emma Pender quoted in Hills, *Struggle for Control*, p. 60.
34. Henry Weaver to Frederick Hill, General Post Office, January 25, 1875, POST 30/288 Part II, BT Archives; British Post Office, "Proposal of British Post Office. Further Discussion with Cable Companies," February 1879, POST 30/361, BT Archives.
35. Ira T. Cohen, "Towards a Theory of State Intervention: The Nationalization of the British Telegraphs," *Social Science History* 4 (1980): p. 177.
36. Ibid., p. 181.
37. William J. Reader, *A History of the Institution of Electrical Engineers, 1871–1971* (London: P. Peregrinus on behalf of the Institution of Electrical Engineers, 1987), p. 20.
38. C. H. B. Patey cited in James Anderson, "International Telegraph Convention, London 1879, Proposal of British Post Office. Comments by Cable Companies Letter by James Anderson to Mr. Patey," POST 30/361 Part I, BT Archives.

39. Ibid.
40. The question of representation was also debated in the *Times* of London in June 1879. Mercator, "International Telegraph Conference: To the Editor of the Times," *Times*, June 13, 1879; Ernest M'Kenna, "The International Telegraph Conference: To the Editor of the Times," *Times*, June 17, 1879; H. R. Meyer, "The International Telegraph Conference," *Times*, June 17, 1879.
41. James Anderson, International Telegraph Convention, London 1879 Proposal of British Post Office. Comments by Cable Companies Letter by James Anderson to Mr. Patey.
42. On the National Telegraph Act and the Butler Amendment, see Richard R. John, *Network Nation: Inventing American Telecommunications* (Cambridge, MA: Belknap Press of Harvard University Press, 2010), pp. 116, 165.
43. James Anderson, "International Telegraph Convention, London 1879 Proposal of British Post Office. Comments by Cable Companies Letter by James Anderson to Mr. Patey."
44. James Anderson et al., "Letter to C.H.B. Patey," January 15, 1878, POST 30/361, BT Archives, p. 1.
45. General Post Office, "Telegraph Conference 1879. Meeting of Representatives of Telegraph Companies in Mr. Patey's Room," February 10, 1879, BT Archives. The only proposition Great Britain made in the end at the London conference was one on language; International Telegraph Union, "Propositions Pour la Conférence Télégraphique Internationale de Londres. Supplément, 1879," POST 30/361, BT Archives, p. 2.
46. "International Telegraph Conference," *Times*, June 23, 1879.
47. "The International Telegraph Conference," *Daily News*, June 7, 1879; British Post Office, "International Telegraph Convention. London 1879. Entertainment of Delegates. Arrangements," January 4–7, 1879, POST 30/361, BT Archives.
48. Emma Pender to Marion des Voeux, July 13, 1879, Emma Pender Papers, CWA.
49. John Pender, "Speech to the Delegates" cited in "International Telegraph Conference," *Times*, June 18, 1879.
50. "International Telegrams," *Daily News*, July 26, 1879.
51. At the 1903 conference, sixty-eight government delegates met sixty-seven private telegraph company delegates. International Telecommunication Union, *Procès-Verbaux: Conference Télégraphique Internationale de Londres. Liste de Participants* (Bern: International Telecommunication Union, 1904).
52. On a different meaning of cable diplomacy, see David P. Nickles, "Submarine Cables and Diplomatic Culture," in *Communications Under the Seas: The Evolving Cable Network and its Implications*, eds. Bernard S. Finn and Daqing Yang (Cambridge, MA: MIT Press, 2009), p. 219.
53. "The French Atlantic Cable," *Glasgow Herald*.
54. Carter, *Cyrus Field*, p. 131.

55. 34th Congress, 3rd Session, House of Representatives, "Submarine Telegraph—Again," in *The Congressional Globe*, ed. U.S. Congress, pp. 586–89 (1857), p. 588.
56. Ibid.
57. Sohui Lee, "Anglophobia," in *Britain and the Americas: Culture, Politics, and History; a Multidisciplinary Encyclopedia*, ed. Will Kaufman (Santa Barbara, CA: ABC-CLIO, 2005), p. 88.
58. John, *Network Nation*, p. 104.
59. 34th Congress, 3rd Session, House of Representatives, "Submarine Telegraph," in *The Congressional Globe*, ed. U.S. Congress, pp. 729–31 (1857), p. 731.
60. Carter, *Cyrus Field*, p. 131.
61. Gillian Cookson, "The Transatlantic Telegraph Cable—Eighth Wonder of the World," *History Today* 50 (2000): p. 50.
62. John S. A. Gordon, *A Thread Across the Ocean: The Heroic Story of the Transatlantic Cable* (New York: Walker, 2002), p. 70.
63. Carter, *Cyrus Field*, p. 124.
64. "Europe: Debate on the Alabama Question in the English House of Commons," *Chicago Tribune*, March 9, 1868.
65. Cyrus Field, "Memoranda 1868," Field Papers, NYPL.
66. Adrian Cook, *The Alabama Claims: American Politics and Anglo-American Relations, 1865–1872* (Ithaca, NY: Cornell University Press, 1975), pp. 167–70; James Graves, *Thirty Six Years in the Telegraphic Service 1852–1888 Being a Brief Autobiography of James Graves MSTE.*, ed. Donard de Cogan (1909/2008), p. 58, https://dandadec.files.wordpress.com/2013/07/technical-autobiography.pdf.
67. Justin McCarthy, ed., *The Settlement of the Alabama Question. The Banquet Given at New York to the British High Commissioners by Cyrus W. Field, 1871* (London 1872), pp. vii–viii.
68. "Benting Chuff: The Public Feeling About the Alabama Claims," *New York Times*, June 9, 1869.
69. Ibid.
70. American Scenic and Historic Preservation Society, "The André Monument Property at Tappan, April 1905," Field Papers, NYPL; "The André Anniversary: To-days Ceremonies at Tarrytown," *New York Times*, September 23, 1880.
71. Dean of Westminster, cited in American Scenic and Historic Preservation Society, "The André Monument Property at Tappan."
72. Washington Heights Century Club, "André's Monument. Anglo-Maniac Field's Homage to a Spy!," 1883, Field Papers, NYPL.
73. "Mr. Field's Pet Memorial: The Third Attempt to Blow Up the André Memorial," *New York Times*, November 5, 1885; "Attempted Destruction of Major André's Monument," *Pall Mall Gazette*, April 1, 1882.
74. Alexander Hamilton, "Cyrus W. Field's Audacity," *New York Times*, December 20, 1885.

75. "Sympathy Made to Order; No Ready Flow of Tears over Andre's Monument. The Amusing Attempt of Cyrus W. Field and His Friends to Manufacture an Indignation Meeting in Tappan," *New York Times*, November 11, 1885.
76. Hamilton, "Field's Audacity."
77. "Mr. Field's Pet Memorial," *New York Times*.
78. Bright, *Submarine Telegraphs*, p. 80.
79. Hamilton, "Field's Audacity."
80. Bright, *Submarine Telegraphs*, p. 330.
81. Ibid., p. xiv.
82. E. Weedon to Henry Weaver, "Extract from *New York Times*," November 7, 1870, Vol.1/Reel 1, Letter Book SI.
83. Bright, *Submarine Telegraphs*, p. 324.
84. The Eastern Telegraph Company's employees called themselves "The Exiles."
85. F. Perry to Managing Director, "E.P. Earle's Letter Grievance," November 30, 1887, Vol. 8/Reel 10, Letter Book SI.
86. E. Weedon, "Extract from Memo to General Manager Dated April 22, '76 'Land Staff,'" August 1885, Vol. 7/Reel 9, Letter Book SI.
87. F. Perry to General Manager, "Alterations in Office," February 26, 1894, Vol. 10/Reel 12, Letter Book SI; E. Weedon to A. M. Mackay, "Furniture," February 23, 1881, Vol. 3/Reel 5, Letter Book SI; E. Weedon to Managing Director, "Furniture Renewal," March 7, 1881, Vol. 3/Reel 5, Letter Book SI.
88. E. Weedon to Dr. Anderson, December 16, 1879, Vol. 3/Reel 5, Letter Book SI.
89. In 1881, the staff founded the Heart's Content Curling Club. E. Weedon to H. Weaver, December 29, 1881, Vol. 4/Reel 6, Letter Book SI. On currency, see E. Weedon to Dr. Anderson, December 11, 1879, Vol. 3/Reel 5, Letter Book SI, and F. Perry to General Superintendent, February 5, 1895, Vol. 11/Reel 13, Letter Book SI.
90. E. Weedon to Managing Director, "Church Grant," September 20, 1883, Vol. 6/Reel 8, Letter Book SI; E. Weedon to H. Weaver, December 28, 1875, Vol. 1/Reel 2, Letter Book SI.
91. In 1877, the school was opened with thirteen students including four "natives." E. Weedon to H. Weaver, January 11, 1877, Vol. 2/Reel 1, Letter Book SI. According to de Cogan, Samuel Seymour Stentaford was the first Newfoundlander employed at Heart's Content. D. de Cogan, "Cable Talk: Relations Between the Heart's Content and Valentia Cable Stations 1866–1886," University of East Anglia, p. 4, https://dandadec.files.wordpress.com/2013/07/valentia-e28093-heart_s-content-relations.pdf.
92. James Graves, *The Autobiography of James Graves*, ed. Donard de Cogan (Electronic version of two early autobiographies, 1880s/2005), p. 44; Bright, *Submarine Telegraphs*, p. 241.
93. Tarrant, *Atlantic Sentinel*, Prologue; Sean T. Cadigan, *Newfoundland and Labrador: A History* (Toronto: University of Toronto Press, 2009), pp. 125, 134.

94. Cadigan, *Newfoundland and Labrador*, p. 141. Also, the railway did not find favor. Surveying in 1881 led to open rioting and abuse of the engineers. Ibid., p. 142.
95. E. Weedon to A .M. Mackay, "Water for the Public," August 7, 1883, Vol. 6/Reel 8, Letter Book SI; E. Weedon to Managing Director, "Drugs & Holiday Allowance," July 1, 1881, Vol. 3/Reel 5, Letter Book SI; E. Weedon to General Manager, February 1, 1884, Vol. 6/Reel 8, Letter Book SI.
96. E. Weedon to Henry Weaver, June 6, 1870, Vol. 1/Reel 1, Letter Book SI; the doctor also gave advice "by telegraph" to any of the landline offices out west, E. Weedon to H. Weaver, "Business Statement," April 3, 1871, Vol. 1/Reel 1, Letter Book SI. On the details of outside practice, further see F. Perry to Dr. Anderson, February 2, 1885, Vol. 7/Reel 9, Letter Book SI.
97. Anglo-American Telegraph Company, "General Orders, Rules and Regulations to Be Observed by the Officers, Clerks and Servants of The Company," August 1880, SI, pp. 3, 10.
98. E. Weedon to Mr. Sullivan, October 27, 1880, Vol. 2/Reel 1, Letter Book SI.
99. E. Weedon to Mr. Sullivan, February 27, 1883, Vol. 6/Reel 8, Letter Book SI.
100. E. Weedon to H. Weaver, "Bonavista Extension," June 12, 1877, Vol. 2/Reel 1, Letter Book SI.
101. Willson Beckles emphasizes the great estrangement between the local population in *The Tenth Island: An Account of Newfoundland* (London: Grant Richards, 1897), p. 45. Also in contemporary historiography on Newfoundland, the cable station remains strangely absent: Moses Harvey, *A Short History of Newfoundland: England's Oldest Colony* (London; Glasgow: W. Collins, 1890), or John D. Rogers, *A Historical Geography of the British Colonies: Newfoundland* (Oxford: Clarendon Press, 1911).
102. D. W. Prowse, *A History of Newfoundland: From the English Colonial and Foreign Records* (London: Eyre and Spottiswoode, 1896), p. 497.
103. "Newfoundland and the Cable," *New York Times*, May 3, 1873.
104. Prowse, *Newfoundland*, p. 498.
105. E. Weedon to H. Weaver, "Bonavista Extension," June 12, 1877, Vol. 2/Reel 1, Letter Book SI.
106. Henry M. Field, *History of the Atlantic Telegraph* (New York: C. Scribner & Co, 1866), p. 16.
107. Prowse, *Newfoundland*, p. 486.
108. Cadigan, *Newfoundland and Labrador*, p. 136.
109. E. Weedon to H. Weaver, "Private Letter on the Coming Election," October 21, 1873, Vol. 1/Reel 1, Letter Book SI.
110. Tarrant, *Atlantic Sentinel*, p. 77.
111. E. Weedon to H. Weaver, "Bonavista Extension," June 12, 1877, Vol. 2/Reel 1, Letter Book SI. For the 1873 election campaign, see E. Weedon to H. Weaver, "Private Letter on the Coming Election," October 21, 1873, Vol. 1/Reel 1, Letter Book SI.

112. Mark McGowan, "Irish Catholics," in *Encyclopedia of Canada's People*, ed. Paul R. Magocsi (Toronto: Publication for the Multicultural History Society of Ontario by Toronto University Press, 1999), p. 741.
113. Donald MacKay, *Flight from Famine: The Coming of the Irish to Canada* (Toronto: Natural Heritage Books, 2009), p. 335.
114. Willeen G. Keough, "Contested Terrains: Ethnic and Gendered Spaces in the Harbour Grace Affray," *The Canadian Historical Review* 90 (2009): p. 30.
115. Ibid., p. 32.
116. Cadigan, *Newfoundland and Labrador*, p. 140.
117. Keough, "Contested Terrains," p. 36.
118. E. Weedon to Cyrus W. Field, September 8, 1877, Vol. 2/Reel 1, Letter Book SI; E. Weedon to H. Weaver, November 26, 1878, Vol. 2/Reel 1, Letter Book SI.
119. E. Weedon to Managing Director, "Church Grant," September 20, 1883, Vol. 6/ Reel 8, Letter Book SI.
120. E. Weedon to H. Weaver, December 31, 1881, Vol. 4/Reel 6, Letter Book SI.
121. Ibid.
122. E. Weedon to Managing Director, "Horan & Wallace," February 18, 1882, Vol. 4/ Reel 6, Letter Book SI.
123. Cadigan, *Newfoundland and Labrador*, p. 143.
124. Eric J. Hobsbawm, *The Age of Empire: 1848–1875* (New York: Vintage Books, 1996), pp. 46–51.
125. Griset and Headrick, "Submarine Telegraph Cables," p. 553.
126. Kunert, *Telegraphen-Seekabel*, pp. 59–67; O. Moll, *Die Unterseekabel in Wort und Bild* (Köln: Westdeutscher Schriftenverein, 1904), pp. 12–14.
127. Hobsbawm, *Age of Empire*, p. 57.
128. Ibid.
129. The Great Ship Company, *The Great Eastern Steam Ship. The Past—The Future*, London, 1858, p. 10.
130. Tiago Saraiva, "Inventing the Technological Nation: The Example of Portugal (1851–1898)," *History and Technology* 23 (2007): pp. 263–73.
131. Hills, *Struggle for Control*, pp. 69–70.
132. Jorma Ahvenainen, *The Far Eastern Telegraphs: The History of Telegraphic Communications Between the Far East, Europe, and America Before the First World War* (Helsinki: Suomalainen Tiedekatemia, 1981), p. 166; Charles Bright, "An All-British or Anglo-American Pacific Cable," in *Imperial Telegraphic Communication*, ed. Charles Bright (London: P.S. King, 1911), p. 6.
133. Hills, *Struggle for Control*, p. 5.
134. Ibid.
135. George Squier, "The Influence of Submarine Cables upon Military and Naval Supremacy," *Proceedings of the U.S. Naval Institute* 26, no. 4 (1901): p. 602.
136. Kunert, *Telegraphen-Seekabel*, p. 202.
137. Ibid. 202, 230–31.

138. Bright, *Submarine Telegraphs*, pp. 137–40.
139. Charles Bright, "Imperial Telegraphic Communication and the 'All-British' Pacific Cable: London Chamber of Commerce, Pamphlet No. 40," in *Imperial Telegraphic Communication*, ed. Charles Bright (London: P.S. King, 1911); Bright, "Pacific Cable."
140. Daniel R. Headrick, *The Tools of Empire: Technology and European Imperialism in the Nineteenth Century* (New York: Oxford University Press, 1988), p. 162.
141. Bill Burns, "1899 Cape Town–St. Helena Cable," accessed July 2, 2011, http://www.atlantic-cable.com/Cables /1899StHelena/index.htm.
142. Headrick, *Tools of Empire*, p. 162.
143. W. D. Alexander, "The Story of the Transpacific Cable: (reprinted from Hawaiian Historical Journal Vol. 18) January 24, 1911," in *Development of Submarine Cable Communications*, ed. B. S. Finn (New York: Arno Press, 1980), p. 50.
144. 41st Congress, 2nd Session, Bill H.R. 1778, April 14, 1870; 41st Congress, 2nd Session. Bill S. 958. May 31, 1870.
145. 42nd Congress, 2nd Session, Bill H.R. 629, December 18, 1871.
146. 43rd Congress, Congressional Record, June 16, 1874.
147. "The Hawaiian Cable," *New York Times*, February 8, 1895.
148. Winseck and Pike, *Communication and Empire*, p. 157.
149. Bowell, *Proceedings of the Colonial Conference, June 28–July 9, 1894* (Ottawa: S. E. Dawson, 1894), pp. 21–30; National Archives Canada, p. 21; Sandford Fleming to Sir Hector Langevin, Minister of Public Works, April 8, 1886, Material relating to a proposed Pacific Cable, 1886–1899, NAEST 17/102, IET Archives, p. 1.
150. John Lamb to Kempe, October 18, 1893, Post 30/803, 6097/98, File 26, BT Archives.
151. Robert Boyce, "Imperial Dreams and National Realities: Britain, Canada and the Struggle for a Pacific Telegraph Cable, 1879–1902," *The English Historical Review* 115 (2000): p. 50.
152. "A Pacific Cable," *The Times*, November 18, 1892, or "The Colonies. The Pacific Cable Conference," *The Times*, June 9, 1896. Similarly Charles Bright, "Pacific Cable. To the Editor of the Times," *The Times*, December 26, 1898.
153. Boyce, "Imperial Dreams," pp. 44–45.
154. Hamilton, Memo, April 29, 1897, Public Record Office, Britain, cited in Winseck and Pike, *Communication and Empire*, p. 165.
155. See the exchange of letters with Canada's High Commissioner in the *London Times* in 1894.
156. "Pacific Cable Plan Blocked?" *New York Times*, April 29, 1899.
157. Boyce, "Imperial Dreams," p. 62.
158. Sandford Fleming to MacDonald, July 6, 1880, Fleming Papers NCA.
159. "Proposed Pacific Cable to Australia," *The Times*, November 24, 1888.
160. *Novoe Vremya* 1898, cited in Bright, "Pacific Cable," p. 7.
161. Boyce, "Imperial Dreams," p. 49.

162. Ibid., p. 66.
163. Winseck and Pike, *Communication and Empire*, pp. 165–67.
164. "Honolulu Cable Project," *New York Times*, July 28, 1892.
165. "The Hawaiian Cable," *Los Angeles Times*, August 24, 1895; "For a Cable to Honolulu," *New York Times*, February 10, 1895.
166. "Hawaiian Cable Prospects," *New York Times*, August 3, 1898.
167. Abram Hewitt to James B. McCreary, House of Representatives, February 15, 1895, Abram Hewitt Papers, CUA.
168. Emphasis mine. Bright, "Imperial Telegraphic Communication," p. 33.
169. Abram Hewitt to John Pender, February 12, Abram Hewitt Papers, CUA.
170. Abram Hewitt to John Pender, February 15, 1895, Abram Hewitt Papers, CUA.
171. Abram Hewitt to John Pender, February 11, 1895, Abram Hewitt Papers, CUA; "For a Pacific Cable," *New York Times*, February 11, 1899.
172. "Corliss Pacific Cable Bill," *New York Times*, June 11, 1902.
173. Winseck and Pike, *Communication and Empire*, p. 170.
174. "Pacific Cable Hearing: House Commerce Committee Considers the Matter. Effect of Wireless," *New York Times*, January 12, 1902.
175. "Corliss Pacific Cable Bill: The House Adopts Rule for Debate and Considers the Measure—Mr. Dalzell Speaks," *New York Times*, June 11, 1902.
176. "Pacific Cable in a Year: Differences with Mackay Company Adjusted," *Washington Post*, November 21, 1902.
177. Winseck and Pike, *Communication and Empire*, p. 170;
178. Carl Willhelm Guilleaume, letter to Reinhold von Sydow, August 6, 1901, Rheinisch-Westfälisches Wirtschaftsarchiv Köln.
179. Ahvenainen, *Far Eastern Telegraphs*, p. 174.
180. Griset and Headrick, "Submarine Telegraph Cables," pp. 566–67.
181. George Ward to O. Moll, July 1, 1910, in BArch R/901/80740.
182. Kaiserlich deutsche Botschaft U.S.A. to Reichskanzler von Bethman Hollweg, October 1, 1910, BArch R/901/80740.
183. Thomas C. Owen, *The Corporation Under Russian Law, 1800–1917: A Study in Tsarist Economic Policy* (Cambridge: Cambridge University Press, 1991), p. 184.
184. Carl Ludwig Löffler, HO 144/308/B5783, National Archives Kew; for Siemens see "Death of Sir William Siemens," *Birmingham Daily Post*, November 24, 1883.
185. Bright, "Imperial Telegraphic Communication," p. 62.

7. THE WIRING OF THE WORLD

1. Cyrus W. Field, *Ocean Telegraphy: The Twenty-Fifth Anniversary of the Organization of the First Company Ever Formed to Lay an Ocean Cable* (New York, 1879), p. 34.
2. Daniel R. Headrick, *The Tools of Empire: Technology and European Imperialism in the Nineteenth Century* (New York: Oxford University Press, 1988), p. 162; Pascal

Griset and Daniel R. Headrick, "Submarine Telegraph Cables: Business and Politics, 1838–1939," *Business History Review* 75 (2001): p. 572.

3. Daniel R. Headrick, *The Invisible Weapon: Telecommunications and International Politics, 1851–1945* (New York: Oxford University Press, 1991), p. 28.
4. Field, *Ocean Telegraphy*, p. 3.
5. Gillian Cookson, "The Transatlantic Telegraph Cable—Eighth Wonder of the World," *History Today* 50 (2000): pp. 44–51.
6. Alfred D. Chandler, Jr., "The Emergence of Managerial Capitalism," *The Business History Review* 58 (1984): p. 496.
7. Henry A. Moriarty to the Editor of the *Standard*, March 8, 1870, Moriarty Letter Book CWA.
8. Contrary to the theory of "gentlemanly capitalism"; P. J. Cain and A. G. Hopkins, "The Political Economy of British Expansion Overseas, 1750–1914," *The Economic History Review* 33 (1980): pp. 463–90.
9. Michael Fry, *The Scottish Empire* (East Linton: Tuckwell Press, 2001).
10. Harold James, *History of Krupp* (Princeton: Princeton University Press, 2012).
11. Jürgen Osterhammel, *Die Verwandlung der Welt: Eine Geschichte des 19. Jahrhunderts* (München: Beck, 2009), chapter XV.
12. Charles H. Parker, *Global Interactions in the Early Modern Age, 1400–1800* (Cambridge: Cambridge University Press, 2010), and David Porter, *Comparative Early Modernities: 1100–1800* (Basingstoke: Palgrave Macmillan, 2012).
13. On early global markets, see Alexander Engel, *Farben der Globalisierung: Die Entstehung moderner Märkte für Farbstoffe 1500–1900* (Frankfurt/Main: Campus Verl., 2009).
14. A. W. Holland, *The Real Atlantic Cable* (London: Nabu Press, 1914), p. 6.
15. Ibid.
16. Bill Glover, "Cable Timeline—1845 to 2012," accessed September 19, 2011, www.atlantic-cable.com.
17. "Our London Correspondence," *Glasgow Herald*, September 2, 1875.
18. Eric J. Hobsbawm, *The Age of Capital: 1848–1875* (New York: Vintage Books, 1996), p. 4.
19. Eric J. Hobsbawm, *Industry and Empire: The Birth of the Industrial Revolution* (New York: The New Press, 1999), p. 156; Alfred D. Chandler, Jr., *The Visible Hand: The Managerial Revolution in American Business* (Cambridge, MA: Belknap Press of Harvard Univ. Press, 2002).
20. John Pender, cited in J. D. Scott, *Siemens Brothers 1858–1958: An Essay in the History of Industry* (London: Weidenfeld & Nicholson, 1958), p. 57.
21. B. S. Finn, "Submarine Telegraphy: A Study in Technical Stagnation," in *Communications Under the Seas: The Evolving Cable Network and its Implications*, eds. Bernard S. Finn and Daqing Yang (Cambridge, MA: MIT Press, 2009); Richard Noakes, "Industrial Research at the Eastern Telegraph Company, 1872–1929," *The British Journal for the History of Science* 47 (2014): pp. 119–46.

22. David Edgerton, *Science, Technology and the British Industrial "Decline": 1870–1970* (Cambridge: Cambridge University Press, 1996), pp. 18, 25.
23. William Siemens, "The Society of Telegraph Engineers: Inaugural Address," in *Journal of the Society of Telegraph Engineers: Including Original Communications on Telegraphy and Electrical Science*, Vol. 1 (London: 1872/1873), p. 22.
24. Hartmut Böhme, "Einleitung," in *Netzwerke: Eine Kulturtechnik der Moderne*, eds. Jürgen Barkhoff, Hartmut Böhme, and Jeanne Riou (Köln: Böhlau, 2004), pp. 21–22.
25. James Anderson, "Submarine Cables in War Time," *Pall Mall Gazette*, October 22, 1884.
26. On the commercialization of news, see Gerald J. Baldasty, *The Commercialization of News in the Nineteenth Century* (Madison, WI: University of Wisconsin Press, 1992).
27. Hobsbawm, *Age of Capital*, p. 60.
28. "Yankee Snacks," *The Newcastle Weekly Courant*, September 12, 1891.
29. A. J. Gillam, *Great Britain versus America Cable Matches 1895–1901* (London: The Chess Player, 1997), p. 7.
30. Abram Hewitt to John Pender, February 11, 1895, Abram Hewitt Papers, CUA.
31. P. M. Kennedy, "Imperial Cable Communications and Strategy, 1870–1914," *The English Historical Review* 86 (1971): pp. 728–52.
32. Headrick, *Tools of Empire*.
33. Griset and Headrick, "Submarine Telegraph Cables," pp. 566–67.
34. J. H. Heaton, "Penny-a-Word Telegrams Throughout the Empire," *Proceedings of the Royal Colonial Institute* 40 (1908–1909): p. 6.
35. Hugh G. Aitken, *Syntony and Spark—the Origins of Radio* (Princeton, NJ: Princeton University Press, 1985); Sungook Hong, *Wireless: From Marconi's Black-Box to the Audion* (Cambridge, MA: MIT Press, 2001).
36. Margot Fuchs, "Anfänge der drahtlosen Telegraphie im Deutschen Reich 1897–1918," in *Vom Flügeltelegraphen zum Internet: Geschichte der modernen Telekommunikation*, eds. Hans-Jürgen Teuteberg and Cornelius Neutsch (Stuttgart: Steiner, 1998), p. 116.
37. Griset and Headrick, "Submarine Telegraph Cables," p. 574.
38. Heidi Tworek, "'The Path to Freedom'? Transocean and German Wireless Telegraphy, 1914–1922," *Historical Social Research. Historische Sozialforschung* 35 (2010): p. 227.
39. Ibid., pp. 215–17.
40. Griset and Headrick, "Submarine Telegraph Cables," p. 574.
41. Kenneth G. Beauchamp, *History of Telegraphy* (London: Institution of Electrical Engineers, 2001), p. 239.
42. Richard R. John, *Network Nation: Inventing American Telecommunications* (Cambridge, MA: Belknap Press of Harvard University Press, 2010), p. 12.

APPENDIX. ACTORS OF GLOBALIZATION

1. This compilation of biographies is based on the *Oxford Dictionary of National Biography*, the *Dictionary of American Biography*, the American National Biography, and scraps and pieces from a variety of archival material.

BIBLIOGRAPHY

ARCHIVAL SOURCES

America Singing Collection, Library of Congress.

Anderson, James, Papers, Cable and Wireless Archive, Porthcurno Telegraph Museum.

Anglo-American Telegraph Company, Papers, Cable and Wireless Archive, Porthcurno Telegraph Museum.

Anglo-American Telegraph Company, Papers, Smithsonian Institution Archives.

Anonymous. Note Book from a Telegraph Construction & Maintenance Company Employee, March 2, 1870–October 23, 1873, Cable and Wireless Archive, Porthcurno Telegraph Museum.

Atlantic Telegraph Company. Papers, Cable and Wireless Archive, Porthcurno Telegraph Museum.

Behm, Charles F. W., Papers, New York Public Library, Manuscript Division.

Board of Trade, Papers, Companies Registration Offices, National Archives London.

Brett, Jacob, Papers, Special Collection Manuscripts, Archives of the Institution of Engineering and Technology.

Briefsammlung der Brüder Siemens, 1842–1892, Siemens Corporate Archives.

Brumidi, Constantino, "The Telegraph, 1862," United States Senate Collection.

Brunel, Isambard Kingdom, Papers, Bristol University Archives.

Cooper, Peter, Papers, Cooper Union Archives.

Direct United States Cable Company, Papers, Cable and Wireless Archive, Porthcurno Telegraph Museum.

Early Activities and Correspondence of the IET, 1871–1929, Papers, Archives of the Institution of Engineering and Technology.

The Eastern Telegraph Company, Limited, Papers, Cable and Wireless Archive, Porthcurno Telegraph Museum.

Field, Cyrus, Papers, New York Public Library, Manuscript Division.

Field, Cyrus, Papers, Smithsonian Institution Archives.

Fleming, Sandford, Papers, National Archives Canada.

Granville, W. P., Diary, Cable and Wireless Archive, Porthcurno Telegraph Museum.

Great Eastern Steamship Company, Papers, Cable and Wireless Archive, Porthcurno Telegraph Museum.

Hewitt, Abram, Papers, Cooper Union Archives.

Kennan, George, Papers, New York Public Library, Manuscript Division.

"L'Arrivée à Paris, par le Tunnel de la Manche, 1884," Sidney M. Edelstein Library, Hebrew University of Jerusalem.

Moriarty, Henry August, Private Letter Book of Captain Henry August Moriarty, Cable and Wireless Archive, Porthcurno Telegraph Museum.

New York, Newfoundland and London Telegraph Company, Papers, Cable and Wireless Archive, Porthcurno Telegraph Museum.

O. Roty, "Medal Commemorating the Opening of the Suez Canal, 1869," Commemorative Medals, National Maritime Museum London.

Pender, Emma, Papers, John Pender Papers, Cable and Wireless Archive, Porthcurno Telegraph Museum.

Pender, John, Papers, Cable and Wireless Archive, Porthcurno Telegraph Museum.

Pictorial Americana, Prints and Photographs Division, Library of Congress.

Postes et Telegraphes, Cables Transatlantiques Anglo-Americains, Archives Nationales France.

Postes et Telegraphes, Compagnie Francaise des Câbles Télégraphiques, Papers, Archives Nationales France.

Postes et Telegraphes, Compagnie Francaise du Télégraphe de Paris a New York, Papers, Archives Nationales France.

Post Office, Papers, British Telecom Archives.

Preece, William, Papers, Archives of the Institution of Engineering and Technology.

Public Record Office, Papers, National Archives London.

R4701 Reichspostministerium, Bundesarchiv Berlin-Lichterfelde.

R901 Auswärtiges Amt, Bundesarchiv Berlin-Lichterfelde.

Spratt, George, Diaries, 1871–1909, Cable and Wireless Archive, Porthcurno Telegraph Museum.

The Telcon Story, National Archive for Electrical Science and Technology, Archives of the Institution of Engineering and Technology.

Telegraphy Books, National Archive for Electrical Science and Technology, Archives of the Institution of Engineering and Technology.

U.S. Congress, Papers, Library of Congress.

U.S. Sanitary Commission, Records/English Branch Archives, New York Public Library, Manuscript Division.

Western Union Collection, Smithsonian Institution Archives.

NEWSPAPERS AND PERIODICALS

Aberdeen Weekly Journal
Belfast News-Letter
Birmingham Daily Post
Boston Daily Globe
Bristol Mercury
Caledonian Mercury (Edinburgh)
Chicago Daily Tribune
Electrician
Era (London)
Freeman's Journal and Daily Commercial Advertiser (Dublin)
Glasgow Herald
Hampshire Telegraph and Sussex Chronicle, etc.
Harper's Weekly. A Journal of Civilization
Journal of the Institution Electrical Engineers (London)
Journal of the Society of Telegraph Engineers (London)
Leeds Mercury
Liverpool Mercury
London Daily News (London)
London Times
Nation
Newcastle Weekly Courant
New York Herald
New York Times
New York Tribune
Observer (London)
Pall Mall Gazette (London)
Paris Herald
The Penny Illustrated Paper and Illustrated Times (London)
Punch
Reynold's Newspaper (London)
The Telegraphist. A Monthly Journal for Postal, Telephone, and Railway Telegraph Clerks (London)
The Trewman's Exeter Flying Post (Exeter)
Washington Post
Western Mail (Cardiff)

PUBLISHED PRIMARY SOURCES

Alexander, W. D. "The Story of the Transpacific Cable: (reprinted from Hawaiian Historical Journal vol. 18) January 24, 1911." In *Development of Submarine Cable Communications*. Edited by B. S. Finn, 51–71. New York: Arno Press, 1980.

American Institute of Electrical Engineers. *Historical Sketch of Its Organization and Work: Founded 1884*. New York: American Institute of Electrical Engineers, 1892.

Anderson James. "Anglo-Mediterranean Telegraphs: To the Editor of the Times." *Times*, September 24, 1868.

——. "Submarine Cables in War Time." *Pall Mall Gazette*, October 22, 1884.

——. "The Telegraph to India: To the Editor of the Times." *Times*, December 7, 1868.

——. "Telegraph Reform, Being a Paper read by Sir James Anderson Before a Congress of the Chambers of Commerce of the British Empire, at the Indian and Colonial Exhibition, South Kensington, on July 6th (and the Discussion upon the Same on July 6th and 7th) 1886, July 6." In *Cables in Time of War Etc*. Edited by James Anderson, 3–26. London: George Tucker, 1886.

——, ed. *Cables in Time of War Etc*. London: George Tucker, 1886.

Anglo-American Telegraphic Code and Cypher Co. *The Anglo-American Telegraphic Code to Cheapen Telegraphy and to Furnish a Complete Cypher*. Adapted to use in general correspondence; included business, social, political, and all other sources of correspondence. New York, 1891.

Anglo-American Telegraph Company, ed. *Letter Books Heart's Content Letter Book from Anglo-American Telegraph Co. Station at Heart's Content, Newfoundland. 1870–1876*. Originals preserved at Heart's Content Cable Museum, under the Historic Resource Division, Department of Provincial Affairs, Government of Newfoundland and Labrador, St. John's Newfoundland. 11 Vols., Reproduced 1973 by B. Finn, Smithsonian Institutions.

Articles of Confederation and Perpetual Union Between the States of New Hampshire, Massachusetts Bay, Rhode Island, and Providence Plantations, Connecticut, New York, New Jersey, Pennsylvania, Delaware, Maryland, Virginia, North Carolina, South Carolina, and Georgia. Williamsburg: Alexander Purdie, 1777.

Bamber, E. F., ed. *The Scientific Works of William Siemens, Civil Engineer: A Collection of Addresses, Lectures, Etc*. III. London: John Murray, Albemarle Street, 1889.

Beckles, Willson. *The Tenth Island: An Account of Newfoundland*. London: Grant Richards, 1897.

Bell, Alexander G. "A Lecture Entitled Researches in Electric Telephony: Delivered Before the Society of Telegraph Engineers October 31, 1877." Edited by Society of Telegraph Engineers. London/New York: 1878.

Blundell, J. W. A. *The Manual of Submarine Telegraph Companies 1872*. London: Author, 1872.

Bowell. *Proceedings of the Colonial Conference, June 28-July 9, 1894*. Printed by order of Parliament. Ottawa: S. E. Dawson, 1894.

Bright, Charles. "An All-British or Anglo-American Pacific Cable." In *Imperial Telegraphic Communication*. Edited by Charles Bright, 4–17. London: P.S. King, 1911.

——. "The Extension of Submarine Telegraphy in a Quarter-Century: Reprinted from Engineering Magazine (December 1898)." In *Development of Submarine Cable Communications*. Edited by B. S. Finn, 417–28. New York: Arno Press, 1980.

——. *Imperial Telegraphic Communication.* Edited by Charles Bright. London: P.S. King, 1911.

——. "Imperial Telegraphic Communication and the 'All-British' Pacific Cable: London Chamber of Commerce, Pamphlet No. 40." In *Imperial Telegraphic Communication.* Edited by Charles Bright, 29–75. London: P.S. King, 1911.

——. *Sir Charles Tilston Bright Civil Engineer: With Which Is Incorporated the Story of the Atlantic Cable, and the First Telegraph to India and the Colonies.* London: Archibald Constable, 1908.

——. *The Story of the Atlantic Cable.* New York: D. Appleton, 1903.

——. *Submarine Telegraphs: Their History, Construction and Working.* London: C. Lockwood, 1898.

——. "Universal Penny-a-word Telegraphy: Letter to the Editor." *Times,* November 18, 1908.

Brown, J. *The Cable and Wireless Communications of the World: A Survey of Present Day Means of International Communication by Cable and Wireless.* London: Sir Isaac Pitman, 1927.

Chandler, Albert B. *A New Code, or Cipher Specially Designed for Important Private Correspondence by Telegraph. Applicable as Well to Correspondence by Mail.* Washington DC, 1869.

Clark, Latimer. "On a Standard Voltaic Battery." *Journal of the Society of Telegraph Engineers* 7 (1878): 85–107.

Cooley, Charles H. *Social Organization: A Study of the Larger Mind.* New York: C. Scribner's, 1909.

Cooper, Peter. *A Sketch of the Early Days and Business Life of Peter Cooper. An Autobiography.* New York, 1877.

Department of Foreign Affairs and Trade, Canberra. *Convention for the Protection of Submarine Telegraph Cables, 1884.* Australian Treaty Series 1, 1901.

Drapner, George. "Notes of Earth Currents on February 4th, 1872: (reprinted from the Times)." *Journal of the Society of Telegraph Engineers* 1, no. 1 (1872): 110–11.

Emerson, Ralph Waldo. "The Young American: A Lecture Read Before the Mercantile Library Association." Boston, February 7, 1844.

Engels, Friedrich. "Herrn Eugen Dührings Umwälzung der Wissenschaft." In *Karl Marx, Friedrich Engels Ausgewählte Werke.* Edited by Mathias Bertram. Digitale Bibliothek 11. Berlin: Directmedia, 2004.

"Europa and America. Report of the Proceedings at an Inauguration Banquet, Given by Mr. Cyrus W. Field, of New York, at the Palace Hotel, Buckingham Gate, on Friday, the 15th April, 1864, in Commemoration of the Renewal by the Atlantic Telegraph Company, (After a Lapse of Six Years) of Their Efforts to Unite Ireland & Newfoundland, by Means of a Submarine Electric Telegraph Cable, 15th April 1864." London, 1864.

Field, Cyrus W. *Ocean Telegraphy: The Twenty-Fifth Anniversary of the Organization of the First Company Ever Formed to Lay an Ocean Cable.* New York, 1879.

Field, David D. "Address at the de Lesseps Banquet: Given at Delmonicos, March 1, 1880." In *Speeches, Arguments and Miscellaneous Papers of David Dudley Field.* Edited by David D. Field and A. P. Sprague, 359–61. New York: D. Appleton, 1884.

——. *Draft Outlines of an International Code.* New York: Diossy, 1872.

——. *An International Code: Address on this Subject, Before the Social Science Association, at Manchester, October 5, 1866.* Edited by YA Pamphlet Collection. New York: W.J. Read, 1867.

Field, David D., and Albéric Rolin. *Projet d'un Code International.* Paris: Gand, 1881.

Field, David D., and A. P. Sprague, eds. *Speeches, Arguments and Miscellaneous Papers of David Dudley Field.* New York: D. Appleton, 1884.

Field, Henry M. *History of the Atlantic Telegraph.* New York: C. Scribner, 1866.

——. *The Life of David Dudley Field.* New York: Charles Scribner's, 1898.

Geistbeck, Michael. *Der Weltverkehr: Telegraphie und Post, Eisenbahnen und Schiffahrt in ihrer Entwicklung dargestellt.* Freiburg: Herbersche Verlagshandlung, 1895.

Goddard, S. A. "The Atlantic Telegraph: To the Editor of the Daily Post." *Birmingham Daily Post*, August 2, 1866.

Golder, Dwight. *Official Cable Code and General Information for European Tourists Including French and German Phrases with English Pronunciation.* New York, 1887.

Gooch, Daniel. *Diaries of Sir Daniel Gooch Baronet: With an Introductory Notice by Sir Theodore Martin K.C.B.* London: Kegan Paul, Trench Trübner, 1892.

Gould, Benjamin A. *Transatlantic Longitude as Determined by the Coast Survey Expedition of 1866: A Report to the Superintendent of the U.S. Coast Survey.* Smithsonian Contributions to Knowledge. Washington, DC: Smithsonian Institution, 1869.

Grant, E. B. *Report on the Western Union Telegraph Company.* New York: The Sun Job Print, Printinghouse Square, 1869.

——. *The Western Union Telegraph Company: Its Past, Present and Future.* New York: Hotchkiss, Burnham, 1883.

Grant, Ulysses. "Monday, December 6, 1869." *Journal of the Senate of the United States of America, 1789–1873* (1869).

Graves, Edward. "Inaugural Address of the New President, Mr. Edward Graves." *Journal of the Society of Telegraph Engineers and Electricians* 17, no. 70 (1888): 4–30.

Graves, James. *The Autobiography of James Graves.* 1880s/2005; Electronic version of two early autobiographies. Transcribed and Edited by Donard de Cogan.

——. "Earth Currents, and the Aurora Borealis of 4th February 1872." *Journal of the Society of Telegraph Engineers* 1, no. 1 (1872): 102–4.

——. "On Curbed Signals on Long Cables." *Journal of the Society of Telegraph Engineers* 8 (1879): 92.

——. *Thirty Six Years in the Telegraphic Service 1852–1888 Being a Brief Autobiography of James Graves MSTE.* 1909/2008. Transcribed by Dominic de Cogan. Edited by Donard de Cogan.

Great Britain, and Joint Committee to Inquire into the Construction of Submarine Telegraph Cables. *Report of the Joint Committee Appointed by the Lords of the Committee of Privy Council for Trade and the Atlantic Telegraph Company to Inquire into the Construction of Submarine Telegraph Cables.* London: G.E. Eyre and W. Spottis-Woode, 1861.

Gurney, Samuel. *Epitome of Proceedings at a Telegraphic Soirée Given by Samuel Gurney, Esq., M.P., At 25, Prince's Gate, Hyde Park, March 26, 1862.* London: Thomas Piper, at the Electrician Office, 1862.

Hale, Chandler. "The Projected Cable-line to the Philippines." *The North American Review* 171, no. 524 (1900): 82–89.

Hamilton, Alexander. "Cyrus W. Field's Audacity." *New York Times*, December 20, 1885.

Harvey, Moses. *A Short History of Newfoundland: England's Oldest Colony.* London: W. Collins, 1890.

Heaton, J. H. "A Cable Post: The Possibilities of Atlantic Submarine Communication." *The North American Review* 160, no. 463 (1895): 659–67.

——. "The Cable Telegraph System of the World." *Arena* 38, no. 214 (September 1907): 226–29.

——. "Imperial Telegraph Cables: To the Editor of the Times." *Times*, March 3, 1902.

——. "Penny-a-Word Telegrams Throughout the Empire." *Proceedings of the Royal Colonial Institute* 40 (1908–1909): 3–37.

——. "The Postal and Telegraphic Communication of the Empire." *Proceedings of the Royal Colonial Institute* 19 (1887–1888): 171–221.

——. "Universal Penny-a-Word Telegrams: Letters to the Editor." *Times*, November 23, 1908.

——. "Wanted, Cheap Imperial Telegraphs." *Times*, July 13, 1899.

——. "The World's Cables and the Cable's Rings." *Financial Review of Reviews*, May 1908.

Hennig, Richard. "Die historische Entwicklung der deutschen Seekabelunternehmungen." *Beiträge zu Geschichte der Technik und Industrie. Jahrbuch des Vereines Deutscher Ingenieure* 1 (1909): 241–50.

Highton, Edward. *The Electric Telegraph: Its History and Progress.* London: John Weale, 1852.

Holland, A. W. *The Real Atlantic Cable.* London: Nabu, 1914.

House of Representatives, 34th Congress. 3rd session. "Submarine Telegraph." In *The Congressional Globe.* Edited by U.S. Congress, 729–31. 1857.

——. "Submarine Telegraph—Again." In *The Congressional Globe.* Edited by U.S. Congress, 586–89. 1857.

House of Representatives, 41st Congress. "Friday, December 10, 1869." In *Journal of the House of Representatives of the United States* (1869–1870).

Institution of Civil Engineers, ed. *The Practical Applications of Electricity. A Series of Lectures Delivered at the Institution of Civil Engineers, Session 1882/3.* London: Institution of Civil Engineers, 1884.

Internationale Telegraphen Union. *Internationale Telegraphen-Conferenz von St. Petersbourg: Vertrag, Dienstreglement und Tarifftabellen.* Bern: ITU, 1875.

International Telecommunication Union. *Procès-Verbaux: Conference Télégraphique Internationale de Londres. Liste de Participants.* Bern: International Telecommunication Union, 1904.

Judson, Isabella F. *Cyrus W. Field: His Life and Work.* New York: Harper & Brothers Publishers, 1896.

Kapp, Ernst. *Grundlinien einer Philosophie der Technik: Zur Entstehungsgeschichte der Cultur aus neuen Gesichtspunkten.* Braunschweig, 1877.

Kennan, George. *Tent Life in Siberia: and Adventures Among the Koraks and Other Tribes in Kamtchatka and Northern Asia.* New York: G.P. Putnam's, 1877.

Knies, Karl. *Der Telegraph als Verkehrsmittel: Über den Nachrichtenverkehr überhaupt.* München: Fischer, 1996 [reprint from 1857].

Lardner, Dionysius. *The Electric Telegraph Popularised: With One Hundred Illustrations.* London: Walton and Maberly, 1855.

Maury, Matthew F. *The Physical Geography of the Sea.* London, 1855.

Maury, Matthew F., and L. Warrington. *Wind and Current Chart of the North Atlantic.* Washington, DC: U.S. Hydrographical Office, 1847/1852.

Maver, William. "Ocean Telegraphy: Extract from the Electrical World [American]." *The Telegraphist. A Monthly Journal for Postal, Telephone and Railway Telegraph Clerks* 2 (June 1884): 87–88.

McCarthy, Justin, ed. "The Settlement of the Alabama Question. The Banquet Given at New York to the British High Commissioners by Cyrus W. Field, 1871." London 1872.

Mercator. "International Telegraph Conference: To the Editor of the Times." *Times,* June 13, 1879.

Meyer, H. R. "The International Telegraph Conference." *Times,* June 17, 1879.

Moll, O. *Die Unterseekabel in Wort und Bild.* Köln: Westdeutscher Schriftenverein, 1904.

Morse, Samuel F. B., and Edward L. Morse. *Samuel F. B. Morse: His Letters and Journals.* New York: Da Capo Press, 1973.

Mullaly, John. *The Laying of the Cable, or the Ocean Telegraph.* New York: D. Appleton, 1858.

Nelson, George H., ed. *A Collection of Letters to Charles William Siemens 1823–1883: With a Foreword by Sir George H. Nelson.* London: The English Electric Company Limited, 1953.

Old Electric. "Introductory." *The Telegraphist. A Monthly Journal for Postal, Telephone and Railway Telegraph Clerks,* August 1, 1884.

Palmer. *Palmer's European Travelers and Telegraph Code.* 1880.

Parkinson, J. C. *The Ocean Telegraph to India.* London: William Blackwood, 1870.

Pastor Rev. J. Bray. "On the Atlantic Cable." *The New York Morning Express,* August 17, 1858.

Pender, John. "Letter to Jay Gould, December 11th: The New American Telegraph Company." *Glasgow Herald*, December 30, 1880.

Pender, Rose. *No Telegraph: Or, a Trip to Our Unconnected Colonies. 1878.* London: Gilbert & Rivington, 1879.

Pole, William. *The Life of Sir William Siemens: Member of Council of the Institution of Civil Engineers.* London: John Murray, Albemarle Street, 1888.

Porter, Adrian. "The Life and Letters of Sir John Henniker Heaton Bt." *Times*, September 9, 1914.

——. *The Life and Letters of Sir John Henniker Heaton Bt.: By His Daughter Mrs. Adrian Porter with Numerous Illustrations.* New York: John Lane, 1916.

Preece, William H. "Minutes of the Ordinary General Meeting and the Annual Meeting Held on December 22, 1880." *Journal of the Society of Telegraph Engineers* 10, no. 35 (1881): 1–3.

——. "The Progress of Telegraphy: 15. February 1883." In *The Practical Applications of Electricity. A Series of Lectures Delivered at the Institution of Civil Engineers, Session 1882/3.* London: Institution of Civil Engineers, 1884.

Proudhon, Pierre-Joseph. *Qu'est-ce que la propriété? Ou recherches sur le principe du droit et du gouvernement.* Paris: A la Librarie de Prévot, 1841.

Prowse, D. W. *A History of Newfoundland: From the English Colonial and Foreign Records.* London: Eyre and Spottiswoode, 1896.

Rawlinson, H. B., and William Boutcher. "The Telegraph to India." *Times*, May 31, 1858.

Renault, Louis. *Etudes sur les rapports internationaux: La Poste et le Télégraphes.* Paris, 1877.

Russel, Edgar, Adolphus W. Greely, and Samuel Reber. *Handbook of Submarine Cables.* Washington, DC: Government Printing Office, 1905.

Russell, William H., and Robert Dudley. *The Atlantic Telegraph by W.H. Russell L.L.D. Illustrated by Robert Dudley: Dedicated by Special Permission to His Royal Highness Albert Edward Prince of Wales.* London: Day & Son, 1865.

Sabine, Robert. *The Electric Telegraph.* London: Virtue Brothers, 1867.

Saward, George. *The Trans-Atlantic Submarine Telegraph: A Brief Narrative of the Principal Incidents in the History of the Atlantic Telegraph Company.* London, 1878.

Scrymser, James A. *Personal Reminiscence of James A. Scrymser: In Times of Peace and War.* New York: Eschenbach, 1915.

Shaffner, T. P. "Communication with America, via the Faröes, Iceland, and Greenland." *Proceedings of the Royal Geographical Society of London* 4, no. 3 (1859–1860): 101–8.

——. *The Telegraph Manual: A Complete History and Description of the Semaphoric, Electric and Magnetic Telegraphs of Europe, Asia, Africa, and America, Ancient and Modern with Six Hundred and Twenty-Five Illustrations.* New York: Pudney & Russell, Publishers, 1859.

Siemens, Werner von. *Lebenserinnerungen von Werner von Siemens: Mit dem Bildniss des Verfassers in Kupferätzung*. Berlin: Julius Springer, 1908.

——, ed. *Wissenschaftliche und technische Arbeiten*. Berlin: Julius Springer, 1889.

Siemens, William. "Address as President of the Society of Telegraph Engineers: Delivered on January 23, 1878." In *The Scientific Works of William Siemens, Civil Engineer: A Collection of Addresses, Lectures, Etc.* Edited by E. F. Bamber, 162–81, III. London: John Murray, 1889.

——. "Deep-Sea Telegraphs." In *The Scientific Works of William Siemens, Civil Engineer: A Collection of Addresses, Lectures, Etc.* Edited by E. F. Bamber, 2–23, III. London: John Murray, 1889.

——. "On Certain Means of Measuring and Regulating Electric Currents." *Proceedings of the Royal Society of London* 20 (1878–1879): 292–97.

——. "On Determining the Depth of the Sea Without the Use of Sounding-Line." *Philosophical Transactions of the Royal Society of London* 166 (1876): 671–92.

——. "The Society of Telegraph Engineers: Inaugural Address." In *Journal of the Society of Telegraph Engineers: Including Original Communications on Telegraphy and Electrical Science* 1 (1872–1873): 19–33.

——. "Testing Electric Cables." In *The Scientific Works of William Siemens, Civil Engineer: A Collection of Addresses, Lectures, Etc.* Edited by E. F. Bamber, 29–48, III. London: John Murray, 1889.

Slater, Robert. *Telegraphic Code to Ensure Secrecy in the Transmission of Telegrams. By Robert Slater, Secretary of the Societé du Cable Transatlantique Francaise, Limited: Second Edition*. London, 1897.

Smith, Willoughby. "Inaugural Address of President for 1883, Willoughby Smith." *Journal of the Society of Telegraph Engineers* 12, no. 46 (1883): 5–26.

——. *The Rise and Extension of Submarine Telegraphy*. New York: Arno Press, 1974.

Squier, George. "The Influence of Submarine Cables upon Military and Naval Supremacy." *Proceedings of the U.S. Naval Institute* 26, no. 4 (1901): 599–622.

Stanley, Henry M. *The Autobiography of Sir Henry Morton Stanley*. Boston: Houghton Mifflin, 1890.

Stevens, J. A. *Report of the Proceedings of a Meeting Called to Further the Enterprise of the Atlantic Telegraph. Held at the Hall of The Chamber of Commerce, New York, March 4, 1863*. New York: John W. Amerman Printer, 1863.

Sumner, Samuel B., and Charles Sumner. *Poems*. New York: The Author Publishing Company, 1877.

The Covenant of the League of Nations with a Commentary Thereon. London: His Majesty's Stationary Office, 1919.

Thomson, William. "The President's Inaugural Address." *Journal of the Society of Telegraph Engineers* 3, no. 7 (1874): 1–21.

Thurn, H. *Die Seekabel unter besonderer Berücksichtigung der deutschen Seekabeltelegraphie: In technischer, handelswirtschaftlicher, verkehrspolitischer und strategischer Beziehung dargestellt*. Leipzig: S. Hirzel, 1909.

Twenty-Seventh Anniversary of the First Atlantic Cable: Mr. Cyrus W. Field's Banquet at the Star and Garter Hotel, Richmond, on August 5th 1885. London: Printed for Private Circulation, 1886.

U.S. Congress, ed. *The Congressional Globe*, 1857.

van Rensselaer, C. *Miscellaneous Sermons, Essays and Addresses: By the Rev. Cortland van Resselaer*. Philadelphia: J.B. Lippincott, 1861.

——. *Signals from the Atlantic Cable: An Address Delivered at the Telegraphic Celebration, September 1st, 1858, in the City Hall, Burlington, New Jersey*. Philadelphia: J. M. Wilson, 1858.

Verne, Jules. *A Floating City*. London: Sampson Low, Marston, 1871.

Walsh, George E. "Cable Cutting in War." *The North American Review* 167, no. 503 (1898): 498–502.

War Department: Office of the Chief of Staff. *Rules of Land Warfare*. Washington, DC: Government Printing Office, 1914.

Watts, John. *The Facts of the Cotton Famine*. London: Simpkin, Marshall, 1866.

Webb, F. C. "On the Practical Operations Connected with the Paying-Out and Repairing of Submarine Telegraph Cables: A Paper Given on February 23, 1858, with Subsequent Discussion." *Minutes of Proceedings of the Institution of Civil Engineers* 17 (1858): 262–366.

Whitehouse, Edward O. W. *Atlantic Telegraph: Letter from a Shareholder to Mr. Whitehouse, and His Reply*. London: Bradbury & Evans, Whitefriars, 1858.

——. *Recent Correspondence between Mr. Wildman Whitehouse and the Atlantic Telegraph Company*. London: Bradbury & Evans, Whitefriars, 1858.

——. *Reply to the Statement of the Directors of the Atlantic Telegraph Company: Published in the "Daily News" of September 20, and "Times" September 22, 1858*. London: Bradbury & Evans, Whitefriars, 1858.

SECONDARY WORKS

Adams, Iestyn. *Brothers Across the Ocean: British Foreign Policy and the Origins of the Anglo-American "Special Relationship" 1900–1905*. London: Tauris Academic Studies, 2005.

Adas, Michael. *Dominance by Design: Technological Imperatives and America's Civilizing Mission*. Cambridge, MA: Belknap Press of Harvard University Press, 2006.

——. *Machines as the Measure of Men: Science, Technology, and Ideologies of Western Dominance*. Ithaca: Cornell University Press, 1989.

Ahvenainen, Jorma. *The European Cable Companies in South America Before the First World War*. Helsinki: Finnish Academy of Science and Letters, 2004.

——. *The Far Eastern Telegraphs: The History of Telegraphic Communications Between the Far East, Europe, and America Before the First World War*. Helsinki: Suomalainen Tiedeakatemia, 1981.

——. "The International Telegraph Union: The Cable Companies and the Governments." In *Communications Under the Seas: The Evolving Cable Network and Its*

Implications. Edited by Bernard S. Finn and Daqing Yang, 61–80. Cambridge, MA: MIT Press, 2009.

——. "The Role of Telegraphs in the 19th Century Revolution of Communication." In *Kommunikationsrevolutionen: Die neuen Medien des 16. und 19. Jahrhunderts*. Edited by Michael North, 73–80. Köln: Böhlau, 1995.

Aitken, Hugh G. *Syntony and Spark—The Origins of Radio*. Princeton, NJ: Princeton University Press, 1985.

Alborn, Timothy L. *Conceiving Companies: Joint-stock Politics in Victorian England*. London: Routledge, 2003.

Aldcroft, D. H. "The Entrepreneur and the British Economy, 1870–1914." *The Economic History Review* 17, no. 1 (1964): 113–34.

Allen, Harry Cranbrook. *Great Britain and the United States: A History of Anglo-American Relations (1783–1952)*. Hamden, CT: Archon Books, 1969.

Alleyne, Mark D. *News Revolution: Political and Economic Decisions About Global Information*. New York: St. Martin's Press, 1997.

Anchordoguy, Marie. "Nippon Telegraph and Telephone Company (NTT) and the Building of a Telecommunications Industry in Japan." *The Business History Review* 75, no. 3 (2001): 507–41.

Anderson, Benedict. *Imagined Communities: Reflections on the Origin and Spread of Nationalism*. London: Verso, 2006.

Appadurai, Arjun, ed. *Globalization*. Durham, NC: Duke University Press, 2001.

Appleyard, Rollo. *The History of the Institution of Electrical Engineers (1871–1931)*. London: The Institution of Electrical Engineers, 1939.

Armytage, W. H. G. *A Social History of Engineering*. London: Faber and Faber, 1961.

Baglehole, K. C. *A Century of Service: A Brief History of Cable and Wireless LTD. 1868–1968*. Essex: Anchor Brendon Limited, 1969.

Baker, Ray S., ed. *Woodrow Wilson and World Settlement: Written from His Personal and Unpublished Material*. New York: Doubleday, Page & Company, 1922.

Baldasty, Gerald J. *The Commercialization of News in the Nineteenth Century*. Madison, WI: University of Wisconsin Press, 1992.

Barkhoff, Jürgen, Hartmut Böhme, and Jeanne Riou, eds. *Netzwerke: Eine Kulturtechnik der Moderne*. Köln: Böhlau, 2004.

Barth, Boris, and Jürgen Osterhammel, eds. *Zivilisierungsmissionen: Imperiale Weltverbesserung seit dem 18. Jahrhundert*. Konstanz: UVK-Verl.-Ges., 2005.

Bayly, Christopher A. *The Birth of the Modern World: 1780–1914*. Malden, MA: Blackwell, 2009.

——. *Empire and Information: Intelligence Gathering and Social Communication in India, 1780–1870*. New Delhi: Cambridge University Press, 2007.

Beachy, Robert, ed. *Women, Business, and Finance in Nineteenth-Century Europe: Rethinking Separate Spheres*. Oxford: Berg, 2006.

Beauchamp, Kenneth G. *History of Telegraphy*. London: Institution of Electrical Engineers, 2001.

Bédarida, François, and A. S. Forster. *A Social History of England 1851–1990*. London: Routledge, 1998.

Bektas, Yakup. "Displaying the American Genius: The Electromagnetic Telegraph in the Wider World." *The British Journal for the History of Science* 34, no. 2 (2001): 199–232.

Berghoff, Hartmut. *Englische Unternehmer 1870—1914: Eine Kollektivbiographie führender Wirtschaftsbürger in Birmingham, Bristol und Manchester*. Göttingen: Vandenhoeck & Ruprecht, 1991.

Berliner, Don. *Airplane Racing: A History, 1909–2008*. Jefferson, NC: McFarland, 2009.

Billington, Ray A., and Martin Ridge. *Westward Expansion: A History of the American Frontier*. Albuquerque, NM: University of New Mexico Press, 2001.

Blank Albrecht. "Ausgewählte Familien und Personen." http://www.albrecht-blank.de/ahnenblan/pafg172.htm.

Blondheim, Menahem. *News over the Wires: The Telegraph and the Flow of Public Information in America, 1844–1897*. Cambridge, MA: Harvard University Press, 1994.

Bluma, Lars, "Progress by Technology? The Utopian Linkage of Telegraphy and the World Fairs, 1851–1880." In *Global Communication Electric: Business, News, and Politics in the World of Telegraphy*. Edited by Michaela Hampf and Simone Müller-Pohl. Frankfurt/Berlin: Campus, 2013.

Boase, G. C., and Rev. Anita McConnell. "Lampson, Sir Curtis Miranda." In *Oxford Dictionary of National Biography*. Edited by H. C. G. Matthew et al. Oxford: Oxford University Press, 2004. http://www.oxforddnb.com/view/article/15957.

Böhme, Hartmut. "Einleitung." In *Netzwerke: Eine Kulturtechnik der Moderne*. Edited by Jürgen Barkhoff, Hartmut Böhme, and Jeanne Riou, 17–36. Köln: Böhlau, 2004.

Boli, John, and George M. Thomas, eds. *Constructing World Culture: International Nongovernmental Organizations Since 1875*. Stanford, CA: Stanford University Press, 1999.

Bonnerjea, B. "The Aurora Borealis in Folk-Lore." *Notes and Queries* 175, no. 7 (1938): 113–14.

Bösch, Frank. *Medialisierte Ereignisse: Performanz Inszenierung und Medien seit dem 18. Jahrhundert*. Frankfurt am Main: Campus-Verl., 2010.

——. *Öffentliche Geheimnisse: Skandale, Politik und Medien in Deutschland und Großbritannien 1880–1914*. München: Oldenbourg, 2009.

Bourdieu, Pierre. "The Forms of Capital." In *Handbook of Theory and Research for the Sociology of Education*. Edited by J. Richardson, 241–58. New York: Greenwood Press, 1986.

Bowler, Peter J., and Iwan R. Morus. *Making Modern Science: A Historical Survey*. Chicago: University of Chicago Press, 2005.

Boyce, Robert. "Imperial Dreams and National Realities: Britain, Canada and the Struggle for a Pacific Telegraph Cable, 1879–1902." *The English Historical Review* 115, no. 460 (2000): 39–70.

——. "Submarine Cables as a Factor in Britain's Ascendancy as a World Power, 1850–1914." In *Kommunikationsrevolutionen: Die neuen Medien des 16. und 19. Jahrhunderts*. Edited by Michael North, 81–99. Köln: Böhlau, 1995.

Boyd-Barrett, Oliver, and Terhi Rantanen, eds. *The Globalization of News*. London: Sage, 1998.

Bridgman, Roger. "Brett, John Watkins (1805–1863)." In *Oxford Dictionary of National Biography*. Edited by H. C. G. Matthew et al. Oxford: Oxford University Press, 2004. http://www.oxforddnb.com/view.

Britton, John A. "The Confusion Provoked by Instantaneous Discussion: The New International Communications Network and the Chilean Crisis of 1891–1892 in the United States." *Technology and Culture* 48 (October 2007): 729–57.

Brooke, David. "Brassey, Thomas (1805–1870)." In *Oxford Dictionary of National Biography*. Edited by H. C. G. Matthew et al. Oxford: Oxford University Press, 2004. http://www.oxforddnb.com/view/article/3289.

Buchanan, Robert A. *The Engineers: A History of the Engineering Profession in Britain, 1750–1914*. London: Kingsley, 1989.

Buchanan, Robert A. "Institutional Proliferation in the British Engineering Profession, 1847–1914." *The Economic History Review* 38, no. 1 (1985): 42–60.

Buchanan, Robert A., and Isambard K. Brunel. *Brunel: The Life and Times of Isambard Kingdom Brunel*. London: Hambledon Continuum, 2008.

Bunch, Bryan H., and Alexander Hellemans. *The History of Science and Technology: A Browser's Guide to the Great Discoveries, Inventions, and the People Who Made Them, from the Dawn of Time to Today*. Boston: Houghton Mifflin, 2004.

Bureau International de l'Union Télégraphique. *L'Union Télégraphique Internationale (1865–1915)*. Bern: Bureau International de l'Union Télégraphique, 1915.

Burns, Bill. "1899 Cape Town–St. Helena Cable." http://www.atlantic-cable.com/Cables/1899StHelena/index.htm (accessed July 2, 2011).

Buschauer, Regine. *Mobile Räume: Medien- und diskursgeschichtliche Studien zur Tele-Kommunikation*. Bielefeld: Transcript-Verl., 2010.

Cadigan, Sean T. *Newfoundland and Labrador: A History*. Toronto: University of Toronto Press, 2009.

Caedel, Martin. "Cobden and Peace." In *Rethinking Nineteenth-Century Liberalism: Richard Cobden Bicentenary Essays*. Edited by Anthony Howe, Simon Morgan, and Richard Cobden, 189–207. Burlington, VT: Ashgate, 2006.

Cain, P. J., and A. G. Hopkins. "The Political Economy of British Expansion Overseas, 1750–1914." *The Economic History Review* 33, no. 4 (1980): 463–90.

Carter, Samuel. *Cyrus Field: Man of Two Worlds*. New York: Putman, 1968.

Cassis, Youssef. *Capitals of Capital: A History of International Financial Centres, 1780–2005*. Cambridge: Cambridge University Press, 2006.

Chandler, Alfred D. Jr. "The Emergence of Managerial Capitalism." *The Business History Review* 58, no. 4 (1984): 473–503.

——. *The Visible Hand: The Managerial Revolution in American Business*. Cambridge, MA: Belknap Press of Harvard University Press, 2002.

Channon, Geoffrey. "Gooch, Sir Daniel." In *Oxford Dictionary of National Biography*. Edited by H. C. G. Matthew et al. Oxford: Oxford University Press, 2004. http://www.oxforddnb.com/view/article/10939 (accessed September 15, 2008).

Chapman, Jane. *Comparative Media History: An Introduction; 1789 to the Present*. Cambridge: Polity Press, 2008.

Chauvin Municipal Office. "The Village of Chauvin: Our History." http://www.villageofchauvin.ca/history.htm (accessed August 9, 2011).

Cheah, Pheng, ed. *Cosmopolitics: Thinking and Feeling Beyond the Nation*. Minneapolis, MN: University of Minnesota Press, 2000.

——. "The Cosmopolitical–Today." In *Cosmopolitics: Thinking and Feeling Beyond the Nation*. Edited by Pheng Cheah, 20–41. Minneapolis, MN: University of Minnesota Press, 2000.

Checkland, Olive, and Sydney Checkland. *Industry and Ethos: Scotland, 1832–1914*. Edinburgh: Edinburgh University Press, 1997.

Choisi, Jeanette. *Wurzeln und Strukturen des Zypernkonfliktes 1878 bis 1990: Ideologischer Nationalismus und Machtbehauptung im Kalkül konkurrierender Eliten*. Stuttgart: Steiner, 1993.

Claeys, Gregory. *Imperial Sceptics: British Critics of Empire, 1850–1920*. Cambridge: Cambridge University Press, 2010.

Clavin, Patricia. "Defining Transnationalism." *Contemporary European History* 14, no. 4 (2005): 421–39.

Coates, Vary T., and Bernard S. Finn. *A Retrospective Technology Assessment: Submarine Telegraphy—The Transatlantic Cable of 1866*. San Francisco: San Francisco Press, 1979.

——. "The Commercial Cable Co. and Their Waterville Station." Paper presented at IEE History of Technology Weekend, Trinity College Dublin, July 1987.

Cohen, Ira T. "Towards a Theory of State Intervention: The Nationalization of the British Telegraphs." *Social Science History* 4, no. 2 (1980): 155–205.

Colley, Linda. *Britons: Forging the Nation; 1707–1837*. New Haven, CT: Yale University Press, 1992.

Combs, Mary B. "CUI BONO? The 1870 British Married Women's Property Act, Bargaining Power, and the Distribution of Resources within Marriage." In *A Special Issue on Women and Wealth*. Edited by Carmen D. Deere and Cheryl R. Doss. *Feminist Economics* 12, 1–2 (2006): 51–83.

Commercial Cable Company. "How Submarine Cables Are Made, Laid, Operated and Repaired." http://atlantic-cable.com/Article/1915CCC/index.htm (accessed March 8, 2011).

Conrad, Christoph. "Vorbemerkung," *Geschichte und Gesellschaft* 28 (2002): 339–42.

Conrad, Sebastian, ed. *Globalgeschichte: Theorien, Ansätze, Themen*. Frankfurt: Campus-Verl., 2007.

——. *Globalisation and the Nation in Imperial Germany*. Cambridge: Cambridge University Press, 2010.

Conrad, Sebastian, and Andreas Eckert. "Globalgeschichte, Globalisierung, multiple Modernen: Zur Geschichtsschreibung der modernen Welt." In *Globalgeschichte: Theorien, Ansätze, Themen.* Edited by Sebastian Conrad, 7–49. Frankfurt: Campus-Verl., 2007.

Conrad, Sebastian, and Sorcha O'Hagan. *German Colonialism: A Short History.* Cambridge: Cambridge University Press, 2011.

Conrad, Sebastian, and Jürgen Osterhammel. *Das Kaiserreich transnational: Deutschland in der Welt 1871–1914.* Göttingen: Vandenhoeck & Ruprecht, 2006.

Cook, Adrian. *The Alabama Claims: American Politics and Anglo-American Relations, 1865–1872.* Ithaca: Cornell University Press, 1975.

Cookson, Gillian. *The Cable: The Wire That Changed the World.* Strout: Tempus Publishing, 2003.

——. "Ruinous Competition: The French Atlantic Telegraph of 1869." *Entreprises et Histoire* December, no. 23 (1999): 93–107.

——. "The Transatlantic Telegraph Cable—Eighth Wonder of the World." *History Today* 50 (2000): 44–51.

Cooper, Frederick. "Networks, Moral Discourse, and History." In *Intervention and Transnationalism in Africa: Global-Local Networks of Power,* 23–46. Cambridge: Cambridge University Press, 2001.

Crawford, Elisabeth. *Nationalism and Internationalism in Science, 1880–1939: Four Studies of the Nobel Population.* Cambridge: Cambridge University Press, 1992.

Crockett, Albert S. *When James Gordon Bennett Was Caliph of Bagdad.* New York: Funk and Wagnalls, 1926.

Cubitt, Geoffrey, ed. *Imagining Nations.* Manchester: Manchester University Press, 1998.

Davis, Lance E., and Robert A. Huttenback. *Mammon and the Pursuit of Empire: The Political Economy of British Imperialism, 1860–1912.* Cambridge: Cambridge University Press, 2010.

Dean, J. N. "Gutta Percha and Balata, with Particular Reference to Their Use in Submarine Cable Manufacture." *Journal of the Royal Society of Arts* 92 (June 1944).

de Cogan, D. "Cable Talk: Relations between the Heart's Content and Valentia Cable Stations 1866–1886." University of East Anglia.

Deere, Carmen D., and Cheryl R. Doss, eds. "A Special Issue on Women and Wealth." Special Issue, *Feminist Economics* 12, 1–2 (2006).

Dejung, Christof, ""Towards a global social history? New studies on the global history of the middle class," *Neue Politische Literatur* 59 (2014): pp. 229–253.

Dictionary of American Biography: Gale Biography in Context. Web. New York: Charles Scribner's, 1936.

duBoff, Richard. "The Telegraph in Nineteenth-Century America: Technology and Monopoly." *Comparative Studies in Society and History* 26, no. 4 (1984): 571–86.

Dugan, James. *The Great Iron Ship.* London: Hamilton, 1953.

Dunn, Gano. *Peter Cooper (1791–1883): "A Mechanic of New York."* New York: Newcomen Society of New York, 1949.

Dupont, Brandon, Alka Gandhi, and Thomas J. Weiss. "The American Invasion of Europe: The Long-Term Rise in Overseas Travel, 1820–2000." *NBER Working Paper Series Working Paper* 13977 (May 2008).

Dwyer, John B. *To Wire the World: Perry M. Collins and the North Pacific Telegraph Expedition*. Westport, CT: Praeger, 2001.

Edgerton, David. "The Contradictions of Techno-Nationalism and Techno-Globalism: A Historical Perspective." *New Global Studies* 1 (2007): 1–32.

——. "From Innovation to Use: Ten Eclectic Theses on the Historiography of Technology." *History and Technology* 16 (1999): 1–26.

——. *Science, Technology and the British Industrial "Decline": 1870–1970*. Cambridge: Cambridge University Press, 1996.

Edgerton, David, and S. M. Horrocks. "British Industrial Research and Development Before 1945." *The Economic History Review* 47, no. 2 (1994): 213–38.

Eisenstadt, Shmuel N. *Multiple Modernities*. New Brunswick, NJ: Transaction Publishers, 2002.

Engel, Alexander. *Farben der Globalisierung: Die Entstehung moderner Märkte für Farbstoffe 1500–1900*. Frankfurt/Main: Campus Verl., 2009.

Feldenkirchen, Wilfried, ed. *Siemens: Von der Werkstatt zum Weltunternehmen*. München: Piper, 2003.

Ferguson, Niall. *Civilization: The West and the Rest*. London: Lane, 2011.

——. *The Ascent of Money*. New York: Penguin Press, 2008.

Finn, B. S., ed. *Development of Submarine Cable Communications*. New York: Arno Press, 1980.

——. "Submarine Telegraphy: A Study in Technical Stagnation." In *Communications Under the Seas: The Evolving Cable Network and Its Implications*. Edited by Bernard S. Finn and Daqing Yang, 9–24. Cambridge, MA: MIT Press, 2009.

Finn, Bernard S., and Daqing Yang, eds. *Communications Under the Seas: The Evolving Cable Network and Its Implications*. Cambridge, MA: MIT Press, 2009.

Finzsch, Norbert, and Ursula Lehmkuhl, eds. *Atlantic Communications: The Media in American and German History from the Seventeenth to the Twentieth Century*. Oxford: Berg, 2004.

Fischer, Claude S. *America Calling: A Social History of the Telephone to 1940*. Berkeley: University of California Press, 1992.

Fischer-Tiné, Harald, and Michael Mann, eds. *Colonialism as Civilizing Mission: Cultural Ideology in British India*. London: Anthem Press, 2004.

Flanagan, John T. "Review: Domestic Manners of the Americans, by Frances Trollope." *American Literature*, March 1951.

Friedman, Michael, and Alfred Nordmann, eds. *The Kantian Legacy in Nineteenth-Century Science*. Cambridge, MA: MIT Press, 2006.

Fry, Michael. *The Scottish Empire*. East Linton: Tuckwell Press, 2001.

Fuchs, Margot. "Anfänge der drahtlosen Telegraphie im Deutschen Reich 1897–1918." In *Vom Flügeltelegraphen zum Internet: Geschichte der modernen Telekommunikation*.

Edited by Hans-Jürgen Teuteberg and Cornelius Neutsch, 113–31. Stuttgart: Steiner, 1998.

Gagen, Wendy. "The Manly Telegrapher: The Fashioning of a Gendered Company Culture in the Eastern and Associated Telegraph Companies." In *Global Communication Electric: Business, News, and Politics in the World of Telegraphy*. Edited by Michaela Hampf and Simone Müller-Pohl. Frankfurt/Berlin: Campus, 2013.

Garratt, G. R. M. *One Hundred Years of Submarine Cables*. London: Her Majesty's Stationary Office, 1950.

Geyer, Michael H., and Charles Bright, "World History in a Global Age." *The American Historical Review* 100, no. 4 (October, 1995): 1034–60.

Geyer, Martin H., and Johannes Paulmann, eds. *The Mechanics of Internationalism: Culture, Society, and Politics from the 1840s to the First World War*. Oxford: Oxford University Press, 2008.

Geyer, Martin H., and Johannes Paulmann. "Introduction: The Mechanics of Internationalism." In *The Mechanics of Internationalism: Culture, Society, and Politics from the 1840s to the First World War*. Edited by Martin H. Geyer and Johannes Paulmann, 1–26. Oxford: Oxford University Press, 2008.

Gillam, A. J. *Great Britain vs. America Cable Matches 1895–1901*. London: The Chess Player, 1997.

Gilmore, Paul. "The Telegraph in Black and White." *ELH* 69, no. 3 (2002): 805–33.

Gispen, Kees. "Der gefesselte Prometheus: Die Ingenieure in Großbritannien und den Vereinigten Staaten 1750–1945." In *Geschichte des Ingenieurs: Ein Beruf in sechs Jahrtausenden*. Edited by Walter Kaiser and Wolfgang König, 127–78. München: Hanser, 2006.

Glover, Bill. "Direct United States Cable Company." http://www.atlantic-cable.com/CableCos/DirectUS/index.htm (accessed July 23, 2011).

Gordon, John S. *A Thread Across the Ocean: The Heroic Story of the Transatlantic Cable*. New York: Walker, 2002.

Görges, Hans. *50 Jahre Elektrotechnischer Verein: Festschrift zum fünfzigjährigen Bestehen des Elektrotechnischen Vereins. 1879–1929*. Berlin: Elektrotechnischer Verein, 1929.

Gould, Lewis L. *America in the Progressive Era, 1890–1914*. Harlow: Longman, 2001.

Gowing, Margaret. "Science, Technology and Education: England in 1870." *Oxford Review of Education* 4, no. 1 (1978): 3–17.

Green, Alan. "Further Evidence in the Defence of Wildman Whitehouse? Porthcurno Research Blog." http://researchatporthcurno.blogspot.com/2012/01/further-evidence-in-defence-of-wildman.html (accessed May 7, 2013).

Grewal, David S. *Network Power: The Social Dynamics of Globalization*. New Haven, CT: Yale University Press, 2008.

Griset, Pascal. *Entreprise, technologie et souveraineté: Les télécommunications transatlantiques de la France, XIXe-XXe siècles*. Paris: Editions Rive Droite, 1996.

Griset, Pascal, and Daniel R. Headrick. "Submarine Telegraph Cables: Business and Politics, 1838–1939." *Business History Review* 75, no. 3 (2001): 543–78.

Grodinsky, Julius. *Jay Gould: His Business Career*. Philadelphia: University of Philadelphia Press, 1957.

Gubbins, David, and Emilio Herrero-Bervera, eds. *Encyclopedia of Geomagnetism and Paleomagnetism*. Dordrecht: Springer Science+Business Media B.V., 2007.

Hahn, Hans-Werner. *Die Industrielle Revolution in Deutschland*. München: Oldenbourg Wissenschaftsverlag, 2011.

Haigh, Kenneth R., and Edward Wilshaw. *Cableships and Submarine Cables*. London: Coles, 1968.

Hampf, Michaela, and Ursula Lehmkuhl, eds. *Radio Welten: Politische soziale und kulturelle Aspekte atlantischer Mediengeschichte vor und während des Zweiten Weltkriegs*. Berlin: LIT-Verl., 2006.

Hampf, Michaela, and Simone Müller-Pohl, eds. *Global Communication Electric: Business, News and Politics in the World of Telegraphy*. Frankfurt/Berlin: Campus, 2013.

Hanlon, Christopher. *America's England: Antebellum Literature and Atlantic Sectionalism*. Oxford: Oxford University Press, 2013.

Harrison, John F. C. *Late Victorian Britain, 1875–1901*. London: Routledge, 2000.

Hartmann, Frank. *Globale Medienkultur: Technik, Geschichte, Theorien*. Wien: WUV, 2006.

Harvey, David. "Between Space and Time: Reflections on the Geographical Imagination." *Annals of the Association of American Geographers* 80, no. 3 (1990): 418–34.

Headrick, Daniel R. "A Double-Edged Sword: Communications and Imperial Control in British India." *Historical Social Research. Historische Sozialforschung* 35, no. 1 (2010): 51–65.

——. *The Invisible Weapon: Telecommunications and International Politics, 1851–1945*. New York: Oxford University Press, 1991.

——. "Review: Entreprise, technologie, et souveraineté: Les télécommunications transatlantiques de la France (XIXe-XXe siecle)." *Technology and Culture* 40, no. 1 (1999): 167–69.

——. *The Tentacles of Progress: Technology Transfer in the Age of Imperialism, 1850–1940*. New York: Oxford University Press, 1988.

——. *The Tools of Empire: Technology and European Imperialism in the Nineteenth Century*. New York: Oxford University Press, 1981.

Hearn, Chester G. *Circuits in the Sea: The Men, the Ships, and the Atlantic Cable*. Westport, CT: Praeger, 2004.

Henry, Nancy, and Cannon Schmitt, eds. *Victorian Investments: New Perspectives on Finance and Culture*. Bloomington, IN: Indiana University Press, 2009.

Herman, Arthur. *The Scottish Enlightenment: The Scots' Invention of the Modern World*. London: Fourth Estate, 2001.

Herren, Madeleine. "Governmental Internationalism and the Beginning of a New World Order in the Late Nineteenth Century." In *The Mechanics of Internationalism: Culture, Society, and Politics from the 1840s to the First World War*. Edited by Martin H. Geyer and Johannes Paulmann, 121–44. Oxford: Oxford University Press, 2008.

——. *Hintertüren zur Macht: Internationalismus und modernisierungsorientierte Außenpolitik in Belgien, der Schweiz und den USA 1865–1914*. München: Oldenbourg, 2000.

Hickman, Mary J., and Bronwen Walter. "Deconstructing Whiteness: Irish Women in Britain." *Feminist Review* 50 (1995): 5–19.

Hillier, Jean, ed. *Habitus: A Sense of Place*. Aldershot: Ashgate, 2008.

Hills, Jill. *The Struggle for Control of Global Communication: The Formative Century*. Urbana, IL: University of Illinois Press, 2002.

Hirst, Paul. "Politics: Territorial or Non-Territorial?" In *Habitus: A Sense of Place*. Edited by Jean Hillier, 68–82. Aldershot: Ashgate, 2008.

Historischer Verein des Kantons Schaffhausen, ed. *Schaffhauser Beiträge zur Geschichte*, 1956.

Hobsbawm, Eric J. *The Age of Capital: 1848–1875*. New York: Vintage Books, 1996.

——. *The Age of Empire: 1875–1914*. New York: Vintage Books, 1989.

——. "The Example of the English Middle Class." In *Bourgeois Society in Nineteenth-Century Europe*. Edited by Jürgen Kocka and Allan Mitchell, 127–50. Oxford: Berg, 1993.

——. *Industry and Empire: The Birth of the Industrial Revolution*. New York: The New Press, 1999.

——. *The Invention of Tradition*. Cambridge: Cambridge University Press, 2009.

Hochfelder, David. *The Telegraph in America, 1832–1920*. Baltimore: Johns Hopkins University Press, 2012.

Hoehling, A. A. *The Jeannette Expedition: An Ill-Fated Journey to the Arctic*. London: Abelard-Schuman, 1969.

Holtorf, Christian. *Das erste transatlantische Telegraphenkabel von 1858 und seine Auswirkungen auf die Vorstellungen von Zeit und Raum*. Berlin: Humboldt-Universität Dissertation, 2009.

Hong, Sungook. *Wireless: From Marconi's Black-Box to the Audion*. Cambridge, MA: MIT Press, 2001.

Horsman, Reginald. *Race and Manifest Destiny: The Origins of American Racial Anglo-Saxonism*. Cambridge, MA: Harvard University Press, 1994.

Hovenkamp, Herbert. *Enterprise and American Law 1836–1937*. Cambridge, MA: Harvard University Press, 1991.

Howard, Michael E. *War and the Liberal Conscience*. New York: Columbia University Press, 2008.

Howe, Anthony, Simon Morgan, and Richard Cobden, eds. *Rethinking Nineteenth-Century Liberalism: Richard Cobden Bicentenary Essays*. Aldershot: Ashgate, 2006.

Hughes, Thomas P. *American Genesis: A Century of Invention and Technological Enthusiasm, 1870–1970*. Chicago: University of Chicago Press, 2004.

Hugill, Peter J. "The Geopolitical Implications of Communication Under the Seas." In *Communications Under the Seas: The Evolving Cable Network and Its Implications*. Edited by Bernard S. Finn and Daqing Yang, 257–77. Cambridge, MA: MIT Press, 2009.

——. *Global Communications Since 1844: Geopolitics and Technology*. Baltimore: John Hopkins University Press, 1999.

Hunt, Bruce J. "The Ohm Is Where the Art Is: British Telegraph Engineers and the Development of Electrical Standards." *Osiris* 2, no. 9 (1994): 48–63.

——. "Preece, Sir William Henry (1834–1913)." In *Oxford Dictionary of National Biography*. Edited by H. C. G. Matthew et al. Oxford: Oxford University Press, 2004. http://www.oxforddnb.com/view /article/35605 (accessed 7 May 2010).

——. "Varley, Cromwell Fleetwood (1828–1883)." In *Oxford Dictionary of National Biography*. Edited by H. C. G. Matthew et al. Oxford: Oxford University Press, 2004. http://www.oxforddnb.com/view /article/28114 (accessed 24 Sept 2008).

Huntington, Samuel P. *The Clash of Civilizations and the Remaking of World Order*. London: Touchstone Books, 1998.

Ignatiev, Noel. *How the Irish Became White*. New York: Routledge, 1995.

Innis, Harold A. *Empire and Communications*. Lanham, MD: Rowman & Littlefield, 2007.

The Institution of Engineering and Technology. "Hertha Ayrton." http://www.theiet .org/about/libarc/archives/biographies/ayrtonh.cfm (accessed May 2, 2010).

International Telecommunications Union. *International Telecommunications Union: Celebrating 130 Years, 1865–1965*. Geneva, Switzerland: International Telecommunications Union, 1995.

Iriye, Akira. *Global Community: The Role of International Organizations in the Making of the Contemporary World*. Berkeley, CA: University of California Press, 2004.

Itkowitz, David C. "Fair Enterprise or Extravagant Speculation: Investment, Speculation, and Gambling in Victorian England," *Victorian Studies* 45 (2002): 121–47.

Jacobson, Harold. "International Institutions for Telecommunications: The ITU's Role." In *The International Law of Communications*. Edited by Edward McWhinney, 51–68. Dobbs Ferry, NY: Sijthoff; Oceana, 1971.

James, Harold. *History of Krupp*. Princeton: Princeton University Press, 2012.

Jansen, Marius B. *The Emergence of Meiji Japan*. Cambridge: Cambridge University Press, 1997.

Jenkins, Philip. *A History of the United States*. Basingstoke: Palgrave Macmillan, 2007.

Jepsen, Thomas C. *My Sisters Telegraphic: Women in the Telegraph Office, 1846–1950*. Athens: Ohio University Press, 2000.

Jezler, Ferdinand. "Johann Conrad Im Thurn." In *Schaffhauser Beiträge zur Geschichte*. Edited by Historischer Verein des Kantons Schaffhausen, 311–6 I. 1956.

John, Richard R. "Field, Cyrus West." In *American National Biography*. http://www.anb .org/articles/10/10-00542.html (accessed September 18, 2008).

——. *Network Nation: Inventing American Telecommunications*. Cambridge, MA: Belknap Press of Harvard University Press, 2010.

"John William Mackay." In *Dictionary of American Biography: Gale Biography In Context*. Web. New York: Charles Scribner's, 1936.

Jordanova, Ludmilla. "Science and Nationhood: Cultures of Imagined Communities." In *Imagining Nations*. Edited by Geoffrey Cubitt, 192–211. Manchester: Manchester University Press, 1998.

Kaelble, Hartmut, ed. *The European Way: European Societies During the 19th and 20th Centuries*. Oxford: Berghahn, 2004.

Kaempffert, Waldemar. "Communication and World Peace." *Proceedings of the American Philosophical Society* 70, no. 3 (1931): 273–84.

Kaiser, Walter, and Wolfgang König, eds. *Geschichte des Ingenieurs: Ein Beruf in sechs Jahrtausenden*. München: Hanser, 2006.

Kant, Immanuel, and Michael Friedman. *Metaphysical Foundations of Natural Science*. Cambridge: Cambridge University Press, 2004.

Kaspersen, Lars B. *Anthony Giddens: An Introduction to a Social Theorist*. Oxford: Blackwell, 2000.

Kaufman, Will, ed. *Britain and the Americas: Culture, Politics, and History; a Multidisciplinary Encyclopedia*. Santa Barbara, CA: ABC-CLIO, 2005.

Kende, Istvan. "History of Peace: Concept and Organization from the Late Middle Ages to the 1870s." *Journal of Peace Research* 26, no. 3 (1989): 233–47.

Kennedy, P. M. "Imperial Cable Communications and Strategy, 1870–1914." *The English Historical Review* 86, no. 341 (1971): 728–52.

Kennedy, William. "Notes on Economic Efficiency in Historical Perspective: The Case of Great Britain, 1870–1914." *Research in Economic History* 9 (1984): 109–41.

Keough, Willeen G. "Contested Terrains: Ethnic and Gendered Spaces in the Harbour Grace Affray." *The Canadian Historical Review* 90, no. 1 (2009): 29–70.

Klein, Carole. *Gramercy Park: An American Bloomsbury*. Baltimore: Johns Hopkins University Press, 1999.

Klein, Maury. *The Life and Legend of Jay Gould*. Baltimore: Johns Hopkins University Press, 1986.

Kocka, Jürgen. "The Middle Classes in Europe." In *The European Way: European Societies During the 19th and 20th Centuries*. Edited by Hartmut Kaelble, 15–43. Oxford: Berghahn, 2004.

Kocka, Jürgen, and Allan Mitchell, eds. *Bourgeois Society in Nineteenth-Century Europe*. Oxford: Berg, 1993.

Koskenniemi, Martti. *The Gentle Civilizer of Nations: The Rise and Fall of International Law 1870–1960*. Cambridge: Cambridge University Press, 2010.

Krajewski, Markus. *Restlosigkeit: Weltprojekte um 1900*. Frankfurt am Main: Fischer-Taschenbuch-Verl., 2006.

Krause, Günther. *Der Internationale Fernmeldeverein*. Franfurt/Berlin: Alfred Metzner Verlag, 1960.

Kropholler, Jan. *Internationales Einheitsrecht: Allg. Lehren*. Tübingen: Mohr, 1975.

Kunert, A. *Telegraphen-Seekabel*. Köln-Mühlheim: Karl Glitscher, 1962.

Laborie, Léonard. *L'Europe mise en réseaux: La France et la coopération internationale dans les postes et les télécommunications (années 1850-années 1950)*. Brussels: Peter Lang, 2010.

Landes, David S. *The Unbound Prometheus: Technological Change and Industrial Development in Western Europe from 1750 to the Present*. Cambridge: Cambridge University Press, 2008.

Laney, Al. *Paris Herald: The Incredible Newspaper*. New York: Greenwood Press Publishers, 1947.

Latour, Bruno. "Die Logistik der Immutable mobiles." In *Mediengeographie: Theorie—Analyse—Diskussion*. Edited by Jörg Döring and Tristan Thielmann, 111–44. Bielefeld: Transcript-Verl., 2009.

——. *Reassembling the Social: An Introduction to Actor-Network-Theory*. Oxford: Oxford University Press, 2007.

Law, John, and John Hassard. *Actor Network Theory and After*. Oxford: Blackwell, 2005.

Ledger, Sally. *The New Woman: Fiction and Feminism at the Fin de Siècle*. Manchester: Manchester University Press, 1997.

Lee, Sohui. "Anglophobia." In *Britain and the Americas: Culture, Politics, and History; a Multidisciplinary Encyclopedia*. Edited by Will Kaufman, 87–90. Santa Barbara, CA: ABC-CLIO, 2005.

Lindquist, Svante. "The Spectacle of Science: An Experiment in 1744 Concerning the Aurora Borealis." *Configurations* 1, no. 1 (1993): 57–94.

Longinow, Michael A. "News Gathering." In *American Journalism: History, Principles, Practices*. Edited by William D. Sloan and Lisa M. Parcell, 144–52. Jefferson, NC: McFarland, 2002.

Loya, Thomas A., and John Boli. "Standardization in the World Polity: Technical Rationality Over Power." In *Constructing World Culture: International Nongovernmental Organizations Since 1875*. Edited by John Boli and George M. Thomas, 169–97. Stanford, CA: Stanford University Press, 1999.

Lynch, A. C. "Bright, Sir Charles Tilston." In *Oxford Dictionary of National Biography*. Edited by H. C. G. Matthew et al. Oxford: Oxford University Press, 2004. http://www.oxforddnb.com/view/article/3415 (accessed September 15, 2008).

MacKay, Donald. *Flight from Famine: The Coming of the Irish to Canada*. Toronto: Natural Heritage Books, 2009.

Magee, Gary B., and Andrew S. Thompson. *Empire and Globalisation: Networks of People, Goods and Capital in the British World, c. 1850–1914*. Cambridge: Cambridge University Press, 2010.

Magocsi, Paul R., ed. *Encyclopedia of Canada's People*. Toronto: Toronto University Press, 1999.

Maier, Charles. "Consigning the Twentieth Century to History: Alternative Narratives for the Modern Era." *American Historical Review* 105, no. 3 (2000): 807–31.

Malatesta, Maria. *Professional Men, Professional Women: The European Professions from the Nineteenth Century until Today*. Los Angeles, CA: Sage, 2011.

Maltby, Josephine, and Janette Rutterford. "'She Possessed Her Own Fortune': Women Investors from the Late Nineteenth to the Early Twentieth Century." *Business History* 48, no. 2 (2006): 220–53.

Maltby, Josephine, and Janette Rutterford. "'The Widow, the Clergyman and the Reckless': Women Investors in England, 1830–1914." In *A Special Issue on Women and Wealth*. Edited by Carmen D. Deere and Cheryl R. Doss. *Feminist Economics* 12, no. 1–2 (2006): 111–38.

Manter, Ethel. *Rocket of the Comstock: The Story of John William Mackay*. Caldwell, ID: Caxton Printers, 1950.

Marland, E. A. "British and American Contributions to Electrical Communication." *The British Journal for the History of Science* 1, no. 1 (June 1962): 31–48.

McCarthy, Diane. "Cable Makes Big Promises for African Internet." CNN.com, July 23, 2009. http://edition.cnn.com/2009/TECH/07/22/seacom.on/index.html (accessed June 11, 2011).

McConnell, Anita. "Glass, Sir Richard Atwood (1820–1873)." In *Oxford Dictionary of National Biography*. Edited by H. C. G. Matthew et al. Oxford: Oxford University Press, 2004. http://www.oxforddnb.com/view/article/10801 (accessed September 24, 2008).

——. "Pender, Sir John." In *Oxford Dictionary of National Biography*. Edited by H. C. G. Matthew et al. Oxford: Oxford University Press, 2004. http://www.oxforddnb.com/view/article/21831 (accessed September 15, 2008).

——. "Smith Willoughby (1828–1891)." In *Oxford Dictionary of National Biography*. Edited by H. C. G. Matthew et al. Oxford: Oxford University Press, 2004. http://www.oxforddnb.com/view/article/25942 (accessed September 15, 2008).

McDonald, Philip B. *A Saga of the Seas: The Story of Cyrus W. Field and the Laying of the First Atlantic Cable*. New York: Wilson-Erickson, 1937.

McGowan, Mark. "Irish Catholics." In *Encyclopedia of Canada's People*. Edited by Paul R. Magocsi, 734–63. Toronto: Toronto University Press, 1999.

McLaughlin, Robert. "Irish Nationalism and Orange Unionism in Canada: A Reappraisal." *Eire-Ireland* 41, no. 3/4 (2006): 80–109.

McLuhan, Marshall. *The Global Village: Transformations in World Life and Media in the 21st Century*. New York: Oxford University Press, 1992.

——. *Understanding Media: The Extensions of Man*. London: Routledge, 2001.

McNeill, William H. "Globalization: Long Term Process or New Era in Human Affairs?" *New Global Studies* 2, no. 1 (2008): 1–9.

McWhinney, Edward, ed. *The International Law of Communications*. Dobbs Ferry, NY: Sijthoff; Oceana, 1971.

Moore, Frank, ed. *Speeches of John Bright, M.P. on the American Question: With an Introduction by Frank Moore*. Boston: Little, Brown and Company, 1865.

Morus, Iwan R. "The Nervous System of Great Britain: Space, Time and the Electric Telegraph in the Victorian Age." *The British Journal for the History of Science* 33, no. 4 (2000): 455–75.

Mowlana, Hamid. *Global Information and World Communication: New Frontiers in International Relations*. London: Sage, 1998.

Müller-Pohl, Simone. "'By Atlantic Telegraph': A Study on Weltcommunication in the 19th Century." *Medien & Zeit* 4 (2010): 40–54.

——. "Working the Nation State: Submarine Cable Actors, Cable Transnationalism and the Governance of the Global Media System, 1858–1914." In *The Nation State and Beyond. Governing Globalization Processes in the Nineteenth and Twentieth Century*. Edited by Isabella Löhr and Roland Wenzlhuemer, 101–3. New York: Springer Press, 2012.

Mwiti, Lee. "East Africa: Sea Cable Ushers in New Internet Era." Allafrica.com, July 23, 2009. allafrica.com (accessed August 3, 2009).

Myers, Philip E. *Caution and Cooperation: The American Civil War in British-American Relations*. Kent, OH: Kent State University Press, 2008.

Nevins, Allan. *Abram S. Hewitt. With Some Account of Peter Cooper*. New York: Octagon Books, 1935.

Nicholls, David. "The Manchester Peace Conference of 1853." *Manchester Region History Review* 5, no. 1 (1991): 11–21.

——. "Richard Cobden and the International Peace Congress Movement, 1848–1853." *Journal of British Studies* 30, no. 4 (1991): 351–76.

Nickles, David P. "Submarine Cables and Diplomatic Culture." In *Communications Under the Seas: The Evolving Cable Network and Its Implications*. Edited by Bernard S. Finn and Daqing Yang, 209–26. Cambridge, MA: MIT Press, 2009.

——. "Telegraph Diplomats: The United States' Relations with France in 1848 and 1870." *Technology and Culture* 40.1 (1999): 1–25.

——. *Under the Wire: How the Telegraph Changed Diplomacy*. Cambridge, MA: Harvard University Press, 2003.

Noakes, Richard. "Cromwell Varley FRS, Electric Discharge and Victorian Spiritualism." *Notes and Records of the Royal Society* 61 (2007): 5–21.

——. "Industrial Research at the Eastern Telegraph Company, 1872–1929." *The British Journal for the History of Science* (2013): 1–28.

North, Michael, ed. *Kommunikationsrevolutionen: Die neuen Medien des 16. und 19. Jahrhunderts*. Köln: Böhlau, 1995.

O'Connor, Richard. *The Scandalous Mr. Bennett*. Garden City, NY: Doubleday, 1962.

Ogilvie, J. S. *Life and Death of Jay Gould, and How He Made His Millions*. New York: Arno Press, 1981.

Oppenheim, L., and Ronald Roxburgh. *International Law: A Treatise*. Clark, NJ: Lawbook Exchange, 2005.

Osterhammel, Jürgen. *Die Verwandlung der Welt: Eine Geschichte des 19. Jahrhunderts*. München: Beck, 2009.

——. *Europe, the "West" and the Civilizing Mission*. The 2005 Annual Lecture. London: German Historical Institute, 2006.

Osterhammel, Jürgen, and Niels P. Petersson. *Globalization: A Short History*. Princeton, NJ: Princeton University Press, 2005.

Ott, Julia. *When Wall Street Met Main Street*. Cambridge, MA: Harvard University Press, 2011.

Owen, Thomas C. *The Corporation Under Russian Law, 1800–1917: A Study in Tsarist Economic Policy*. Cambridge: Cambridge University Press, 1991.

Palmer, Michael, Oliver Boyd-Barrett, and Terhi Rantanen. "Global Financial News." In *The Globalization of News*. Edited by Oliver Boyd-Barrett and Theri Rantanen, 61–78. London: Sage, 1998.

Paolino, Ernest N. *The Foundation of the American Empire: William Henry Seward and U.S. Foreign Policy*. Ithaca: Cornell University Press, 1973.

Parker, Charles H. *Global Interactions in the Early Modern Age, 1400–1800*. Cambridge: Cambridge University Press, 2010.

Parker, Franklin. *George Peabody: A Biography*. Nashville: Vanderbilt University Press, 1995.

Parkman, Francis, and David Levin. *The Oregon Trail*. Harmondsworth: Penguin Books, 1983.

Petroski, Henry. *Remaking the World: Adventures in Engineering*. New York: Vintage Books, 1999.

Pfitzner, Helmuth. *Seekabel und Funktelegraphie: Im überseeischen Schnellnachrichtenwesen*. Leipzig: Curt Böttger Verlag, 1931.

Pike, Robert M., and Dwayne R. Winseck. "The Politics of Global Media Reform, 1907–1923." *Media, Culture & Society* 26 (2004): 643–75.

Pittock, Murray. *The Invention of Scotland: The Stuart Myth and the Scottish Identity, 1638 to the Present*. London: Routledge, 1991.

Pollard, A. F. "Clark, (Josiah) Latimer (1822–1898)." In *Oxford Dictionary of National Biography*. Edited by H. C. G. Matthew et al. Oxford: Oxford University Press, 2004. http://www.oxforddnb.com/view /article/5469 (accessed September 15, 2008).

Pomeranz, Kenneth. *The Great Divergence: China, Europe, and the Making of the Modern World Economy*. Princeton, NJ: Princeton University Press, 2000.

Porter, David. *Comparative Early Modernities: 1100–1800*. Basingstoke: Palgrave Macmillan, 2012.

Potter, Simon J. *News and the British World: The Emergence of an Imperial Press System, 1876–1922*. Oxford: Clarendon Press, 2003.

Pound, Arthur. *The Telephone Idea: Fifty Years After*. New York: Greenberg, 1926.

Preda, Alex. "The Rise of the Popular Investor: Financial Knowledge and Investing in England and France, 1840–1880." *The Sociological Quarterly* 42, no. 2 (2001): 205–32.

——. "Socio-Technical Agency in Financial Markets: The Case of the Stock Ticker." *Social Studies of Science* 36, no. 5 (2006): 753–82.

Quinn, Carol. "Dickinson, Telegraphy, and the Aurora Borealis." *The Emily Dickinson Journal* 13, no. 2 (2004): 58–78.

Rantanen, Terhi. *When News Was New*. Malden, MA: Wiley-Blackwell, 2009.

Reader, William J. *A History of the Institution of Electrical Engineers, 1871–1971*. London: P. Peregrinus on behalf of the Institution of Electrical Engineers, 1987.

Reed, M. C. *Investment in Railways in Britain, 1820–1844: A Study in the Development of Capital Market*. Oxford: Oxford University Press, 1975.

Rees, Jim. *The Life of Capt. Robert Halpin. Arklow, Co. Wicklow*. Ireland: Dee-Jay Publications, 1992.

Rens, Jean-Guy. *The Invisible Empire: A History of the Telecommunications Industry in Canada 1846–1956*. Montréal: McGill-Queen's University Press, 2001.

Richardson, Angelique, and Chris Willis. *The New Woman in Fiction and Fact: Fin-de-siècle Feminisms*. New York: Palgrave, 2002.

Richardson, J., ed. *Handbook of Theory and Research for the Sociology of Education*. New York: Greenwood Press, 1986.

Robb, George. "Ladies of the Ticker: Women, Investment, and Fraud in England and America 1850–1930." In *Victorian Investments: New Perspectives on Finance and Culture*. Edited by Nancy Henry and Cannon Schmitt, 120–42. Bloomington, IN: Indiana University Press, 2009.

——. *White-Collar Crime in Modern England. Financial Fraud and Business Morality, 1845–1929*. Cambridge: Cambridge University Press, 1992.

Robbins, Bruce. "Introduction Part I: Actually Existing Cosmopolitanism." In *Cosmopolitics: Thinking and Feeling Beyond the Nation*. Edited by Pheng Cheah, 1–19. Minneapolis, MN: University of Minnesota Press, 2000.

Robertson, Charles L. *The International Herald Tribune: The First Hundred Years*. New York: Columbia University Press, 1987.

Roediger, David R. *The Wages of Whiteness: Race and the Making of the American Working Class*. London: Verso, 1991.

Rogers, John D. *A Historical Geography of the British Colonies: Newfoundland*. Oxford: Clarendon Press, 1911.

Rosenberg, Emily S., ed. *A World Connecting: 1870–1945*. Cambridge, MA: Belknap Press of Harvard University Press, 2012.

The Royal Swedish Academy of Science. "The Nobel Prize in Physics, 2009—Press Release." http://nobelprize.org/nobel_prizes/physics/laureates/2009/press.html (accessed June 11, 2011).

Sacks, David, Oswyn Murray, and Lisa R. Brody, eds. *Encyclopedia of the Ancient Greek World*. New York: Facts On File, 2005.

Saraiva, Tiago. "Inventing the Technological Nation: The Example of Portugal (1851–1898)." *History and Technology* 23, no. 3 (2007): 263–73.

Schimmelfennig, Frank. *Internationale Politik*. Paderborn: Schöningh, 2010.

Schivelbusch, Wolfgang. *Geschichte der Eisenbahnreise: Zur Industrialisierung von Raum und Zeit im 19. Jahrhundert*. Wien/München: Fischer Taschenbuch Verlag, 1977.

Scholz, Franz. *Krieg und Seekabel: Eine völkerrechtliche Studie*. Berlin: Verlag von Franz Vahlen, 1904.

Schudson, Michael. *Discovering the News: A Social History of American Newspapers*. New York: Basic Books, 1998.

Schwarzlose, Richard A. *The Nation's Newsbrokers. The Rush to Institution, from 1865 to 1920*. Evanston, IL: Northwestern University Press, 1990.

Scott, J. D. *Siemens Brothers 1858–1958: An Essay in the History of Industry*. London: Weidenfeld & Nicholson, 1958.

Seitz, Don C. *The James Gordon Bennetts—Father and Son, Proprietors of the New York Herald*. New York: University of Michigan Library, 1974.

Shakespeare, William. *A Midsummer Night's Dream*. Stuttgart: Reclam, 2011.

Shami, Seteney. "Prehistories of Globalization: Circassian Identity in Motion." In *Globalization*. Edited by Arjun Appadurai, 220–50. Durham, NC: Duke University Press, 2001.

Shrivastava, K. M. *News Agencies from Pigeon to Internet*. Elgin, IL: New Dawn Press, 2007.

Siemens, Georg. *History of the House of Siemens*. Freiburg: Alber, 1957.

Silverman, Kenneth. *Lightning Man: The Accursed Life of Samuel F. B. Morse*. New York: Knopf, 2003.

Sloan, William D., and Lisa M. Parcell, eds. *American Journalism: History, Principles, Practices*. Jefferson, NC: McFarland, 2002.

Slocum, Victor. *Castaway Boats*. Dobbs Ferry, NY: Sheridan House, 2001.

Smith, Adam. *The Wealth of Nations*. New York: Bantam Dell, 2003.

Spear, A. M. "Canning, Sir Samuel (1823–1908)." In *Oxford Dictionary of National Biography*. Edited by H. C. G. Matthew et al. Oxford: Oxford University Press, 2004. http://www.oxforddnb.com/view /article/32279 (accessed September 15, 2008).

Spencer, David R. *The Yellow Journalism: The Press and America's Emergence as a World Power*. Evanston, IL: Northwestern University Press, 2007.

Sprenger, Florian. "Between the Ends of a Wire: Electricity, Instantaneity and the World of Telegraphy." In *Global Communication Electric: Business, News and Politics in the World of Telegraphy*. Edited by Michaela Hampf and Simone Müller-Pohl, 355–81. Frankfurt/Berlin: Campus, 2013.

Standage, Tom. *The Victorian Internet: The Remarkable Story of the Telegraph and the Nineteenth Century's Online Pioneers*. London: Phoenix, 2003.

Stichweh, Rudolf. "The Genesis and Development of a Global Public Sphere: Revised version of a paper first published in Development 46, 2003, 26–29." (2006): 1–11.

Storey, Graham. *Reuters' Century, 1851–1951*. London: Parrish, 1951.

Stowe, William W. *Going Abroad: European Travel in Nineteenth-Century American Culture*. Princeton, NJ: Princeton University Press, 1994.

Takai, Ronald. *Iron Cages: Race and Culture in 19th-Century America*. New York: Oxford University Press, 1979.

Tarrant, Donald R. *Atlantic Sentinel: Newfoundland's Role in Transatlantic Cable Communications*. St. John's, Newfoundland: Flanker Press, 1999.

Taylor, Dorceta E. *The Environment and the People in American Cities, 1600–1900s: Disorder, Inequality, and Social Change*. Durham, NC: Duke University Press, 2009.

Telegraph Construction and Maintenance Company. *The Telcon Story 1850–1950*. London: Telegraph Construction and Maintenance Company, 1950.

Travis, Anthony S. "Engineering and Politics: The Channel Tunnel in the 1880s." *Technology and Culture* 32, no. 3 (1991): 461–97.

Trevelyan, George. *The Life of John Bright*. Boston: Houghton Mifflin, 1913.

Tully John, "A Victorian Ecological Disaster: Imperialism, the Telegraph, and Gutta-Percha." *Journal of World History* 20 (2009): 559–79.

Tworek, Heidi. "'The Path to Freedom'? Transocean and German Wireless Telegraphy, 1914–1922." *Historical Social Research. Historische Sozialforschung* 35, no. 1 (2010): 209–33.

Tyrrell, Ian R. "What Is Transnational History?" http://iantyrrell.wordpress.com/what-is-transnational-history/ (accessed September 8, 2011).

U.S. Senate. "Telegraph. Catalogue of Fine Art." http://www.senate.gov/artandhistory/art/artifact/Painting_33_00019.htm#bio (accessed September 18, 2011).

Vertovec, Steven. "Transnationalism and Identity." *Journal of Ethnic and Migration Studies* 27 (2001): 573–82.

Villard, Henry S. *Blue Ribbon of the Air: The Gordon Bennett Races.* Washington, DC: Smithsonian Institution Press, 1987.

Virilio, Paul, and Philip Beitchman, trans. *The Aesthetics of Disappearance.* Los Angeles: Semiotext(e), 2009.

Wake, Jehanne. "Kleinwort, Alexander Frederick Henry (1815–1886)." In *Oxford Dictionary of National Biography.* Edited by H. C. G. Matthew et al. Oxford: Oxford University Press, 2004. http://www.oxforddnb.com/view/article/48912 (accessed November 26, 2010).

Wayland, John W. *The Pathfinder of the Seas: The Life of Matthew Fontaine Maury.* Richmond: Garrett & Massie, 1930.

Weiher, Sigfrid von. *Die englischen Siemens-Werke und das Siemens-Überseegeschäft in der zweiten Hälfte des 19. Jahrhunderts.* Berlin: Duncker u. Humblot, 1990.

Wellman, Barry, and Wenhong Chen. "The Global Digital Divide—Within and Between Countries." *IT & Society* 1, no. 7 (2004): 18–25.

Wenzlhuemer, Roland. *Connecting the Nineteenth-Century World: The Telegraph and Globalization.* Cambridge: Cambridge University Press, 2013.

——. "The Dematerialization of Telecommunication: Communication Centres and Peripheries in Europe and the World, 1850–1920." *Journal of Global History* 2 (2007): 345–72.

——. "Editorial: Telecommunication and Globalization in the Nineteenth Century." *Historical Social Research. Historische Sozialforschung* 35, no. 1 (2010): 7–18.

Wilke, Jürgen. "The Telegraph and Transatlantic Communication Relations." In *Atlantic Communications: The Media in American and German History from the Seventeenth to the Twentieth Century.* Edited by Norbert Finzsch and Ursula Lehmkuhl, 107–34. Oxford: Berg, 2004.

Wilkins, Mira. *The History of Foreign Investment in the United States to 1914.* Cambridge, MA: Harvard University Press, 1989.

Winder, Gordon M. "Imagining World Citizenship in the Networked Newspaper: La Nación Reports the Assassination at Sarajevo, 1914." *Historical Social Research. Historische Sozialforschung* 35, no. 1 (2010): 140–66.

Winkler, Jonathan R. *Nexus: Strategic Communications and American Security in World War I.* Cambridge, MA: Harvard University Press, 2008.

Winseck, Dwayne R., and Robert M. Pike. *Communication and Empire: Media, Markets, and Globalization, 1860–1930*. Durham: Duke University Press, 2007.

Yang, Daqing. "Submarine Cables and the Two Japanese Empires." In *Communications Under the Seas: The Evolving Cable Network and Its Implications*. Edited by Bernard S. Finn and Daqing Yang, 227–56. Cambridge, MA: MIT Press, 2009.

——. *Technology of Empire: Telecommunications and Japanese Imperialism 1930–1945*. Cambridge, MA: Harvard University Press, 2003.

Zweig, Stefan. *Decisive Moments in History: Twelve Historical Miniatures*. Riverside, CA: Ariadne Press, 1999.

INDEX

CPSIA information can be obtained
at www.ICGtesting.com
Printed in the USA
LVOW11*0900020817
543398LV00003B/5/P